AF788702

A Collection of Dreams about the Future of Technology Education

International Technology Education Studies

Series Editor

Marc J. de Vries (*Delft University of Technology, The Netherlands*)

Editorial Board

John R. Dakers (*Delft University of Technology, The Netherlands*)
Jonas Hallström (*Linköping University, Sweden*)
Philip Reed (*Old Dominion University, USA*)

VOLUME 21

The titles published in this series are listed at *brill.com/ites*

A Collection of Dreams about the Future of Technology Education

A Festschrift in Honour of Marc J. de Vries

Edited by

John R. Dakers

BRILL

LEIDEN | BOSTON

Cover illustration: iStock.com/Peach_iStock

All chapters in this book have undergone peer review.

The Library of Congress Cataloging-in-Publication Data is available online at https://catalog.loc.gov

Typeface for the Latin, Greek, and Cyrillic scripts: "Brill". See and download: brill.com/brill-typeface.

ISSN 1879-8748
ISBN 978-90-04-71092-4 (paperback)
ISBN 978-90-04-71093-1 (hardback)
ISBN 978-90-04-71094-8 (e-book)
DOI 10.1163/9789004710948

Copyright 2025 by John R. Dakers. Published by Koninklijke Brill BV, Plantijnstraat 2, 2321 JC Leiden, The Netherlands.
Koninklijke Brill BV incorporates the imprints Brill, Brill Nijhoff, Brill Schöningh, Brill Fink, Brill mentis, Brill Wageningen Academic, Vandenhoeck & Ruprecht, Böhlau and V&R unipress.
Koninklijke Brill BV reserves the right to protect this publication against unauthorized use. Requests for re-use and/or translations must be addressed to Koninklijke Brill BV via brill.com or copyright.com.
For more information: info@brill.com.

This book is printed on acid-free paper and produced in a sustainable manner.

For Marc J. de Vries
And all those who study, learn and research technology education worldwide.
Also, in memory of William (Bill) Dugger and Jacques Ginestié

Contents

Acknowledgements

I want to thank the contributing authors for taking time out of their busy schedules to produce these wonderful chapters. Their splendid contributions in this book go a long way in challenging the status quo. I am in awe of their wisdom and generosity. I would also like to acknowledge Wendy's unstinting support, which has, as ever, been made manifest in so many ways.

I also want to thank the many thousands of dedicated educators worldwide who will, ultimately, be responsible for delivering the dreams outlined in this book.

Thanks must also go to John Bennet from the Publisher for his help, support and guidance in completing this book. I want also to thank the enormous support given to me by Brill Publishers in general, and particularly that of Christine Hededam and Vignesh Vivekanandan, who have guided me through the process.

Finally, I must thank Marc de Vries for his helpful discussions with me regarding this collection and for his support and guidance over the years on many matters related to my research.

Figures and Tables

Figures

Tables

Notes on Contributors

Piet Ankiewicz
is an Emeritus Professor of Technology Education in the Faculty of Education at the University of Johannesburg. He considers himself as a theoretical educationalist who specialises in studying and developing theoretical models and philosophical frameworks to understand and explain the tenets of STEM – science, technology, engineering, and mathematics – education. His research interests include the affordances of the philosophy of technology for technology classroom pedagogy, teacher education, indigenous technology knowledge systems, and STEM education. He also has an interest in students' attitudes towards technology. He has been rated by the National Research Foundation as an established researcher with international recognition. He is a member of the Editorial Board of the *International Journal of Technology and Design Education*.

Frank Banks
is an Emeritus Professor in Teacher Education at The Open University in the UK where he authored the first open and distance initial teacher education course in Technology, and later directed the innovative on-line initial teacher education programme. He was Director, International Development in Teacher Education, responsible for the University's strategic direction of major education projects in sub-Saharan Africa, Bangladesh, and India. Frank has worked as a schoolteacher of science, technology, engineering science and mathematics in different secondary high schools in England and in Wales. He has authored or edited 32 books/handbooks for teachers, over one hundred academic papers, and has acted as a consultant in the professional development of teachers to Egyptian, South African and Argentinean government agencies, UNESCO, and the World Bank.

Dennis Cheek
is Dean of, the School of Entrepreneurship & Humanities and university-wide Professor of Creativity, Innovation & Entrepreneurship at Universitas Ciputra, Surabaya, Indonesia. Since 2023, he is a Visiting Professor of Innovation & Entrepreneurship at IÉSEG School of Management in France. Cheek has been a CEO, CFO, CLO, and serial entrepreneur with experience at regional, national, and international levels. He taught in middle and high schools in three nations, held leadership positions in the NY and RI state education departments, and was a vice president at the John Templeton and Ewing Marion Kauffman foundations. Professor Cheek consulted with 56 institutions of higher education

in 12 nations, and taught in 7 bachelor's, 12 master's, and 8 doctoral programs including Penn State Great Valley, University of Pennsylvania (Wharton & GSE), and the National University of Singapore. With PhDs from Penn State (curriculum & instruction/science education) and Durham University (theology & technology), he has authored or edited over 850 publications. Dennis undertook post-doctoral studies at the Paris School of Business. He is a Fellow of AAAS (STEM education) and the Royal Society of Arts. He was admitted into the Society of Ordained Scientists in 1999, an ecumenical religious order within the global Anglican communion.

Hongwu Dai

is an assistant professor of Foreign Languages at Yunnan Normal University in Kunming, China. She earned a BA in English education from Central China Normal University and an MA in applied linguistics from La Trobe University. She is currently a Ph.D. candidate in organisational leadership at Eastern University in the USA. She has published two books in China (*English teaching skills and techniques and Study on reading children's English picture books*). Hongwu served as production translator into English for the award-winning documentary "*The Inheritance of the Green Field*". She has won numerous university teaching awards. Most recently, she co-authored papers presented at the AECT Summer Research Symposia and published them in the AECT Communications and Technology: Issues and Innovations series by Springer in *Transdisciplinary learning experience design: Futures, synergies, and innovations* (in press, 2024) and *Towards inclusive learning design: Social justice, equity, and community* (2023) as well as papers in the 43rd, 44th, and 45th AECT International Convention Annual Proceedings volumes. One of these papers, "Exploring systems thinking in school principals' educational technology leadership in Yunnan Province, PRC", was selected to be further expanded by the authors and appeared in *Distance Learning* (2023).

John R. Dakers

is a researcher and author on the subject of (ethno)technological literacy. His first book was published in 2006, entitled *Defining Technological Literacy*, later updated and published as a revised second edition, published in 2014, by Palgrave MacMillan. He also had another book published in 2014 entitled *New Frontiers in Technological Literacy* by Palgrave MacMillan. He has co-edited two books published by Brill (formerly Sense) entitled *Analysing Best Practices in Technology Education* with Marc de Vries, Rod Custer and Gene Martin in 2007 and *Reflections on Technology for Educational Practitioners with* Marc de Vries and Jonas Hallstöm in 2019. John has been an invited speaker to many events in different countries, including Denmark, England, The Netherlands, Finland, Sweden, France and the United States. He continues to write about technology education and technological literacy but now emphasises the philosophy of Deleuze and Guattari. His latest book is *A Nomadic Pedagogy About Technology: Teaching the Ongoing Process of Becoming Technologically Literate*, published by Brill in 2022–23.

Christian Detweiler
studies emerging technologies and the design practices and technological cultures from which these emerge. His work focuses on developing methods that help designers analyse technologies and account for values in design. He has a BA in Media Studies from the University of Amsterdam and an MSc in Media Technology from Leiden University. He did his PhD on accounting for values in design at the Ethics/Philosophy of Technology and Interactive Intelligence research groups at Delft University of Technology. He is the head of the Philosophy and Professional Practice Research Group at The Hague University of Applied Sciences (https://www.thehagueuniversity.com/research/research-groups/details/philosophy-and-professional-practice). He also works as an educator in a User Experience Design program, where he explores critical practices with design students.

Andrew Doyle
is Lecturer in Technology Education at the University of Waikato, where he teaches across a broad range of initial teacher education programs. Before New Zealand, Andrew worked in or studied technology education in the Irish, Swedish, and English national contexts. In 2020 he defended his doctoral thesis at the Department of Learning in Engineering Sciences at KTH Royal Institute of Technology, Stockholm. His research is focused on better understanding the nature of technology for education, including teachers' representations of the purpose(s) of teaching technology.

Wendy Fox-Turnbull
is Associate Dean of Academic Education at the University of Waikato, Hamilton, New Zealand.

Jonas Hallström
is Professor of Technology Education at the Department of Behavioural Sciences and Learning, Linköping University, Sweden. His research concerns historical, philosophical, and educational aspects of technology, with a

particular focus on technological systems, computational thinking, artificial intelligence, modelling, and design. His most recent book is *Programming and Computational Thinking in Technology Education: Swedish and International Perspectives* on Brill (2024, co-edited with Marc J. de Vries).

Alison Hardy
is an associate professor at Nottingham Trent University's Institute of Education, where she teaches and supervises research students. Alison's research focuses on investigating the values attributed to design and technology (D&T) education by different stakeholders.

The other main focus of Alison's work is shaping the D&T subject through her leadership of Archer Exchange, a group of leading D&T experts who shape the subject for the modern day through consultancy, training, research, and evaluation. She works closely with teachers to build and sustain the D&T research community, encouraging them to share their research and contribute to shaping the direction of future D&T research. In 2022, Alison led the Building Research Capacity and Engagement (BRaCE) project, involving D&T teachers in the PATT conference to further these goals.

Alison has edited "Learning to Teach Design and Technology" and "Debates in Design and Technology Education," and shares her ideas through her website, dralisonhardy.com, and the "Talking D&T" podcast. She serves on editorial boards, advises the Department for Education on the D&T curriculum, and has been a trustee for the D&T Association. Alison's passion for education and commitment to the D&T community make her a dedicated professional in the field.

Eva Hartell
currently serves as the head of research in Haninge municipality and works as a researcher both at the KTH Royal Institute of Technology in Sweden and the University of Bergen in Norway.

With a doctorate in educational assessment from KTH in 2015, her primary focus lies in bridging teaching and learning by embedding formative assessment with or without the support of digital technologies. She collaborates closely with teachers and schools in various national and international practitioner-based research and innovation projects, funded by organisations like the European Union and the British Council. She shares her expertise widely through publishing– including international handbooks on digital learning and STEM education and teacher magazines and through supervision, workshops, and keynotes.

Her passion for evidence-informed education originates from her background as a distinguished K-12 STEM teacher, providing the foundation for her ongoing mission to connect educational research with school practices.

Bio https://www.kth.se/profile/ehartell

K-ULF project https://www.kth.se/en/k-ulf.

Social media X (former Twitter) @EvaHartell and Instagram @evahartell & @tillit_till_it.

Pasi Ikonen
is a university teacher in Technology Education. He has been working as a University Teacher at the University of Jyväskylä since 2003. Over his career, he has been involved in several national and international in-service training programs. Mr. Ikonen also has experience in international scientific and teacher professional development projects in several countries. His main research interests are teacher education and technology education.

Henk Jochemsen
(PhD) is, among others, a professor emeritus of Christian philosophy at Wageningen University & Research (2008–2018). From 2002–2006 he was the chair of the research group Ethics of Care at the Ede Christian University of Applied Sciences, of which Johan Hegeman was the assistant chair. In this function, he worked on the topic of this contribution.

Alister Jones
was appointed Deputy Vice-Chancellor of the University of Waikato in 2011. He is a former Dean of Education and was previously a Research Professor and Director of the Wilf Malcolm Institute of Educational Research. He has managed and directed research projects that have informed policy, curriculum, and teacher development in New Zealand and internationally. His main areas of research are curriculum, teaching, learning and assessment and he has over 300 publications. In 2000, he was awarded the NZ Science and Technology Medal for his significant contribution to the development of technology education both nationally and internationally. His current portfolio includes international engagement and delivery with a strong focus on the Asia-Pacific regions, including the development of off-shore programmes and campuses.

Hanna Kauppinen
is a university teacher in Multi-material Craft Education in the Department of Teacher Education at the University of Jyväskylä. She worked as a teacher

of textile crafts, and multi-material crafts and as a class teacher in a comprehensive school before coming to the university in 2023. Her research is in its early stages, but she is interested in researching learning through crafts and the future of crafts in primary education.

Richard Kimbell

his specialist interest is in design learning and its interaction with assessment. He founded the Technology Education Research Unit (TERU) at Goldsmiths University of London in 1990, and over 25 years ran research projects for research councils (eg ESRC, NSF [USA]), for industry (eg LEGO, BP), for government departments (eg DfES, DfID), as well as for professional and charitable organisations (eg Engineering Council, Royal Society of Arts, Design Museum). Kimbell has supervised 14 PhDs to successful completion and acted as external examiner for 26 others in universities all over the world. He has published widely in the field of technology education, including six single-authored books, many chapters for edited compilations, as well as reports commissioned by UK Government Departments, the Congress of the United States, UNESCO and NATO. He has written and presented television programmes and regularly lectures internationally. He has been a consultant to the National Academy of Engineering and the National Science Foundation in the USA, and a Visiting Professor at the University of British Columbia, Canada; the University of Stockholm, Sweden; Edith Cowan University, Perth Australia, Texas A&M University, USA; and the University of the Shannon, Ireland.

Roel Kuiper

studied history and philosophy at the Vrije Universiteit Amsterdam. His expertise is in the history and philosophy of the antirevolutionary and neo-Calvinist movements in the 19th and 20th centuries and also in the Reformational Philosophy of the school of Herman Dooyeweerd.

In 1998 he was appointed professor of Christian Philosophy at the Erasmus University in Rotterdam. In 2014 he became a professor at the Theological University Kampen|Utrecht, where he was also the president of the board in the years 2015–2022. He is currently leading a Center of Expertise about *Education and Identity*.

Main publications

Roel Kuiper published widely in the areas of philosophy, political thought and education. In 2009 Roel Kuiper published *Moreel Kapitaal* (*Moral Capital*), a philosophical approach to modern society and its problems of cohesion in a globalised world.

Later he published books on Christian politics (2013), Public interests (2014), Identity (2015) and Christian education. His latest book is about Citizenship (2023).

Other occupations

In the years 2007–2019 he was a member of the Dutch Senate for the Christian Union. He chaired several committees and boards in the Netherlands. He is also an advisor of the Dutch Department of Education and its Minister in the area of primary and secondary education.

Mike Martin

is a senior lecturer in Teacher Education at Liverpool John Moores University. Mike has been involved in Design and Technology for 30 years as a teacher and teacher educator across primary, secondary and masters level programmes. Mike has regularly presented at international conferences and written book chapters with a particular focus on values and sustainability. His doctoral study used phenomenology to explore the development of student teachers' subject knowledge. More recently his attention has turned to the philosophy of technology and our relationships with technology as makers and users. When not sketching, tinkering, 3D printing or exploring the use of sustainable materials for making at home he can be found dinghy sailing.

Carl Mitcham

is a philosopher of engineering and technology, Professor Emeritus of Humanities, Arts, and Social Sciences at the Colorado School of Mines and Visiting International Professor of Philosophy of Technology at Renmin University of China.

Sonja Niiranen

is a university lecturer in Engineering Pedagogy. She received her master's and doctoral degrees in education in 2006 and 2016, respectively. She worked for 13 years as a technical craft and technology education teacher in the Department of Teacher Education at the University of Jyväskylä. She is currently working in the Faculty of Education and Culture at Tampere University as part of a PGL (Professional Growth and Learning) research group and delivering courses on university pedagogy for the academic staff members at the university.

Charlotta Nordlöf

is an associate senior lecturer at Linköping University, Sweden. She holds a PhD in Technology education and has a background as a technology teacher in upper secondary school. Her research interests are focused on educational and didactic aspects of technology education, with a particular emphasis on teachers' views of technology education. Her thesis examined teachers' attitudes towards technology education and their views on technological knowledge. Currently, she is involved in research projects on authentic STEM education and AI literacy in primary school.

Aki Rasinen
is a university lecturer in Technology Education, Emeritus. He has studied to become a technical subject teacher, BA, MA, LicEd, and PhD. He has worked as a primary school and high school teacher, university lecturer, project manager, curriculum developer, and researcher mainly in Finland but also nine years in Southern African countries. His research interests are technology education curriculum; girls' involvement in technology studies; problem-solving, creativity and innovation; STEAM, and holistic learning.

Philip A. Reed
PhD, DTE, is Professor in the Darden College of Education and Professional Studies at Old Dominion University in Norfolk, Virginia. He has served as a teacher, teacher educator, and administrator in secondary and post-secondary education for over thirty years. Dr. Reed was the 2020–2021 President of the International Technology and Engineering Educators Association (ITEEA). His research focuses on curriculum development and implementation in workforce education.

Timo Rissanen
is a university teacher in Technology Education in the Department of Teacher Education at the University of Jyväskylä. Before coming to the university, he worked as a comprehensive school teacher. His main interest is developing new ways of learning craft and technology education in collaboration with the school teachers. In the field of research, he is interested in curriculums and also learning by doing in the field of craft and technology education.

John M. Ritz
Professor Emeriti, Old Dominion University, Norfolk, VA, USA, has contributed to the technology education profession for over 50 years. He has taught technological literacy, curriculum development, and research methods at the university. He has conducted academic and sponsored research projects with local, national, and international participants and published papers

and provided presentations on his findings. Two of his most cited articles include *Research Needs for Technology Education: An International Perspective* and *STEM and Technology Education: An International State-of-the-Art*. He is a member of the Academy of Fellows for the International Technology and Engineering Educators Association and the Academy of Fellows for the Technical Foundation of America.

Hilda Ruth Beaumont

has worked in education for some 50 years, including 15 years as a teacher, and for most of that time was known as David Barlex. David became an acknowledged leader in design & technology education, curriculum design and curriculum materials development through working as the director of the Nuffield Design and Technology Projects. Recently David identified disruptive technologies relevant to the school curriculum and ways in which young people may be engaged with these. His publication with Frank Banks *Teaching STEM in the secondary school* was very well received with a second edition being published in 2020. His most recent publication with Torben Steeg is *Design and Technology in your School Principles for Curriculum, Pedagogy and Assessment*, published by Routedge in 2024. Since 2021 David has found that he is much, much happier and at ease with himself if dressed and presenting as female. Hence, she is now living full time as HildaRuth and continuing with both academic writing and consultancy under the name of Hilda Ruth Beaumont as an Honorary Senior Research Associate at the Institute of Education UCL.

Marion Rutland

(PhD) is a former Honorary Research Fellow and Principal Lecturer in Design and Technology at the School of Education, University of Roehampton, London. Her roles at the University included Design and Technology Curriculum Area Co-ordinator, Course Leader for the PGCE Design and Technology Secondary Programme, Curriculum Leader for the Primary Design and Technology, Design and Technology Pathway Leader and Tutor for the MA Education (Design and Technology) Programme and PhD Supervisor and External PhD External Examiner. Prior to this she taught food-related subjects for 20 years in a range of secondary schools in the UK and Australia and was an Advisory Teacher for Information Communication Technology in London. She was a President of the National Association of Teachers of Home Economics and Technology (NATHE). Her search interests include food education in schools, curriculum development, information technology (IT), creativity and teaching and learning in design and technology.

Elwin Savelsbergh
is a professor of Curriculum Development at the Utrecht University of Applied Sciences and associate professor at Utrecht University. Key themes in his research program are (technological) literacy, education for sustainable development, citizenship, and pedagogical content knowledge.

Alice Schut
her research interests revolve around understanding and stimulating creative processes in a design context. She did her PhD on design fixation and design feedback at the Science Education and Communication research group at Delft University of Technology. She is currently a researcher at the Philosophy & Professional Practice research group at The Hague University of Applied Sciences. Here, she researches the creative (im)possibilities of designing with generative AI. Additionally, she works as a lead educator in the international User Experience Design Program.

David Spendlove
is Associate Dean for the Faculty of Humanities and Professor of Education at the University of Manchester. Prior to entering Higher Education David was a successful teacher and school leader working nationally on numerous projects. Since moving into higher education, David has worked nationally and internationally on research, policy and curriculum development. He has developed research in a number of areas, including design and technology, creativity, teaching and learning, teacher professionalism and the politics of teacher education.

Kendall N. Starkweather
PhD, CAE, DTE is the Executive Director/CEO, Emeritus of the International Technology & Engineering Educators Association.

Steve Keirl
has worked in the Universities of Tasmania, South Australia and Goldsmiths, London over a period of 25 years. His teaching and research have embraced Design and Technology curriculum theory, philosophy of technology, philosophy of design, ethics, and critical-creative pedagogical practices for advancing technological literacy for democratic life. He promotes critical approaches to understanding our human-existential relationships with technologies and he calls for much deeper education about these relationships. Steve believes that the question of just what it means 'to be human', and to what extent we value the concept, needs urgent attention.

Underpinning this position is a need for serious ethical questioning of our designed world and how it comes to 'be' and Steve calls for resistance to (technological) determinism and its ability to deny hope and human agency towards preferred futures. He also resists capitalist-driven curricula models which marginalise creativity, criticism, the humanities and sustainability.

Maarten Verkerk
(1953) studied chemistry at Utrecht University. In 1982, he obtained his PhD in materials science at the University of Twente. In 2004, he defended his second PhD *Trust and Power on the Shop Floor* (Eburon, 2004) at Maastricht University.

After his first PhD, he conducted fundamental research (Philips Research), managed factories domestically and internationally (Royal Philips), led a psychiatric hospital (Vijverdal, Maastricht), and managed an innovation network in healthcare (VitaValley, Ede).

From 2004 to 2017, he was a professor of Christian Philosophy at the Eindhoven University of Technology. In the period from 2008 to 2019, he held the same position at Maastricht University.

Maarten was a member of the Senate from 2019–2023. Nowadays, he is an unpaid chairman of various foundations and committees.

Maarten has written many articles in national and international journals. He has authored, co-authored, and edited many books. Amongst others, *Breakthrough. From Innovation to Impact* (Owls Foundation, 2014), *Philosophy of Technology. An introduction for technology and business students* (Routledge, 2016), *Ecclesiastes for Managers* (Dordt College Press, 2018), *The Normative Nature of Social Practices and Ethics in Professional Environments* (IGI Global, 2019), *Process Intensification* (De Gruyter, 2020) and *Interdisciplinary and Social Nature of Engineering Practices* (Springer, 2022).

Maarten van der Sanden
PhD, runs his own design workshop called LUS (Dutch for loop, and refers to 'looped thinking' as a part of design thinking and doing in complexity). He has worked for the Delft University of Technology for 25 years as team lead of science communication at the university's corporate office and as an associate professor in Communication Design for Innovation, which was a part of the Department of Science Education and Communication led by Marc de Vries. As for his workshop Maarten is fascinated by how professionals learn and communicate together (and make decision thereof) under deep uncertainty and sometimes high pressure, concerning transformative changes in which technology and society are intertwined. He designs processes and tools to support

this kind of decision making. Marc and Maarten edited a book in 2016 in which they unveiled the synergy between science education and communication.

Gerald van Dijk
is a senior lecturer and researcher at the University of Applied Sciences Utrecht. His research interests include design- and maker education, language-sensitive STEM education, education for sustainable development (ESD) and teacher education.

CHAPTER 1

Introduction to This Festschrift for Marc J. de Vries

A Book Full of Dreams about the Future of Technology Education

John R. Dakers

1 Prologue

In 1984, a research project was begun in The Netherlands to determine the attitude towards and concepts regarding technology held by students aged 12–15. An instrument was developed, and a large-scale sample was taken. The results of the study were so striking that the decision was taken to present these at several international conferences and to promote the international extension of the research. One of the important findings was the following:

> Technology is very important in our society. Technology influences all aspects of society (economy, politics, social relations, the way we work, live, and travel.
>
> Technology, however, has had hardly any influence on the curriculum of general education. This is not a good situation. General education should prepare pupils for living in a technical society.

The quotation above was written in 1986 by the then 28-year-old, Marc J. deVries. It formed part of his contribution to the first-ever international Pupils Attitudes Towards Technology (PATT) workshop, as it was then known, held at Eindhoven University of Technology in the Netherlands. Participants attended from Austria, Canada, France, Kenya, Poland, Sweden, the United Kingdom, and the Netherlands. The proceedings entitled *What do Girls and Boys Think of Technology*? were edited by Jan H Raat and Marc de Vries.

In the following year, Dr. William E. Dugger from Virginia Tech attended a conference of people working with PATT and suggested replicating the PATT study in the United States. Consequently, in the Spring of 1988, Dr Marc deVries worked with Drs. E. Allen Bame and William E Dugger to develop a PATT instrument for use in the United States.[1] A Journal article entitled "Pupils' Attitudes Toward Technology-PATT-USA" was published in the *Journal of Technology Studies*, in 1993 (Bame, Dugger, de Vries, & McBee, 1993). And the rest, as they say, is history.

© JOHN R. DAKERS, 2025 | DOI:10.1163/9789004710948_001

2 Marc's Contribution to Technology Education

The contribution of Marc deVries to technology education is wide-ranging, edifying, inclusive, inspiring, remarkable, or, put simply, colossal.

On December 14, 2023, Marc was installed as a senator for the SGP (Reformed Political Party), the oldest political party in the Netherlands. He succeeded Diederik van Dijk, who had become a member of the House of Representatives. That illustrates the remarkable bit. Going back a few years (many years actually), illustrates all the other bits. Marc studied experimental physics at the Vrije Universiteit Amsterdam (Free University Amsterdam) (M.Sc. 1982) and specialised in educational research in physics education. He received his PhD at the Technische Universiteit Eindhoven (Eindhoven University of Technology) in 1988 for a dissertation on *Technology in Physics Education.* This study implied a move towards a new specialisation: educational research in technology education. In that context, he organised international conferences under the title of Pupils' Attitudes Towards Technology (PATT). This name is related to the part of his PhD research regarding the ideas that pupils of 12–13 years of age have about technology. This series of conferences continues to this day. In 1988, he worked on setting up a new teacher education programme for technology teachers at the Pedagogische Technologische Hogeschool (Pedagogical Technological College) in Eindhoven. Later, he became Head of the department for this program. His involvement in the international developments in technology education caused him to be invited for presentations at international conferences and advisory work for organisations such as OECD, Unesco, and the Council of Europe, as well as for universities in various countries (In Europe, Australia, Chile, Hong Kong, Israel, Taiwan, South-Africa and the USA).

In 1990, Marc moved to Eindhoven University of Technology to become an assistant professor in the philosophy and methodology of technology, with a specialisation in design methodology, under Professor Andries Sarlemijn. In 1992, he organised an international conference on 'Design Methodology and Relationships with Science' in the context of the NATO Scientific Affairs Programme, together with Professor Nigel Cross (Open University England) and Professor D.P. Grant (California Polytechnic University), both editors of international scholarly journals in the field of design methodology.

From 1995 to 2000, he worked for the Stichting Historie der Techniek (Foundation for the History of Technology) in the Netherlands, to write the history of the Philips Natuurkundig Laboratorium (nowadays part of Phillips Research).

His move back to philosophy coincided with the appointment of Professor Anthonie Meijers as the successor of Andries Sarlemijn, who sadly died in 1996. Since then, his field of specialisation has been philosophical research into the nature of technological knowledge. This field is related to his previous fields of specialisation: transfer of technological knowledge (educational research in technology education); use of technological knowledge in design processes (design methodology); and development of technological knowledge (history of technology). In 2002, he co-organised an international conference entitled 'Technological Knowledge: Philosophical Reflections'.

Meanwhile, his work in technology education continued. Marc is currently the editor-in-chief of the International Journal for Technology and Design Education (now published by SpringerNature) and the co-author of a series of Dutch textbooks for technology education in lower secondary education (the title of the series is "Techno*logisch*", i.e. Techno*logical*). He co-founded the Dutch Association of Technology Teachers (Dutch abbreviation: VeDoTech).

Marc is both a Professor of Christian Philosophy of technology and a Professor of Science Education at Delft University of Technology (TUDelft). He was an affiliate Professor of philosophy and pedagogy of technology and engineering in education at the KHT Royal Institute of Technology in Stockholm, Sweden, until 2019.

3 Awards

- The Laureate Citation Award of the International Honorary for Professions in Technology 'Epsilon Pi Tau', in 1998.
- The Special Recognition Award of the International Technology Education Association (ITEA) in 1991.
- Silvius-Wolansky Award (ITEA) for the publication of the book, Analyzing Best Practices in Technology Education, 2008.
- Academy of Fellows Membership (ITEA), 2010.
- Member of Honor Dutch Technology Teachers' Association (VeDoTech), 2010.
- Council of Technology Teacher Education Outstanding Research Award 2011 (with Ammeret Rossouw and Michael Hacker).
- International Development and Support of Design and Technology Award (DATA), 2012.
- Technical Foundation of America (TFA) Fellow Award, Technical Foundation of America, 2015.

4 Other Professional Achievements

- Member of the editorial board of the International Journal for Design and Technology Education, and since 1999, Editor-in-chief.

- Editor-in-chief of the International Technology Education Studies book series (Sense Publishers).
- Member of the editorial board; Journal for Technology Education.
- Member of the editorial board; Design and Technology Education: An International Journal.
- Member of the editorial board; Philosophia Reformata (the international journal for reformational philosophy).
- Chairman of the PATT-Foundation (international platform for technology education).
- Co-organiser of conferences PATT-1 to PATT-40 and three NATO ARW conferences, co-editor of the proceedings of these conferences, co-organiser of 7th International IOSTE Symposium, co-organiser of Second Jerusalem International Science and Technology Education Conference.
- Organiser of the international conference 'Technological Knowledge: Philosophical Reflections', Boxmeer 2002.
- Co-founder of the Dutch Association of Technology Teachers (plus secretary for a period of four years and a member of the editorial board of its journal Terugkoppeling)
- Advisory work for OECD (project Science Math Technology Education, the Curriculum Redefined), European Commission (DG XII: TSER Programme), Technology Enhancement Project (UK), ORT-STEP (South Africa), ORT-Israel, general reporter for science and technology teacher education conference Council of Europe, external expert for ATEP project Fontys Hogescholen, member committee on Assessing Technological Literacy of the National Academy of Engineering (USA).
- Invited lecturer about technology education and philosophy of technology at many (inter-)national conferences and institutes.

Marc has published, both independently as well as collaboratively with others, over 70 refereed journal articles, over 125 chapters in books, many edited collections as well as several monographs. He has written and edited more academic papers than, frankly, I can count. Some books of note include;

- Handbook of Technology Education (Springer International Handbooks of Education) 1st ed. 2018.
- Philosophy of Technology: An Introduction for Technology and Business Students 1st Edition 2015, with Maarten J Verkerk, Jan Hoogland and Jan van der Stoep.
- Innovation Research in Technology and Engineering Management: A Philosophical Approach (ISSN) 1st Edition. 2021.

- The Normative Nature of Social Practices and Ethics in Professional Environments (Advances in Religious and Cultural Studies) 1st Edition. 2019, with Henk Jochemsen.
- Teaching about Technology: An Introduction to the Philosophy of Technology for Non-philosophers (Contemporary Issues in Technology Education) 2016.
- Norms in Technology (Philosophy of Engineering and Technology Book 9) 2012, with Sven Ove Hansson and Anthonie W M Meijers.
- International Handbook of Research and Development in Technology Education 2009, with Alister Jones.
- Shaping Concepts of Technology: From Philosophical Perspective to Mental Images. 1997 with Arley Tamir.

Marc has supervised many post-graduate PhD students, and as far as I am aware, he has attended and spoken at every PATT conference, including the ones held in the USA at the Annual ITEEA conferences. The most recent PATT conference, the 41st, was held in the school of Education and Science, Nanjing Normal University, Nanjing, China, which also falls under the rubric of remarkable!

Marc J. de Vries is a professor of Reformational philosophy, an educator, writer, researcher, and philosopher of technology.

For someone who has given so much to technology education and the philosophy of technology over the years, it has been my privilege and honour to organise and edit this Festschrift dedicated to Marc. It has also been my great privilege to work with so many wonderful colleagues from around the world, who helped make this work possible.

5 Editing a Festschrift

Never having contributed to, or edited a Festschrift before, this has been an interesting adventure for me, full of ups and downs, but mostly ups. The principal 'down' began almost immediately after I had signed the contract with Brill Publishers. What I had to do seemed relatively simple. This was to get in touch with everyone whose names I had written down on an extensive list that I had been working on for weeks. These were people that I knew had known and worked with Marc over the years, and there were many of them. I simply had to email them and ask them if they would like to contribute. I did not anticipate any problems that might arise during this initial foray into the preparation of a Festschrift. What could possibly go wrong?

Well, it turned out that like me, most of those named on my list had retired or left their universities and were, as a result, very difficult, if not impossible, to contact. For those that I did not manage to contact, I offer my sincere apologies. Those that I did manage to contact, however, set into motion the genesis of the 'ups' side of this adventure. Everyone contacted replied with a resounding 'Yes', together with 'Thank you for asking me' and 'It would be an honour to contribute to this Festschrift for Marc'. I know that had I managed to contact the many other colleagues on my list, I would undoubtedly have needed to negotiate a second volume with Brill.

It was in September 2023, that I initiated a conversation with John Bennett from Brill Publishers about the possible publication of a Festschrift for Marc de Vries. I am delighted to say that, on behalf of the publishers, he was very enthusiastic about this proposal. Subsequently, a formal book contract was offered to take this project forward.

Following the signing of the contract, I had a long conversation with Marc, seeking his guidance on how he would like me to take this project forward. He was keen that it should become a collective project in which contributors would write about their own conceptualisations of the future progress of technology education. We concluded that it should, for the most part, be a prospective volume, one that involved thinking through technology, as Carl Mitcham advocates. Marc, being Marc, was less enthusiastic about colleagues extolling his virtues.

6 Defining a Festchrift

My research on the definition of a Festschrift led to the unveiling of many variations. However, the one that most impressed me, was that provided by the late American sociologist Irving Louis Horowitz, who defined a Festschrift as "a call to further work, effort, and energy, a call to the improvement of learning, of a discipline, a science, an artistic vision, or an intellectual position" (Horowitz, 2017, p. 237). This was exactly what Marc wanted.

6.1 *The Call for Contributions*

Over the years, research into technology education has advocated for many modifications in its delivery. These have ranged widely, from, at one end, the abolition of domestic science and technical education delivered as traditional gender-specific subjects, to the introduction of technological literacy and more recently, the international push for STEM/STEAM education, at the other. With the possible exception of food and textile technology, technology education continues to retain a vocational orientation, one seeking to stimulate student interest in, and consequent uptake of, engineering-orientated occupations. Despite the many calls for change in the research literature, technology education as delivered in the classroom continues, stubbornly, to frustrate its evolution into a subject more in alignment with the technologically textured world that we now inhabit. This is especially evident in light of the exponential speed at which technology is evolving today. This is unfortunate because the evolutional aspects of technology have consequences for humans, This, was, for example clearly revealed by Klaus Schwab, the founder and executive chairman of the World Economic Forum. Technologies, he argues, whether considered good or bad, impact on society. Technology is crucially, for Schwab, standing at a crossroads;

> If we miss this window of opportunity to shape new technologies in ways that promote the common good, enhance human dignity and protect the environment, there is a good chance that the challenges we experience today will only be exacerbated, as narrow interests and biased systems further entrench inequalities and compromise the rights of people in every country. (Schwab, 2018, p. x)

In terms of technology education, we are also standing at a crossroads. This will remain problematic for as long as policymakers and their like, continue to ignore this aspect of education about technology. This was indeed made manifest some time ago in the *International Handbook of Technology Education: Reviewing the Past Twenty Years,* edited by Marc J. de Vries and Ilja Mottier, published in 2006 by Sense Publishers, in which Marc lamented that:

> We need to work on a technology education practice that convinces governments, school administrators and school boards, teachers, parents, and pupils that teaching and learning about technology is not only fun but also a necessary component of every (future) citizen's general education. (de Vries, 2006, p. 9)

Sadly, 12 years later in the 2018 *Springer Handbook on Technology Education* Marc lamented yet again:

> [h]opefully in due time, there will be a second International Handbook of Technology Education with a new chapter in the history of technology education that will show that technology education has been able to overcome the hurdles of survival and flourishes in many countries worldwide. (de Vries, 2018a, p. 83)

I am aware that it is not only Marc who holds these views. I have spoken with many colleagues who are equally frustrated with Governments, policymakers and curriculum developers, who continue to turn a blind eye to many of the research-led innovations that promote a technology education that is more in tune with the technologically textured world we now occupy. To that end, I made available, in my call for chapters, a prospective category seeking contributions, that could be written under the aegis of excluding, ignoring, or at least minimising the type of external, hierarchical forces that have, hitherto, resulted in the stultification of creative and innovative conceptual thinking about what technology education might become. This was to free up any authors who wished to write chapters about a future technology education that would serve to meet the needs of what Klause Schwab calls the fourth industrial revolution, one in which emerging technologies – from artificial intelligence to biotechnologies, advanced materials to quantum computing – are driving radical shifts in the way we all live (Schwab, 2018, p. IX). This would, moreover, involve a technology education paradigm that engages young people in conversations about what Kurzweil refers to as the Singularity, an era in which humans transcend Biology (Kurzweil, 2008). The singularity for Kurzweil is "a future period during which the pace of technological change will be so rapid, its impact so deep, that human life will be irreversibly transformed" (Kurzweil, 2008, p. 7).

This 'prospective category' is manifest in the majority of chapters now presented in this book. Whilst they are not all in agreement, they do acknowledge collectively, that a radical change in the delivery of technology education for the future is much needed. They are not alone in this thinking. None other than Bob Dylan has identified this as a problem;

> It's peculiar and unnerving in a way to see so many young people walking around with cell phones and iPods in their ears and so wrapped up in media and video games ... It's a shame to see them so tuned out to real life. Of course, they are free to do that, as if that's got anything to do with freedom. The cost of liberty is high, and young people should understand that before they start spending their lives with all those gadgets. (Bob Dylan, quoted in Kahn, 2010, p. 84)

The majority of chapters have been written in a style chosen by each individual author, based upon their degree of willingness to express their own ideas on the creation of a technology-education curricular paradigm more suited to the 21st century: one that transcends the pre-established mores that continue to dictate policy today. These chapters have served to liberate the thinking of colleagues from their dissatisfaction with the current state of affairs as they strive towards a better one. They have given their vision of technology education, free from the usual constraints and hindrances. Essentially, the chapters provide a free-thinking, ideal and futuristic vision of what each of the authors believes technology education could/should comprise (whilst remaining within the bounds of what may be possible in the real world!).

The chapters have been subject to my editorial review but with a light touch. I have changed the spelling to English U.K. for all chapters and carried out some minor changes, for the sake of clarity, where authors have English U.K. as a second language.

What makes this collection unique is the variety of established published experts it includes- from the fields of technology education, the philosophy of technology and the Reformational philosophy of technology. These contributors, who have all worked with Marc in some form, have provided their vision of what technology education might become, or why they think issues about ethics and norms are important areas for learning about technology. These visions are many and varied. Some authors, for example, argue that STEM education is the way forward. Others may prefer STEAM education, while others again may wish sustainability to be the core philosophy. Some prefer to claim that the impact of technology on global warming should be given a prominent place in the technology education curriculum. Some advocate for technology education to be linked to industry, whilst others (like me for example), argue against this perspective. Some place emphasis on the need for a more Christian perspective on the colossal impact of technology on societies around the world. There is much more than can be listed here. A editor, however, I have resisted being prescriptive; I am aware that many colleagues have strong views and differing perspectives on what constitutes a technology education curriculum for the future. As such I have chosen not to classify or organise the chapters into specific groupings. I have, rather, placed them in no particular order at all. I recognise that as editor, I will have had some influence on the placement of the chapters, even if only subliminally, but I intended to place them randomly. To that end, I hope that readers will choose to read the chapters in whatever way they wish. There is, consequently, no formal beginning or end with only two exceptions, this introductory chapter and Marc's closing chapter.

I would just add that this is a Festschrift that celebrates the contribution of Marc de Vries to the field. As a result, I asked contributors to enjoy writing their chapters and that they should, therefore, adopt a style that would shine a light through the dusty corridors of serious academe!

7 Why Is the Philosophy of Technology Important for Marc de Vries?

Whilst it is well known that Marc has been and continues to be, a strong advocate for technology education in general, his primary focus has always been an analytical perspective on the philosophy of technology. His influences are many and varied, but three major influences relate particularly to a philosophy of technology that calls attention to modern engineering. They are represented in the theories of the Dutch Christian philosopher Herman Dooyeweerd, along with Hendrik van Riessen and Egbert Schuurman. In the four books, discussed below, two are monographs in which Marc discusses Dooyeweerd and Riessen in much more detail and I would recommend these books to those who may not have encountered them. The two others are co-authored with colleagues of Marc.

In his book entitled *Teaching About Technology: An Introduction to the Philosophy of Technology for Non-Philosophers* Marc discusses numerous philosophical aspects suitable for introduction into the pedagogy for teaching technology education. He argues, however, that the type of pedagogy employed will be radically different from that which is normally utilised:

> To help learners acquire an understanding of this aspect [of technology], they need to be stimulated to reflect on this aspect themselves and to develop their own perspectives. Role play and discussion groups can be appropriate strategies for that.

In their book; *Philosophy of Technology: An Introduction for Technology and Business Students*, Marc and his colleagues set out a comprehensive account of the link between Philosophy, Technology and Globalisation. In addition, they give 14 informative portraits of philosophers of technology. One particular and important question they ask is how philosophical reflection can be useful to an engineer.

In answering this question, they differentiate between three functions. The first relates to having an analytical role, the second involves taking a critical look at the reality surrounding the engineering project and the third necessitates an ethical dimension. In answer to the ethical question, they answer thus:

> A critical analysis of culture and society automatically leads to the question 'Well, then how' (should it be seen)? The directional function of philosophy is of great significance for technology. It concerns questions like 'What is a good apparatus?', 'What is the influence of technology on human beings and society?', 'When may we apply a certain technique and when not?', 'What are the effects of new techniques on the environment?', and 'To what extent can technology solve problems in our society?' All

> these types of questions are encountered in the ethics of technology. The directional function of philosophy is closely linked to the analytical and critical functions. The ethics of technology can only be properly developed on the basis of the analysis of technology and in dialogue with the criticism of technology. Thus, the ethics of technology will also have to consider seriously the question to what extent modern technology endangers the freedom of Western humanity. It can only have a meaningful reflection by extensively analysing modern technology and identifying the (possible) social implications in a critical manner. In this way only will the ethics of technology be able to express itself normatively on the development of modern technology. (Verkerk et al., 2016, p. 6)

Innovation Research in Technology and Engineering Management: A Philosophical Approach, published in 2021, offers a more in-depth philosophical account of a technology educational approach that asks thought-provoking questions relating to technological understanding like: "[t]echnological and social developments are closely intertwined. There is an influence of technology on society and vice versa. This mutual relationship is something that needs careful consideration in product development" (de Vries, 2021, p. 28). Citing Apostel, Marc introduces the concept of differing worldviews and how they can affect the different perspectives that individuals may have concerning technology.

Apostel, who was a Belgian philosopher, was a great advocate for a fusion of the exact sciences and the humanities. This was in direct contrast to C.P. Snow's two cultures argument. In discussing the importance of worldviews relating to technology, Marc cites Apostel's concept of 'worldview' as following six philosophical subdisciplines:

1. What is? This ontological question leads to a model of reality as a whole.
2. Where does it all come from? This explanation question yields a model of the past.
3. Where are we going? This prediction question results in a model of the future.
4. What is good and what is evil? This axiological question is answered in a theory of values.
5. How should we act? This praxeological question leads to a theory of actions.
6. What is true and what is false? This epistemological question is answered in a theory of knowledge. (de Vries, 2021, p. 17)

Finally, in their book *Reflections on Technology for Educational Practitioners: Philosophers of Technology Inspiring Technology Education*, Marc and his colleagues put together a book, the content of which is clearly expressed in the title. The book unpacks the concept of what Donald Schön refers to as 'Reflective Practitioners. What the four books discussed bring to light is that the philosophy of technology serves to enhance technology education by engaging students in the wider context of the impact technology has on society.

What becomes evident from the four books mentioned above is that for Marc, technology education for the future must be more than simply a vocationally oriented, craft-based subject domain with a specific emphasis on engineering subjects, all in the interest of the economy. This is revealed in much more detail in the books. This is in line with the arguments that have frequently been put forward by educationalists. The following, for example, represent just some who have been arguing against the vocationalisation of education for years: (Rousseau, 1991; Semetsky, 2006; Freire, 1996; Freinet, 1967; Dewey, 1916).

The imagined liaison between school and industry in the interests of economic wealth production is fallacious and tantamount to sophistry. The technical education I experienced in the 1960s has changed very little between then and now. The products I fabricated in woodwork, metalwork or technical drawing at that time, differ very little from those fabricated by my sixteen-year-old granddaughter during the past few years. Computer-aided design and manufacture (CAD-CAM) may be new, but the emphasis continues to be placed on hand tool production and the reproduction of the same technical drawings. What has changed exponentially in the world outside school, however, is the evolution of technology and the consequential affect this is having on the human/technology relationship (See Bob Dylan above.)

Marc offers many examples in his books to help clarify some of the philosophical issues under discussion. One of his examples that features in several books, chapters and papers is the dual aspect of technology. In his now-famous example, he discusses the dual nature of a screwdriver which, although designed to fix or remove screws, might be used to open a tin of paint or be used as a weapon. Other examples that complement Marc's philosophy can be found, for example in Ray Kurzweil's book *The Singularity Is Near: When Humans Transcend Biology*. This deals with Artificial Intelligence and the future of humanity. For Kurzweil, the singularity is;

> The future period during which the pace of technological change will be so rapid, its impact so deep, that human life will be irreversibly transformed. Although neither utopian nor dystopian, this epoch will transform the concepts that we rely on to give meaning to our lives, from business models to the cycle of human life, including death itself. … the pace of change of our human-created technology is accelerating at an exponential pace. (Kurzweil, 2008, pp. 7–8)

The evolution of technology over time, from the use of simple craft tools through the production of artefacts by hand, to the development of machine fabrication, coupled, as they have become, with design, has formed the basis of technology education as we know it. However, with the advent of digital technologies and robots, culminating now in the concept of A.I., we must surely make those responsible for technology education take note. This is evidenced by many of the chapters in this volume.

Another growing body of research from a variety of cross-sectional, longitudinal and empirical studies, is beginning, moreover to indicate that, since the release of the first iPhone in 2007, and the introduction of social media, there has been a marked increase in mental distress, self-injurious behaviour and suicide among young people with the effects appearing to be greatest among girls" (Abi-Jaoude et al., 2020, Key Points 1). This demonstrates a significant impact, not just on the social aspects of the human/technology relationship, but on the health and well-being leading, in some cases, to a threat to the lives of the youth of today. The researchers argue:

> There is a need for public awareness campaigns and social policy initiatives that promote nurturing home and school environments that foster resilience as youth navigate the challenges of adolescence in today's world. (Abi-Jaoude et al., 2020, Key Points 5)

Marc de Vries has been a staunch campaigner for the need to change the focus of technology education. The sheer volume of research and writing he has produced over the years is testimony to that. His philosophy is that school-based technology education must include a systematic reflection on technology. Following on from the arguments above and those that follow in the chapters of this Festschrift, we can find, in Marc's own words, the need for an empirical turn in the delivery of technology education:

> In 2016 Kroes and Meijers proposed a next turn for the philosophy of technology: an axiological turn. Axiology is the sub-domain in philosophy that is concerned with values. … Kroes and Meijers' proposal is to enhance [the role of values]. They are motivated by the fact that by taking this 'turn,' the philosophy of technology can become (even) more relevant in social debates about new technologies. A fundamental reflection on the nature of the values that are at stake in new technologies could be a valuable support for the development of such technologies but also for responsible use and for policy making with respect to such technologies. This would be different from an 'applied' turn in which

> philosophers would only deal with practical issues concerning values … which can hardly be overestimated. (de Vries, 2018, p. 15)

Evoking the poetry of Sophocles' *Antigone*, Berkowitz analyses the second choral *ode to man*. Written in 441 BCE, *Antigone* gives us a prescient account of the complex characteristics that human beings exhibit in their relationship with the natural world they inhabit, as they continue to evolve technologically.

> Manifold are the deinon (wonders and horrors), but nothing towers more deinon (wondrous and horrifying) than man! (Sophoclese in Berkowitz, 2022, p. 133)

The Greek term 'deinon' does not translate easily into English. It has two, simultaneously opposing meanings, denoting 'man' as being both wondrous and horrifying. In *Antigone*, Sophocles tells us that the technology man invents enables him to sail the seas, trap birds and other animals, build cities and includes many other wondrous powers. Man is deinon. Man is wondrous.

In so doing, however, man also exerts his extraordinary powers over the world by exploiting its resources; he can be cruel to others and can engage in senseless wars, all of which slowly but inexorably extinguish his humanity. "Man's technological mastery endangers that part of man that man himself does not control, his humanity as a created and limited being, a being who struggles to live in and master a world beyond his control" (Berkowitz, 2022, p. 133). Man is deinon. Man is horrific.

There exists a hidden agenda that tends to be overlooked in the rhetoric surrounding technology education. As Sophocles writes in his third strophe of the *Ode to Man* and Berkowitz reframes thus; "if man tames the earth so as to free himself from toil and uncertainty, what then is left of the mystery that is being human? Because man as *deinon* is always-overcoming, he loses his human fallibility and human freedom. In other words, to overcome all obstacles is to be a god, not a man" (Berkowitz, 2022, p. 134)(italics in original).

Technology education then, like all education, must not be prosaic, but must be transformative, as Marc suggests. It must be about much more than learning declarative or procedural knowledge. In this technologically textured world, it must, however, also be about understanding the affective and axiological dimensions of technology. It must include consideration of humankind, culture and society. It must be concerned about the human/technology relationship. It must care about the impact of new and emerging technologies on the youth of today.

The coevolution of humans and their technologies has reached its most profound manifestation in the exponentially evolving concept of artificial intelligence (AI). An example of the incredible power of this is as follows.

There is a television show that appears in many countries around the world called (the country's name)'s Got Talent. In the 'America's Got Talent' show in 2022, a company called Metaphysic ai, digitally remastered a live singer to appear as one of the judges of the show. This was done in real-time. It can be viewed on YouTube. In another round, the same format was used to turn another singer into Elvis Presley. This technology is known as hyperreality or deepfake. Whilst its use on the television show was wonderful and amusing, it has a more horrific and sinister potential, as Stanford University's IT department concluded.

> In the past few years, artificial intelligence technology has crossed a threshold with the capability to make people look and sound like other people. A "deepfake" is fabricated hyper-realistic digital media, including video, image, and audio content. Not only has this technology created confusion, scepticism, and the spread of misinformation, but deepfakes also pose a threat to privacy and security.
>
> With the ability to convincingly impersonate anyone, cybercriminals can orchestrate phishing scams or identity theft operations with alarming precision. In a recent incident,[2] cybercriminals posed as a company's chief financial officer and other colleagues in a Zoom meeting. The elaborate scam led to the loss of $25 million. (Newsroom, 2024)

In the chapters that follow, many offer perspectives that support Marc de Vries' philosophical viewpoint. Several chapters simply offer justified praise to Marc for his inspiration, his influence, his guidance, his breadth of knowledge and his humanity, which I welcome and support. I hope that those who contributed to this Festschrift enjoy the book. I hope that those who read this Festschrift enjoy the experience. I especially hope that Marc de Vries, to whom this Festschrift is dedicated, enjoys the many contributions that celebrate his contribution to the field.

Notes

1 The material regarding the USA was taken from a PATT-USA first report of findings, dated October 25th 1989 by E. Allen Bame and William E Dugger.

2 See CNNN report at https://edition.cnn.com/2024/02/04/asia/deepfake-cfo-scam-hong-kong-intl-hnk/index.html

References

Abi-Jaoude, E., Naylor, K. T., & Pignatiello, A. (2020). Smartphones, social media use and youth mental health. *Canadian Medical Association Journal* (*CMAJ*), *192*(6), 136–141.

Aoki, T. T., Pinar, W. F., & Irwin, R. L. (2004). *Curriculum in a new key: The collected works of Ted T. Aoki*. Routledge.

Bame, E. A., Dugger, W. E., de Vries, M. J., & McBee, J. (1993). Pupils' Attitudes Towards Technology – PATT-USA. *The Journal of Technology Studies, 19*(1), 40–48.

Berkowitz, R. (2022). Singularity and the human condition. In R. Berkowitz (Ed.), *The perils of invention: Lying, technology, and the human condition* (pp. 131–151). Black Rose Books.

Dakers, J. R., Hallström, J., & de Vries, M. J. (2019). *Reflections on technology for educational practitioners: Philosophers of technology inspiring technology education*. Brill Publishers. Leiden.

Dakers, J. R. (2022). *A nomadic pedagogy about technology: Teaching the ongoing process of becoming ethnotechnologically literate*. Brill Publishers.

de Vries, M. J. (2018). Philosophy of technology: Themes and topics. In M. J. de Vries (Ed.), *Handbook of Technology Education* (pp. 7–16). Springer.

de Vries, M. J. (2021). *Innovation research in technology and engineering management: A philosophical approach* (Kindle Edition). Taylor & Francis.

de Vries, M. J. (2005). *Teaching about technology: An introduction to the philosophy of technology for non-philosophers*. Springer.

Dewey, J. (1916). Democracy and Education. In J. Boydston (Ed.), *John Dewey: The middle works*. Illinois University Press.

Freinet, C. (1967). *Education through work: A model for child center learning* (J. Sivell, Trans.). The Edwin Mellon Press.

Freire, P. (1996). *Pedagogy of the oppressed*. (M. B. Ramos, Trans.). Penguin.

Horowitz, I. L. (2017). *Communicating ideas: The politics of scholarly publishing*. Routledge.

Kahn, R. (2010). *Critical pedagogy, ecoliteracy, & planetary crisis: Tthe ecopedagogy movement*. Peter Lang.

Kurzweil, R. (2008). *The singularity is near: When humans transcend biology*. Duckworth Overlook.

Newsroom, Stanford. (2024). Stanford University IT. 22 02. Accessed April 06, 2024, from https://uit.stanford.edu/news/dangers-deepfake-what-watch#:~:text=Not%20only%20has%20this%20technology,theft%20operations%20with%20alarming%20precision

Rousseau, J. J. (1991). *Emile or on education* (A. Bloom, Trans.). Penguin Books.

Sanders, M. (2009). STEM, STEM Education, STEMmania. *The Technology Teacher*, 20–26.

Schwab, K. (2018). *Shaping the future: A guide to building a better future*. Penguin.
Semetsky, I. (2006). *Deleuze, education and becoming*. Sense
Verkerk, M., Hoogland, J., Stoep, J., & de Vries, M. (2016). *Philosophy of technology: An introduction for technology and business students* (Kindle Edition). Taylor and Francis.

CHAPTER 2

A Most Unusual Philosopher

Carl Mitcham

Marc de Vries is a most unusual philosopher of technology. I mean this in the most complementary sense. Most often, philosophers speak only to other philosophers. In the case of philosophers of technology, when they open up the black in-group box, they are most commonly interested in drawing the attention of public policymakers. Marc is the only philosopher of technology to write *An Introduction to the Philosophy of Technology for Non-Philosophers* for educators and the general public. His *Philosophy of Technology: An Introduction for Technology and Business Students* extends his reach out to an even more unusual audience.

I first met Marc sometime in the early 2000s, probably at one of the Netherlands meetings of the Society for Philosophy and Technology. I think we initially found a common cause in our concern for the religious and theological dimensions of technology. I had been in correspondence with Egbert Schuurman, one of Marc's teachers, from the early 1980s, after discovering in the 1970s the work of his mentor Heindrick von Riessen.

Then Marc honoured me with an invitation to give the first Van Riessen Lecture in 2006. Although Marc's and my religious questionings of technology were quite different, dialogue with him was formative in my own continuing attempt to engage with what I think is a fundamental dimension of any attempt to truly grapple with technology. I thus remain grateful to Marc for his own leadership in this area.

© CARL MITCHAM, 2025 | DOI:10.1163/9789004710948_002

CHAPTER 3

A Vision of Technology Education; A USA Perspective

John M. Ritz

1 Introduction

One's perspectives are developed through interactions with their environment. Many have differing views because of these interactions. This researcher has developed a world view of life and education, particularly toward technology education. This view will be explained through this writing, and then it will be used to provide a vision of what type of technology education programmes may emerge for future generations of school-aged learners.

2 Background

Early experiences with technology came from what one saw or experienced in their local environments. One could see workers constructing buildings and repairing utility systems and people travelling by bicycles, cars, and trains. Fascination with technology came easy, e.g., watching a worker install a rotational antenna on the roof so a family could view multiple stations of TV at night. In the 1950s, citizens listened to the radio and purchased records to bring contemporary music into their homes. Technology-related presents were received on holidays. As a post-WWII baby, I grew up during a technological revolution.

For middle grades youth in the USA during the 1960s, the first experiences with technology education were gained through study in industrial-based unit shops, e.g., woodworking, drafting, and metalworking. Students learned the names and usage of tools and made projects following teacher-designed patterns. In high school, there were elective courses in various unit shops. During the last two years of high school drafting, the class was able to solve problems and design products. This was similar to tasks assigned to engineers, as stated by teachers.

Teacher preparation programmes during the 1960s continued to focus on unit labs. Teacher preparation programmes followed a hands-on application

© JOHN M. RITZ, 2025 | DOI:10.1163/9789004710948_003

of tools and machines to replicate or design products. The teacher preparation programmes developed depth of knowledge and skill with increased types of technologies. Design courses were required and also were available as electives. I took a design class and chose a second one as an elective.

One of the most influential courses in an undergraduate teacher preparation programme was student teaching. This author had the opportunity to student-teach at an experimental programme that was funded through the US Office of Education through its ES '70 *Programmes* (*An Educational System for the '70s*) along with grant funding from the Ford Foundation (Cochran, 1970). It was located in Ft. Lauderdale, Florida, and is known as NOVA Schools (NOVA from Innovation). The instructional strategy employed at this K-12 school was individualised instruction. Students from the county had to apply and be academically admitted to the school. All instruction was individualised (faculty wrote learning activity packages, my first attempt at publishing). After completing student teaching, I was hired at NOVA. Technology education at NOVA was located in the Department of Technical Sciences. Today some in science education might have these same notions to call our subject this as we practiced at this school in the 1970s.

While at NOVA, the author learned about the *American Industry Project* developed at the University of Wisconsin-Stout. NOVA integrated threads of this programme into its Technical Science curriculum. The *American Industry* curriculum framework was developed using a taxonomical analysis of the 13 elements of the industry, i.e., communication, transportation, finance, property, research, procurement, relationships, marketing, management, production, materials, processes, and energy (Cochran, 1970).

During the 1960s, the U.S. Department of Education had futuristic leadership and provided access to funding for developing research-based programmes for the future development of programmes for industrial and technology education. The *American Industry Project* (1963) at Stout was one such programme (Cochran, 1970). This funding included support for the *Industrial Arts Curriculum Project* (1965) at The Ohio State University, where courses and instructional materials were systematically developed for teaching construction and manufacturing technology (Towers et al., 1966).

The Maryland Plan developed by Donald Maley (1973) and other programmes that focused on the broader potential field of technology education were under study by such USA leaders as William Warner (1947), Delmar Olson (1963), and Paul DeVore (1972). DeVore developed the *Programme for the Study of Technology* at West Virginia University which led to graduate degrees that focused particularly on the study of technology. DeVore and his doctoral students developed taxonomical structures for the content of communication, production, and transportation technological systems.

Following successful teaching at NOVA and a drafted stint in the US Army, I chose to pursue my master's degree at Stout and study the *American Industry Project*. This education taught me how to gain knowledge of the content for the study of industry and technology and how to plan innovative programs through curriculum design. While at Stout, I also had the opportunity to direct a curriculum development project for the state of Wisconsin to assist in modernising their state curriculum. The work we did introduced new technological systems into the state curriculum framework. What our centre developed in instructional materials was field-tested, evaluated, and implemented in Wisconsin schools.

Because of my interests in curriculum development and the study of technological literacy, and the want to become a university professor, I sought a doctoral program that was leading edge in the study of technology. West Virginia University provided this opportunity. I completed many courses in curriculum development and also in the history of technology, technology and culture, each of the technical systems of production, communication, and transportation systems, and programme development in each of the systems of production, communication, and transportation technology. I studied all technological systems (which most graduates did not), because I wanted to be educated in all, so I could develop a curriculum for the K-12 educational systems. At West Virginia, we developed taxonomical content for production, communication, and transportation systems and transformed this knowledge into instructional programmes.

With this knowledge, I had to rationalise how best to structure content and student-friendly activities for the study of technology. While pursuing the doctoral degree, each graduate student in the programme was assigned to a school-based experimental project – graduate research assistantships. My position required me to work with a team of three teachers in a suburban school district in the southwestern section of West Virginia. The school system contracted with the *Programme for the Study of Technology* at West Virginia University to develop a middle school unified arts programme for eighth-grade students. This entailed content specific knowledge and activities and integrated themes between the subjects. The unified arts programme included the subjects of art, consumer and family science, and technology education. The content for technology education focused on the systems of production, communication, and transportation technology. The linking themes were unifying experiences where a product was designed and mass-produced using knowledge from the three subject areas, the design and creation of an individual product using knowledge and skill from the three subjects, and the research and construction of a futuristic model of the local community. This West Virginia project was designed and overseen for a three-year period. It was the basis of my dissertation. It also helped me plan and analyse the benefits and difficulties of teaching integrative STEM. I will describe my thoughts on STEM later in this review.

Finishing the doctoral degree, I started my university teaching. The programme that I entered at Old Dominion University had already established undergraduate major courses in communication, construction, manufacturing, and transportation technology systems. During the following years, this faculty developed four course sequences in each of the technological systems of communication, production, and energy and transportation. We also had courses in design, inventions and innovations, and technology and society for our technology education majors. The faculty was also insightful and achieved further success by lobbying the general studies committee at the university to have the science requirements transformed to science and technology requirements. We had approved a freshman and junior level course for general education on technological literacy – *Technology in Your World* and *Technology and Society*. Numerous sections of these courses are offered and enrolled with undergraduate students. It was my belief that technological literacy should become a university general education outcome.

In addition to working with local schools to modernise their technology education programmes, grades 6–12, our faculty worked with the state of Virginia to transform the overall technology education curriculum, moving from industry-based practices to technological literacy concept formation. Virginia established curriculum packages for Exploring Technology (grade 6 or 7), Inventions and Innovations (grade 7 or 8), and Technological Systems (grade 8 or 9). At the high school level, curriculum packages were developed for three additional courses which emphasised Technology Foundations, Technology Transfer, and Technology Assessment. The Foundations course was structured around the technology concepts of materials, energy, and information as reported by de Vries (1987). Transfer included early concepts of integrative STEM (science, technology, and mathematics). Assessment focused on the impacts of technology on humans, society, and the environment (Virginia Council on Technology Education for the 21st Century, 1992).

In the USA, under the direction of William E. Dugger and the International Technology and Engineering Educators Association (ITEEA), they developed *Standards for Technological Literacy* (2007, 2002, 2000). In addition, ITEEA developed courses for K-12 students that were standards-based and age-appropriate. This packaged curriculum is titled *Engineering by Design*. ITEEA

TABLE 3.1 Engineering by design course titles

Grade	Titles	Length
6	*Exploring Technology*	18 weeks
7	*Invention and Innovation*	18 weeks
8	*Technological Systems*	18 weeks
9	*Foundations of Technology and Engineering*	36 weeks
10–12	*Technological Design*	36 weeks
11–12	*Advanced Design Applications*	36 weeks
11–12	*Advanced Technological Applications*	36 weeks
11–12	*Engineering Design* (*Capstone*)	36 weeks

has developed instructional packages for K-6 grade students that vary from one to 12 weeks and specific technology and engineering standards-based courses for middle and high schools (see Table 3.1).

Engineering by Design is a standard-based curriculum designed to provide technological and engineering literacy in the USA and other countries who choose to use it. As the reader may have noticed, the first four courses in *Engineering by Design* have the same titles and similar content to those in the Virginia technology curriculum. Courses may be purchased through ITEEA. ITEEA has formed US state partnerships which enable teachers to be trained and have access to these instructional materials for a price. Updating of these courses is on a continuous improvement cycle.

With this background, the reader can see the perspectives the author has developed related to technology and engineering education and where the USA is currently positioned for the future direction of technology and engineering education in its schools. This is the ideal world and many schools still include programs that are based on industrial practices. Most schools employ modern industrial practices using state-of-the-art technologies such as CAD, CNC, 3D printing, and other computer-controlled programmable machines. However, the author is a pragmatic idealist. There are ideals that should direct the study of technology and engineering, but one must be practical about how others will implement what they feel most comfortable for putting technology and engineering into practice. Following will be an explanation of my views on technological literacy, where STEM and engineering should be planned into the curriculum, and some thoughts on the future direction of the technology education profession.

3 Technological Literacy

To function effectively in society, citizens need to possess knowledge of the technology around them. They need to understand some technologies at the macro level, while they need to be more familiar with other specific technologies required for life and work pursuits. Yes, it is good to understand what STEM cell research is and how it might someday improve your life (macro-level). It is also important that one become more familiar with a computing system (e.g., which system is most appropriate to purchase, how to load paper into the printing system, and how to change a printer cartridge to continue to have quality output), and how to purchase the correct size of an automobile battery and how to safely install it. These are literacies and technological literacies.

To function in a society, one develops an understanding of spoken words, reading and writing, and general mathematics (general literacies). In the US Workforce Investment Act of 1998, literacy is defined as "an individual's ability to read, write, speak in English, compute and solve problems at levels of proficiency necessary to function on the job, in the family of the individual and in society" (p. 131). In addition to these general literacies, some believe all need:

> A new form of literacy – a technological literacy … This is a vital necessity if citizens are to participate in assessing and determining the relationship of technological systems to human needs. To function in this role requires that all citizens be conversant in the language of technological systems and comprehend basic concepts of the dynamics of the interrelated systems for all levels of society. (DeVore, 1980, p. 338)

Technological literacy is defined as "the ability to use, manage, understand, and assess technology" (ITEEA, 2000, p. 242). However, in practice, technology's study has at times been focused on developing technical expertise, instead of how useful or pertinent the technologies can be (Ginestié, 2008).

To "understand, use, assess, and manage technology" (ITEEA, 2000, p. 242) is much different than to develop expertise in a few technologies, such as robotics and machining. According to Pearson and Young (2002),

> Technological literacy is not the same as technical competency. Technically trained people have a high level of knowledge and skill related to one or more specific technologies or technical areas . . . a technologically literate person would not necessarily require extensive technical skills. Technological literacy is more of a capacity to understand the broader technological world rather than an ability to work with specific pieces of it. (pp. 21–22)

However, tradition has led educators who teach technology and engineering to teach technical expertise. This may be in part because a technology teacher is given a laboratory with a variety of tools within its confines. It is natural for educators to teach these technologies when they are given these new tools (Ritz, 2011). Might it be a systematic technology means-or-end problem that new technology creates?

Because much of the world continues to experience new technologies and changing economic situations, and the education system is almost void in explaining these developments and how or if they should be used for the betterment of society, these knowledges and abilities should eventually become the focus of education through technological literacy programs. According to DeVore (1972), "It is self-evident that we can control only that which we know about and understand in behavioural terms" (p. 8).

Pearson and Young (2002) stated that "technological literacy – an understanding of the nature and history of technology, a basic hands-on capability related to technology, and an ability to think critically about technological development – is essential for people living in a modern nation" (pp. 11–12). Such people have knowledge of technology and are capable of using it effectively to accomplish various tasks. They can think critically about technological issues and act accordingly. Technological literate people would possess knowledge, ways of thinking and acting, and capabilities that assist them as they interact with the technology found in their environments. These traits of technological literate individuals are cited from Pearson and Young (2002, p. 17) and include the following:

4 Knowledge

- Recognises the pervasiveness of technology in everyday life.
- Understands basic engineering concepts and terms, such as systems, constraints, and trade-offs.
- Is familiar with the nature and limitations of the engineering design process.
- Knows some of the way technology shapes human history and people shape technology.
- Knows that all technologies entail risk, some that can be anticipated and some that cannot.
- Appreciates that the development and use of technology involve trade-offs and a balance of costs and benefits.
- Understands that technology reflects the values and culture of society.

5 Ways of Thinking and Acting

- Asks pertinent questions, of self and others, regarding the benefits and risks of technologies.
- Seeks information about new technologies.
- Participates, when appropriate, in decisions about the development and use of technology.

6 Capabilities

- Has a range of hands-on skills, such as using a computer for word processing and surfing the Internet and operating a variety of home and office appliances.
- Can identify and fix simple mechanical or technological problems at home or work.
- Can apply basic mathematical concepts related to probability, scale, and estimation to make informed judgments about technological risks and benefits. (Pearson & Young, 2002, p. 17)

These knowledge and skills align with those specified in *Standards for Technological Literacy: Content for the Study of Technology* (ITEEA, 2007, 2002, 2000).

7 Content for the Study of Technology

Amassing the work already undertaken in the USA, ITEEA, through the research and developmental work of William E. Dugger, Thomas Loveland and Marie Hoepfl, and many others in the technology and engineering professions, developed *Standards for Technological Literacy* (2007, 2002, 2000) and *Standards for Technological and Engineering Literacy* (2020). ITEEA's *STEM Center for Teaching and Learning* used this knowledge to develop standards-based courses presented earlier as *Engineering by Design*. These materials are contemporary and are continuously revised. I believe these courses will remain the basis for the preparation of American youth in the study of technology and engineering literacy. However, this would be the ideal pathway to achieving technological and engineering literacy. Pragmatically, some teachers will use these materials, while others will continue the teaching of the newest technological tools and systems available to them.

8 STEM and Engineering

For countries to remain economically competitive, their citizens need to foster traits of invention, innovation, and the developmental abilities for creating new products and technologies. To do this, it requires an educated workforce who can invent and develop new products and services. Study at the National Science Foundation (2001) found that educational systems were needed that focused on the subjects of Science, Technology, Engineering, and Mathematics (STEM) (Salinger & Zuga, 2009). Some add Healthcare occupations and the Arts to this acronym. With the need to enhance STEM education, the technology and science education professional associations have cultivated this type of study. Technology education has added engineering design into its knowledge realm. Some believe it is essential to develop curriculum and instructional activities that integrate STEM, i.e., iSTEM. It is believed this will enhance student learning by strengthening the application of the principles of the four subjects, just as scientists, technologists, engineers, and mathematicians do. Others believe that the STEM subjects should be taught in depth as separate subjects. This group does not want to integrate, since they believe that instructional time is needed to develop expertise in their individual specific subject.

It is difficult to integrate subjects that are taught in separate classrooms/laboratories. From practice, I have found that integrative STEM can be sufficiently taught at the elementary level, where a teacher or teacher team can develop and use integrative subjects to design lessons or activities. At the middle grades level, themes can be used where the individual three STEM subject teachers can design projects where the three classes work to find a solution and develop a product, e.g., energy efficiency, agriculture, electric car feasibility, moon housing. These ideas are many. I suggest themes at the middle grades level, since it is difficult to have integration throughout the entire school year. Such practices would require teachers to continuously plan together. If one teacher is not into team-teaching, it is difficult to make this work. If a teacher leaves the school, then the team would have to convince the new teacher to volunteer and invest in the planning, development, and implementation of integrative themes.

At the high school level, science, technology and engineering, and mathematics teachers can use technology and engineering based projects to enhance student interest and depth of learning in their individual classes. Major journal articles and conference presentations reflect on possible integrative STEM units of instruction and activities for each of the individual STEM subjects. It is easier for one to develop subject specific integrative STEM units of instruction, including activities, by using open source AI products. All one needs to do is enter parameters into the search engines and the lessons will self-generate. You can list which individual standards (science, technology and engineering, and mathematics) you would like to use, the particular topic you want the lesson to focus (e.g., energy efficiency, advanced materials, coding, robotics), and the type of product you would like the student to design and make. The more parameters you list, the more robust the Open AI lesson will become. If you do not get what you want, just further explain your parameters in the Open AI platform. I believe you will see educators of technology and engineering integrating science and mathematics content into their curriculum and lessons using Open AI products.

9 Future of the Profession

Ritz and Martin combined three research studies they undertook in the USA. The resulting work was a chapter in the Springer *Handbook of Technology Education* edited by de Vries (2018). This chapter was titled *Visions of the Technology Education Profession by Technology Teacher Educators*. The first of the three studies, *Technology Education Teacher Preparation Curriculum Study* (2018), found faculty teams (N = 28) at USA technology education teacher preparation institutions based their curriculum on The General Study of Technology (M = 3.54), Engineering Design (M = 3.46), and Technological Systems (M = 3.43). (See the study for an explanation of these concepts.) These results were based on a 4-point scale. In addition, teacher educators continue to explore STEM and STEM integration, with this curriculum/instructional approach (M = 3.00). Exploration of STEM in technology teacher preparation will continue into the future.

The second Ritz and Martin study (2013), *Perceptions of New Doctoral Graduates on the Future of the Profession*, surveyed PhD graduates from technology education programmes for a period of up to 5 years immediately preceding their graduation. This study group was N = 34. When asked what the content taught in a formalised K-12 technology and engineering programme should be, there was a general consensus that it should be based on technological literacy, STEM integration, and engineering design. Each of these three foci received more than 50% recommendations.

When asked about the focus of instructional strategies to be used in a quality technology and engineering program, 50% or more of the participants believed that project-based, design-based, and contextual learning were important strategies. There was no one particular K-12 audience for technology and engineering education found through this study (e.g., elementary, middle school, high school).

Participants were asked their views of what the future of their profession would look like. This study found that participants believed national professional associations (64%) and teacher education institutions (70%) will continue to provide professional development and activities for the technology and engineering professions. In addition, participants believed they will be members of the International Technology and Engineering Educators Association (ITEEA) (75%), American Society for Engineering Education (ASEE) (69%), STEM associations (56%), and Council on Technology and Engineering Teacher Education (CTETE) (57%). Participants also plan to attend conferences for continued professional development, including ITEEA (79%), ASEE (62%), and state-level conferences (57%). For the USA population, PATT conferences are not a priority. Finally, participants planned to publish in the *Journal for Technology Education* (86%) and *Technology and Engineering Teacher* (73%).

In their third study, Martin, Ritz, and Kosloski (2014) surveyed current students enrolled in doctoral programmes studying technology education (N = 34). It was found that participants felt the content for K-12 technology and engineering programs should be based on the constructs of technological literacy (n = 21; 62%), technology/engineering design (n = 23; 68%), and STEM integration (n = 27; 80%). Participants also felt that instructional strategies for K-12 technology/engineering programs should be project-based (n = 24; 71%), design-based (n = 28; 82%), and contextual learning (n = 23; 68%).

With the declining number of universities offering a degree in technology and engineering teacher preparation, the researchers sought to determine who would provide professional development activities in the future. Respondents believed teacher education institutions (n = 26; 78.8%), professional associations (n = 23; 69.7%), and distance learning providers (n = 18; 69.7%) would do this. This was prior to COVID and the extensive use of electronic means for all communication and teaching.

The final question to this survey asked participants about the future of the technology and engineering profession. This study's respondents believed the profession will be integrated into a broader STEM organization (n = 30; 88.2%) in the near future. This researcher sees this trend beginning to take hold with the direction that the International Technology and Engineering Educator Association and its journal, *Technology and Engineering Educator: Bringing STEM to Life*. Reviewing the association's website, one will see *STEM Connections, Integrative STEM Educational Methods Certificate, STEMinars, STEM Schools of Excellence*, and its upcoming international conference – *Technology & Engineering: The Soul of STEM*.

10 Summary

The technological revolution has caused academics to transform the content of technology and engineering to be passed along through schooling and the media to youth and citizens. With its roots in the 20th century, the concepts of technology education have matured in the 21st century. I believe this school subject will continue to grow as new technologies and systems are created. Much is shaped by the education which professionals receive, the professional contacts they make, and the literature they read – interactions with their environments. The educators who work in this profession are fortunate to be exposed to research and ideas that have been developed for this school subject. Our profession is fortunate that each generation births new researchers and curriculum developers who study and develop this school subject. The profession has been fortunate, and I believe it to be in good hands for a future generation of leaders.

Technological literacy has been defined and programmes have been developed to educate the masses. We have established ideals for what can be delivered through the study of technology and engineering. We know that some will teach to these ideals, while others will approximate our beliefs as they best see our school subject in their own minds. As individuals explore the integration of math and science concepts into their units of instruction and project designs and their analysis, their integration will improve students' understanding of technology and engineering design. Our profession can better deliver the knowledge needed for a more stable culture and the basis for future workers.

Professionals have plotted the path for technology and engineering education. Others will continue this work. We are fortunate for their research and sharing of these ideals, so our subject can be better delivered. The profession needs to continually political with government officials and school systems leaders to show the need and results of technology and engineering literacy programming.

References

Cochran, L. H. (1970). *Innovative programs in industrial education.* McKnight & McKnight.

DeVore, P. W. (1972). *Education in a technological society: Access to tools.* West Virginia University.

DeVore, P. W. (1980). *Technology an introduction.* Davis Publications, Inc.

Ginestié, J. (2008). *The cultural transmission of artifacts, skills and knowledge.* Sense Publishers.

International Technology and Engineering Educators Association (ITEA/ITEEA). (2007, 2002, 2000). *Standards for technology and engineering literacy: Content for the study of technology* / (2020). *Standards for technological and engineering literacy: The role of technology and engineering in STEM education.* ITEA/ITEEA.

Maley, D. (1973). *The Maryland plan*. Benziger, Bruce & Glencoe, Inc.

Martin, G., Ritz, J., & Kosloski, M. (2014). Technology and engineering doctoral students' perceptions of their profession. *Journal of Technology Studies, 40*(2), 82–95.

Olson, D. W. (1963). *Industrial arts and technology*. Prentice-Hall, Inc.

Pearson, G., & Young, A. T. (2002). *Technically speaking: Why all Americans need to know more about technology*. National Academies Press.

Ritz, J. (2011). A focus on technological literacy in higher education. *Journal of Technology Studies, 37*(1). http://scholar.lib.vt.edu/ejournals/JOTS/v37/v37n1/ritz.html

Ritz, J., & Martin, G. (2013). Perceptions of new doctoral graduates on the future of the profession. *Journal of Technology Studies, 39*(2), 65–78.

Ritz, J., & Martin, G. (2017). Visions of the technology education profession by technology teacher educators. In M. J. de Vries (Ed.), *Handbook of Technology Education* (pp. 699–712). Springer International Handbooks of Education. https://doi.org/10.1007/978-3-319-38889-2_51-1

Salinger, G., & Zuga, K. (2009). Background and history of the STEM movement. In International Technology and Engineering Educators Association (ITEEA) (Ed.), *The overlooked STEM imperative: Technology and engineering* (pp. 4–9). ITEEA.

Towers, E. R., Lux, D. G., & Ray, W. E. (1966). *A rationale and structure for industrial arts subject matter*. The Ohio State University Research Foundation.

Virginia Council on Technology Education for the 21st Century. (1992). *The technology education curriculum K-12*. The Virginia Vocational Curriculum and Resource Center.

Warner, W. E. (1947). *A curriculum to reflect technology*. Industrial Arts Center, The Ohio State University.

Workforce Investment Act. (1998, August 7). *U.S. Public Law,* 105–220.

CHAPTER 4

The Role of Technology Education in Facing the World of Tomorrow

A Perspective based on Knowledge Traditions

Charlotta Nordlöf

1 Introduction

As a technology educator, it is both appealing and straightforward to relate technology teaching to current events in society, news reporting, and other aspects of the surrounding world. Technology, in the form of subject content, is constantly present in society, incorporating both major and minor phenomena that we encounter through media.

When I teach technology teacher students, I sometimes bring the morning paper to the lesson to illustrate to the students that one can find inspiration for technology education in everyday life. It is not difficult to find news related to technology in the paper, and I do this to inspire the students and to help them realise that technology is a subject that is exciting to teach. I do a workshop where the students pick an article from the media and practice highlighting different takes on technology education based on the curriculum. We discuss different perspectives of technology teaching.

Approaching technology education in this way, relating to current issues or engaging themes, is one of the reasons why I enjoy teaching technology, and I want my student teachers to realise this. I like to take this approach to technology education and present it as a school subject where teachers can be creative and focus on the context to make technology education even more meaningful, engaging, and interesting.

Recently, I have begun to think that I need to take this view of news as inspiration for technology education and turn it upside down. Examples of technology education still pop up in every other article in my morning paper, but first and foremost, the knowledge gained from technology education is necessary to be able to understand the world of today and tomorrow. In other words, articles can be an inspiration for technology teaching, but above all, knowledge from technology education is needed to understand and interpret the news reporting in depth. As technology becomes more complex, citizens need

© CHARLOTTA NORDLÖF, 2025 | DOI:10.1163/9789004710948_004

to have technological knowledge to understand the technological and social developments that are reported in the news every day. Our world is technologically intense and has been like that for a long time. Almost 20 years ago de Vries stated that technology education has an important role to play:

> We need to work on a technology education practice that convinces governments, school administrators and school boards, teachers, parents, and pupils that teaching and learning about technology is not only fun but also a necessary component of every (future) citizen's general education. (de Vries, 2006, p. 9)

This kind of general education is sometimes considered as being technological literacy. Technological literacy can be described as an overall, multidimensional understanding of technology, including using technology, understanding technology, and being aware of technology and its relation to society and the environment (Williams, 2017). Technological literacy is necessary to be able to keep up with the progress in the surrounding world, and consequently, technology education is essential for facing the reality of tomorrow. A "technologically literate person dares to question existing solutions to practical problems and identify new problems" (de Vries, 2012, p. 95). Technological literacy can be argued to be the main aim of technology education (Rossouw et al., 2011), and additionally as a base for democratic participation (Williams, 2009).

We educate students to become citizens and problem solvers of tomorrow. The challenges that we face can rarely be disconnected from technology as technology most often plays a significant role. Many of the problems that need to be solved today and the challenges we will face in the future are complex and include technological aspects.

Another aspect of technology education, which I also discuss with the students, is that technology education might feel a bit challenging. Technology education in one classroom can be very different from technology education in another classroom, even though the teaching is based on the same curriculum. Both in teacher education and in school, we like to start with questions like "What is technology?", and that points to something crucial. We need to know what technology is to teach it and to understand it. Further, when comparing technology education syllabuses internationally, we notice differences, but at the same time, we see common denominators: there is a common base for technology education, even though it is difficult to describe it in words. One aspect of that common base could be found in the philosophy of technology (de Vries, 2016) and another in design and problem-solving (de Vries, 2012).

2 Three Categories of Knowledge in Technology Education

To be able to deal with complex challenges and issues from a technology education perspective, I would like to start with the subject of technology. More specifically, I will discuss knowledge in technology education and start at the very roots of the subject. Technology education can roughly be considered as aspects of technology based on crafts, engineering science, and socio-ethical views. This way of understanding and explaining knowledge in technology education is based on a framework called a three-part heuristic framework, which is based on knowledge traditions (Nordlöf et al., 2022). Knowledge traditions are the combined knowledge and skills that have developed within an organised community, where similar tasks are performed, and tips and information, among others are exchanged. The framework can, for example, be used to compare technology education in different countries (Norström, 2022) and for teachers to plan, evaluate, and support collegial discussions.

In literature, technological knowledge is an area discussed from different angles and with different classification systems (e.g. Houkes, 2009). For instance, technological knowledge has been classified for engineering designers, based on the design of integrated circuits. The classification system is based on the dual nature of technical artefacts, as a functional and physical nature of knowledge (de Vries, 2003). Within the engineering community, knowledge is traditionally based on disciplines, e.g. electrical engineering or computer science engineering. In technology education, the model of technology created by Mitcham (1994) is often referred to when explaining technology and technology education (e.g. de Vries, 2016; Svenningsson, 2019). Mitcham describes technology to be fourfold: object, activity, knowledge, and volition. In his model, knowledge is explained as knowledge to produce or knowledge to use technology, not as knowledge in and about technology from a broader perspective.

With these examples in mind, why do we need another classification system? The three-part heuristic framework is designed for technology education and has a structure and foci of its own, compared to other classification systems, since the foundation is knowledge traditions. Therefore, it might be useful in technology education and in understanding the role of technology education in the world of tomorrow.

Table 4.1 presents an overview of the framework. It was developed heuristically by analysing and reflecting on technology education and studying different philosophical and theoretical approaches to technology.

The framework for technology education consists of three categories of technological knowledge, forming an epistemological tripod. The categories

TABLE 4.1 An overview of the three-part heuristic framework for technology education, based on knowledge traditions

	The three legs of the epistemological tripod of technology education		
	Technical skills	**Technological scientific knowledge**	**Socio-ethical technical understanding**
Short description of technological knowledge (tradition)	The first technological knowledge, mastered by humans. Skill or ability. The main focus is to make things work, not why they work. Knowledge *in* technology. (Craftmanship tradition)	Knowledge gained using a general scientific approach, but in a technological context. Understanding why things work is of the greatest importance. Knowledge *in* technology. (Engineering tradition)	Discussing and relating technology to different aspects such as the environment, society and humans. Knowledge *about* technology and its relationship with the human world. (Humanities and social sciences tradition)
Main justification method	Experience	Methods from the technological and natural sciences.	Methods from the humanities and the social sciences.
Example from technology education	Knowledge of how to build, cut and glue cardboard models.	Knowledge of how materials are structured and their properties.	Knowledge of how computers have changed the way we communicate or how society's infrastructure is designed.
Example from professional activities	Craftwork of a blacksmith.	Mechanical calculations of the strength of a bridge.	How a new railway line will affect the everyday life of the local community.

are based on knowledge traditions: (1) technical skills, (2) technological scientific knowledge, and (3) socio-ethical technical understanding. The categories are not mutually exclusive, and their limits are fuzzy (c.f. Håkanson, 2010).

The category of technical skills is based on the craftsmanship knowledge tradition and handicraft professions. The skills are mainly justified by experience, often expressed by actions rather than words. Technological scientific knowledge is based on the engineering knowledge tradition. It is an academic knowledge type, mainly appearing from engineering science and natural science traditions, and is often represented using mathematical models. The third category, Socio-ethical technical understanding, deals with the relationship between technology and the human world. It diverse from the two previous categories since it rather is knowledge about technology than in technology. It studies technology from an external perspective. This category is based on the humanities and social sciences knowledge tradition.

3 The Framework in a Teaching Example

When teaching technology, a particular subject's content will appear differently depending on the perspective from which the content is taught. Taking bridges as an example of subject content in technology education, three different teaching situations could be illustrated as follows:

Classroom 1: The teacher has instructed the students to move the benches in the classroom apart. The students are now making bridges between the benches by folding A4 sheets of paper in different ways. In the next lesson, students will try to make bridges in a similar way using ice-cream sticks and glue.

Classroom 2: The teacher gives a briefing and demonstrates different ways of constructing bridges. The students compare different designs, and the teacher draws arrows to symbolise forces and talks about concepts such as trusses and arches. The students try to make simple triangles and rectangles out of straws to experience the difference between different truss designs.

Classroom 3: The teacher shows the students a map and an aerial photograph and starts with the question: What would our town look like without bridges? Later, the teacher introduces the concepts of infrastructure and technological systems and touches on questions such as why bridges and other links can be targets in war.

Applying the technical knowledge framework to the examples, the teacher in Classroom 1 is mainly concerned with technical skills when the pupils make bridges out of different materials. The teacher focuses on letting the pupils try out different solutions and construction materials. In classroom 2, there is technological scientific knowledge in the form of knowledge about how bridges are constructed, and how different solutions are named and construed. In the third classroom, the emphasis is on socio-ethical technical understanding, as the relationship between technology and society and the importance of bridges for the functioning of society is discussed.

In reality, the categories of knowledge are often integrated in different combinations, but these examples are written to illustrate the categories—the three different types of knowledge in technology education emerge. Further, for students to develop a socio-ethical technical understanding, a knowledge base of technological scientific knowledge is needed.

4 A Future Technology Education Curriculum

Today's technology education includes knowledge from all three knowledge traditions. However, around the world, technology curricula have different foci and emphasise different parts. The Finnish and Scottish technology subjects are examples of curricula dominated by *Technical skills*. In the USA, *Technological scientific knowledge* is present in the standards. *Socio-ethical understanding* is most visible in the Australian and Swedish curricula (Norström, 2022). Additionally, the blend of these three traditions varies between classrooms and from one teacher to another. When looking back, the category technical skills was the initial technological knowledge needed by humans, to fulfil basic human needs. However, when glancing into the future, my notion is that socio-ethical technical understanding will be increasingly essential since the challenges we are facing are complex. Students need to be equipped with technological literacy (as introduced earlier).

As I write this chapter, the media is full of information about the ongoing United Nations Climate Change Conference in Dubai, where the main issues to be negotiated are fossil fuels in energy systems. Such problems are "wicked" (Rittel & Webber, 1973) and, therefore, need to be addressed in a multidimensional approach, both concerning other school subjects and concerning technology as a multifaceted school subject (Dusek, 2006).

Meanwhile, news reporting focuses on how experts believe AI will affect the job market in the future. To have any chance of understanding AI, you need to know something (or a lot) about technology. It is not enough to understand AI from a single perspective; a multidimensional understanding is required to grasp different layers and dimensions. Furthermore, implementing AI will lead to new ways of looking at knowledge (c.f. Stolpe & Hallström, 2024; Tuomi, 2023).

Even though it is difficult to predict the future, I have chosen two examples of technological areas that, based on what we know today, will be important in the future.

5 Example One: Artificial Intelligence (AI)

Artificial intelligence is one example of what is considered a new technology and a technology that will be crucial in the future. Therefore, AI will probably be one example of technology that needs to be processed in technology education in the future. AI is already present in our everyday life. When I enter my car, my phone tells me that the drive home will take 18 minutes and that there are no traffic disruptions. I am used to messages like that, but I don't give it that much attention. I don't necessarily consider it as AI (compared to the AI effect, McCorduck & Cfe, 2004). To understand that the information is provided by AI, it is necessary to know something about AI's construction and operation.

Today's students are part of a generation that will take AI for granted. They will encounter AI in all aspects of their lives. In healthcare, at school, in their leisure time, both on the sofa in front of the TV and on social media. Technologically literate citizens need to be critical of sources, challenge and question information, and critically evaluate AI (c.f. Stolpe & Hallström, 2024). A technology curriculum of tomorrow needs to give students knowledge and tools. That is knowledge in the form of socio-ethical technical understanding in the framework described earlier. It is knowledge about AI and how AI relates to the human world. Even though every citizen won't need to be able to develop AI applications, students need a solid base of knowledge of how AI works. To interpret news related to AI, individuals with a basic understanding of how an algorithm functions and with some understanding of how AI makes decisions, will have an advantage in evaluating and critically assessing both pros and cons as well as fake news. Therefore, technology education also needs to provide students with technological scientific knowledge when it comes to AI.

6 Example Two: Sustainable Development and Problem-Solving

Humanity is facing big and complex problems. These include energy supply, food production, and transport solutions. The common denominator is that these are not problems that can be solved in a limited, smaller context, but are of a larger nature, spanning time and place. These are examples of problems that do not have a single solution but can be described as wicked problems

(Lönngren, 2017; Rittel & Webber, 1973). In technology education, it is not unusual for students to solve technical problems in a limited context, often in the form of developing an artefact. This is a good way of practising problem-solving skills. However, we also need to expose students to challenges that prepare them to become technologically literate citizens and to practice solving more complex problems (cf. Lind et al., 2022). Problems that can be viewed from different angles. By including complex problems in technology education, we can also make the lessons more realistic and meaningful for students by putting what they learn in technology lessons into context. Students need to have a basic understanding of what a technical system is and how systems are interconnected and affect other systems – this can be thought of as knowledge based on technological scientific knowledge in the framework. Further, students need to know how different technological decision affect the environment and the future, and to do that they need knowledge based on socio-ethical technical understanding.

Note that neither of the above examples includes much of the framework's first category, *technical skills*.

7 Discussing Technology Education in the Future

Technology education in Sweden today should include all three knowledge categories in the framework. The teacher interprets the curriculum and decides how to allocate teaching time. Still, the teacher can be affected by factors such as school traditions, habits and culture that make technology education look a certain way. Both students and colleagues may expect certain content or may have expectations of what technology education should be. Such a situation can be limiting for the technology teacher, who may feel that it is difficult to include the full breadth and content of the subject when expectations and traditions are strong. In a study where teachers met in focus groups and discussed knowledge and content in technology education the results revealed that "to build and have fun" – is what the students expect to do in technology class (Nordlöf et al., 2022).

One result of the Swedish School Inspectorate's (2014) review of technology education is "unreflective making", which means that the students do tasks but 'have no idea why they have done it or what they have learned' (Skolinspektionen, 2014, p. 24). The dominant element in teaching was various forms of making, such as building, creating, testing, and constructing, while elements such as reflection and discussion were missing. The making often consumes a lot of time and a large part of the teaching hours in the technology subject (Fahrman, 2021; Nordlöf, 2022). One explanation for the findings regarding the "unreflective making" may be the traditions and culture of technology education, which means that both pupils and teachers have expectations of certain types of content and working methods. These expectations can be challenging to change. Another reason, also mentioned in the report, might be that some teachers may find it difficult to put into words what students should learn.

For various reasons, some of the focus of the subject has been directed toward making, both government documents and curricula formed by traditions. For future curricula, I believe that knowledge from the category of socio-ethical technical understanding will need more space. As I have described in the examples above, knowledge based on how technology affects the world around us and our society, the environment, and human beings will become increasingly important as technology becomes more and more complex. For students to be able to deal with the relationship between technology and the human world, they, therefore, also need a basis, something to relate to. To be able to acquire a socio-ethical technical understanding, some technological scientific knowledge is also necessary, as this knowledge is needed to be able to evaluate, discuss, and argue about different perspectives and alternatives.

Students need to be able to take a holistic and systems view of technology. They need to be aware that technology is not a set of separate artefacts. Technology needs to be placed in a context that also gives meaning to the subject (see Svenningsson et al., 2022). Teachers, therefore, need to take all this into account when planning and implementing technology education.

However, do the challenges of the future mean that old and traditional knowledge will lose its relevance? Not, that is not my point. Technology is a broad subject, and the relationship between technology and people is crucial for the future. There will still be traditional technological knowledge, which is also a solid basis for technological literacy. Being able to write a technical report, make technical drawings, know some basic programming, and understand the design process are examples of timeless technological knowledge (Norström et al., 2023). For those of us who teach the technology teachers of the future or who teach students who are citizens of the future, it is a challenge to choose to teach content and examples that are relevant today and will be relevant in the future. The examples need to be broad and bigger than themselves.

I hope that in the future, when a citizen reads the morning paper (probably not a printed one), she will feel that the knowledge she has gained from technology education is useful in acting in and understanding the world. She and her former classmates will have a general understanding and knowledge that technological phenomena are rarely separate from the world around them and that the choices they make will affect others in one way or another. This will help interpret news and make decisions both in everyday life and in larger issues. If a person lacks technological literacy, how can they make informed decisions and participate in democratic processes such as voting in referendums and elections?

References

de Vries, M. J. (2003). The nature of technological knowledge: Extending empirically informed studies into what engineers know. *Techne: Research in philosophy and technology*, *6*(3), 117–130.

de Vries, M. J., & Mottier, I. (2006). *International handbook of technology education: Reviewing the past twenty years* (Vol. 1). Sense Publishers.

de Vries, M.J. (2012). Teaching for scientific and technological literacy: An international comparison. In U. Pfenning & O. Renn (Eds.), *Wissenschafts-und Technikbildung auf dem Prüfstand* (pp. 93–110). Nomos Verlag.

de Vries, M. J. (2016). *Teaching about technology: An introduction to the philosophy of technology for non-philosophers*. Springer.

Dusek, V. (2006). *Philosophy of technology: An introduction* (Vol. 90). Blackwell Oxford.

Fahrman, B. (2021). *To know a subject - Teachers' views about the subject of technology: How the subject of technology is described and approached by teachers in the lower secondary school* [Licentiate thesis, KTH Royal Institute of Technology]. https://www.diva-portal.org/smash/get/diva2:1557396/FULLTEXT01.pdf

Houkes, W. (2009). The nature of technological knowledge. In *Philosophy of technology and engineering sciences* (pp. 309–350). Elsevier.

Håkanson, L. (2010). The firm as an epistemic community: The knowledge-based view revisited. *Industrial and Corporate Change*, *19*(6), 1801–1828.

Lind, J., Pelger, S., & Jakobsson, A. (2022). Students' knowledge of emerging technology and sustainability through a design activity in technology education. *International Journal of Technology and Design Education*, *32*(1), 243–266. https://doi.org/10.1007/s10798-020-09604-y

Lönngren, J. (2017). *Wicked problems in engineering education: preparing future engineers to work for sustainability* [Thesis, Chalmers University of Technology]. https://publications.lib.chalmers.se/records/fulltext/250857/250857.pdf

McCorduck, P., & Cfe, C. (2004). *Machines who think: A personal inquiry into the history and prospects of artificial intelligence*. CRC Press.

Mitcham, C. (1994). *Thinking through technology: The path between engineering and philosophy*. University of Chicago Press.

Nordlöf, C. (2022). *Lärares transformering av teknikämnet: Om lärares attityder till ämnet och syn på teknisk kunskap i teknikundervisningen* [Teachers' transformation of the technology subject: On teachers' attitudes to the school subject and views of technological knowledge in technology education] [Thesis, Linköping University Electronic Press].

Nordlöf, C., Höst, G., & Hallström, J. (2022). Technology teachers' talk about knowledge: From uncertainty to technology education competence. *Research in Science & Technological Education*, 1–21.

Nordlöf, C., Norström, P., Höst, G., & Hallström, J. (2022). Towards a three-part heuristic framework for technology education. *International Journal of Technology and Design Education*, *32*, 1583–1604. https://doi.org/https://doi.org/10.1007/s10798-021-09664-8

Norström, P. (2022). Technological systems in national standards and curricula. In *Teaching and learning about technological systems* (pp. 35–51). Springer.

Norström, P., Engström, S., & Fahrman, B. (2023). Timeless, socially relevant engineering knowledge and skills for future technology education. In S. Davies, M. McLain, A. Hardy, D. & Morrison-Love (Eds.), *The 40th International Pupils' Attitudes Towards Technology Conference Proceedings 2023*.

Rittel, H. W. J., & Webber, M. M. (1973). Dilemmas in a general theory of planning. *Policy Sciences*, *4*(2), 155–169. https://doi.org/10.1007/BF01405730

Rossouw, A., Hacker, M., & de Vries, M. J. (2011). Concepts and contexts in engineering and technology education: An international and interdisciplinary Delphi study. *International Journal of Technology and Design Education*, *21*, 409–424.

Skolinspektionen [The Swedish National School Inspectorate]. (2014). *Gör det osynliga synligt!* [Make the invisible visible]. Skolinspektionen http://www.skolinspektionen.se/Documents/publikationssok/granskningsrapporter/kvalitetsgranskningar/2014/teknik/kvalgr-teknik-slutrapport.pdf

Stolpe, K., & Hallström, J. (2024). Artificial intelligence literacy for technology education. *Computers and Education Open*, 6, 100159. https://doi.org/10.1016/j.caeo.2024.100159

Svenningsson, J. (2019). Carl Mitcham: Descriptions of technology. In J. R. Dakers, J. Hallström, & M. J. de Vries (Eds.), *Reflections on technology for educational practitioners: Philosophers of technology inspiring technology education*. Brill Academic Publishers.

Svenningsson, J., Höst, G., Hultén, M., & Hallström, J. (2022). Students' attitudes toward technology: Exploring the relationship among affective, cognitive and behavioral components of the attitude construct. *International Journal of Technology and Design Education*, *32*(3), 1531–1551.

Tuomi, I. (2023). Beyond mastery: Toward a broader understanding of AI in education. *International Journal of Artificial Intelligence in Education*. https://doi.org/10.1007/s40593-023-00343-4

Williams, P. J. (2009). Technological literacy: A multiliteracies approach for democracy. *International Journal of Technology and Design Education, 19*(3), 237–254.
Williams, P. J. (2017). Critique as a disposition. In *Critique in design and technology education* (pp. 135–152). Springer.

CHAPTER 5

From Capitalist Vision to Curriculum Crisis

Navigating the Politics of Design and Technology in England's Education System

David Spendlove

1 Introduction

The aim of this chapter is to examine the relatively short history of political influence that has shaped design and technology in England's education system. Examination of the political alignment of the subject remains conspicuously underexplored and under theorised and the intricate interplay of political forces that have shaped both the content and structure of the subject is thus often overlooked and consequently unspoken. I will therefore endeavour to explore the impact of political influence on design and technology education by delving into the intricate political ecosystem that has previously shaped, and now ultimately threatens, the existence and sustainability of the subject.

Attention is therefore directed toward the politics that established and conferred a 'status' on design and technology. This will encompass the broader political ecosystem that has sought to develop, sustain, and more recently undermine its standing within the curriculum. In parallel, an examination of the political influence within design and technology will illustrate the increasing incompatibility of the subject within the complexity of turbulent socio-political, cultural, environmental and economic challenges. Thus, the prevailing populist manifestations of design and technology, that are premised on a western economic mode, reliant upon sustained growth with finite resources and fuelled by exploitation of the cognitive flaws of consumerism and consumption, are shown to be unsustainable and undesirable. The subsequent economic and political influences both within and upon the subject are significant and intricate, yet the discourse surrounding these themes remains surprisingly limited. This becomes increasingly concerning given the rapid demise of the subject within the context of the new 21st century politics of polarisation, post-truth and populism (Naím, 2022).

Within this new political ecosystem, an educated public, able to identify and critique such propaganda, has never been more urgently needed in order to challenge the alternative facts produced by financially distorted and

© DAVID SPENDLOVE, 2025 | DOI:10.1163/9789004710948_005

corporatised social and mainstream media. These foundational educational requirements are, therefore, prerequisites that can and should exist within the school curriculum in multiple and diverse ways. This will enable scrutiny of the prevailing political landscape with its current ideologies, powers and influences. In particular, as we emerge as a society from a period of prolonged disruption and isolation, caused by a global pandemic, into a fractured world of wars, civil unrest, political turbulence, increasing disparities and environmental concerns, the need for a politically informed design and technology educated public has never been greater.

In unpacking the requirements for a future version of design and technology and in order to identify a way forward along with an alternative conception and contribution of the subject, it is necessary to understand how this part of the curriculum, perhaps unwittingly, became a medium for unsustainable, anti-democratic, and exploitative capitalist ideologies. Therefore, a central theme of this chapter is that the prevailing complexities of society today offers a compelling rationale for developing a version of design and technology education that is better equipped and better able to respond to these new challenges. It is, however, acknowledged, that England's fragmented education system is currently far removed from one that is likely to accommodate a subject that aims to create a media-literate, politically informed, economically astute, environmentally aware, design and technologically critical and literate general public.

2 Design and Technology's Political Entanglement with the Neoliberal Curriculum

The manifestation of neoliberalism exists in diverse forms and aims. Fundamentally, however, it represents a profound transition from authoritarian state socialism to liberal market capitalism which accelerated in the aftermath of the collapse of the Soviet Union in the late 1980s. This free-market, capitalist, radical experiment in modernising the economy, civil society and enterprise culture has spread within England to include increased privatisation and market-oriented reform of education with an emphasis on school autonomy and parental choice, alongside market-led mechanisms such as league and performance tables. As such, design and technology emerged at a critical political moment and central to the premise of this chapter, is the assertion that the subject was fundamentally shaped by, and promoted on, the basis of the prevailing neoliberal and capitalist ideologies of the time.

The subsequent neoliberal version of design and technology, emerging from this period which I will refer to as version 1.0, was uncritically adopted, and was a beneficiary of, capitalist ideologies. My position is that the uncritical sustaining of these ideologies, through design and technology education, is no longer justifiable or sustainable. This applies particularly to the indefensible and unnecessary promotion of 'consumption and production' pedagogies that mask contrived processes premised on artificial modes of 'solving problems.' My unequivocal stance is therefore that the 1.0 paradigm is no longer justifiable or acceptable. Consequently, an alternative, designated as 2.0, is proposed. This version explicitly recognises and engages with alternative political and economic dimensions that can authentically become embedded in future iterations of design and technology education and that are in closer alignment with a more harmonious and equitable society.

The conceptual shift to 2.0, therefore, offers a framework for interrogating neoliberal assumptions about civil liberties, political and economic power, society, poverty, media, consumption, and wealth. In essence, 2.0 poses a direct challenge to the unspoken capitalist, consumption-driven, and exploitative narratives that guide everyday actions within the 1.0 paradigm. It is, however, essential to recognise that version 1.0 has previously been the beneficiary of political favour, enjoying growth and popularity, buoyed by the ambitions of previous governments, whilst conversely, the subject has now experienced a decade of neglect and spectacular implosions under the same political ideology. In proposing an alternative 2.0 model, it is, therefore, crucial to examine the intricate political ecosystem, allegiances and consequences of seeking to secure a future, alternative mode of the subject. In particular, there is an imperative to recognise the inherent tensions and nuances involved in navigating between political forces and authentic educational paradigms.

Since the 1988 Education Reform Act (1988), when design and technology was established as a national curriculum subject, the educational infrastructure in England has undergone significant transformations, marked by a shift away from local democratic governance to semi-independent schools in the form of academies. Academisation, modelled on the US public charter school system, has resulted in around 80% of state secondary schools becoming academies, while the conversion rate for primary schools is around 40%. The long-term vision is for all state schools to join multi-academy trusts (MATs) based on the premise of enhancing standards through autonomy, competition, marketisation, and corporatisation, often facilitated through sponsorship and support from businesses, universities, or faith groups.

Such rapid policy reforms, from all political powers, across the last half century, driven by a desire to impose market principles, have resulted in a disjointed school system lacking in local and regional coordination and governance. Moreover, despite political claims that the ongoing fragmentation addresses regional disparities, the attainment gap between disadvantaged and advantaged students has widened, further perpetuating a class-based education system with private school alumni and elite universities sustaining a 'governing elite' that has reproduced itself, and the inequities in society, generation after generation (Beckett, 2017).

Accordingly, the more recent emphasis, driven by successive conservative governments, on a neo-traditional and neoliberal education has fuelled a 'culture war' between progressive and traditional approaches to teaching. This has resulted in the erosion of teacher agency, the narrowing of the curriculum and a shift towards knowledge-rich instruction at the expense of breadth and diversity of understanding. This narrow and constrained view of the curriculum, accompanied by the marginalising of teacher agency, increased academisation and corporatisation of the school system, has given rise to a new education ecosystem in which the current form of design and technology education has become increasingly stifled, marginalised and ultimately threatened.

Understanding the current plight of design and technology education in England, therefore, requires a historical perspective that recognises the emergence of this subject from the 1988 Education Reform Act, a pivotal moment marked by the influence of the Conservative party leader and Prime Minister Margaret Thatcher and the principles of capitalism that shaped her era. From this moment, design and technology, perhaps unintentionally, became entwined with the political ideology of Thatcherism, a period marked by policies that exacerbated social inequality and promoted individual entrepreneurship while diminishing the state's role in the economy by privatizing state-owned industries.

3 Navigating the Legacy of Thatcherism

Margaret Thatcher, the Conservative prime minister from 1979 to 1990, who had also previously served as the Secretary of State for Education from 1970 to 1974, initiated a radical neoliberal reform of education provision whilst in power. This period of far-reaching reform saw contradictory free market principles emerge alongside increased centralisation through the establishment of a National Curriculum, in which English, mathematics, and science were designated as core subjects. Simultaneously, seven foundation subjects were introduced, with 'Technology' being the sole new addition. Notably, technology education was subdivided into two components: design and technology (D&T) and information technology (IT). This laid the groundwork for a new form of design and technology education that aligned with the free-market principles of innovation and entrepreneurship, focusing on designing and making capabilities.

Central to these developments was the Secretary of State for Education Kenneth Baker, a strong advocate for 'technology education.' He envisioned the subject as part of the 'Fourth Industrial Revolution' and as a means of addressing the loss of millions of jobs in the manufacturing industries (Tomlinson, 2021) during the early period of Thatcher's tenure. However, Baker's prioritisation of the needs of industry over the broader opportunities of the subject created a trade-off in which design and technology succumbed to the influence of industry and politics. This ultimately shaped the subject's trajectory and location within the curriculum.

Whilst the Education Reform Act of 1988 granted design and technology a dedicated place on the curriculum, it also instigated additional structural reforms which further increased the adoption and status of the subject. City Technology Colleges and the specialist schools programme, including the establishment of the Technology Colleges programme, provided significant funding to schools where 'Technology' education played a central role. Consequently, within a relatively short time, design and technology had emerged as a political instrument, an unwitting pawn in the neoliberal politics of the time, and integral to structural reforms. The subject, with its limited history, leveraged the Conservative government's initiatives while garnering newfound status and substantial investment. However, this apparent 'success' masked the underlying tensions that would later surface and subsequently make the subject unsustainable

Despite the design and technology community seizing what appeared to be an opportunity, the translation of the subject into reality in the curriculum along with the willingness of schools to acquiesce to political influence has ultimately resulted in short-lived benefits. Complicity within the community, combined with a willingness to be manipulated by the needs of industry and political pressures, have ultimately hindered the establishment of a clear identity and purpose for the subject. Accordingly, design and technology proved more vulnerable than other curriculum subjects as it rode a capitalist wave, uncritically aligning itself with the promotion of industry, production, and consumption. Such willingness to acquiesce ultimately conferred an artificial status on the subject that necessitated a rapid upskilling of the teaching workforce to implement poorly conceived curriculum reforms. Unfortunately, this approach has, in hindsight, proven detrimental. The subject continues its struggle to navigate its identity and purpose, naively contributing to social inequity whilst being insufficiently responsive to the challenges of climate change and environmental damage.

Whilst it may have been difficult to anticipate the crisis of global warming (Letcher, 2021) and the entrenched structural inequalities (Coffey, et al. 2020) that have emerged since the early 1990s, design and technology, approximately 35 years after its formal introduction, finds itself enculturating and normalising a capitalistic mindset, fostering consumption and production under the guise of problem-solving, product design, and manufacturing. Thus, in the midst of human-made climate change, the national curriculum for the subject continues to promote the design and manufacture of 'high-quality prototypes and products for a wide range of users' (DfE, 2013) with minimal consideration of the economic, social, or environmental implications.

As we approach the end of the first quarter of the 21st century, an examination of the evolution of design and technology reveals a subject born out of Thatcher's capitalist vision that silently persists in its everyday practices. Over the last three decades, the theme of 'production', typically focused on artifacts, has remained a constant that is superficially justified, both as a means of aspiration and a mode of assessment. This paradigm, endorsed by the contrived concepts of 'needs and wants' (DfE, 2013), anchors design and technology in the excesses of unethical, hierarchical, neoliberal principles. It thrives on the evolutionary flaws of gratuitous accumulation and consumption.

4 The Systemic Erosion of Design and Technology by Wilful Neglect

Since the inception of Design and Technology, England's political landscape has been controlled largely by two parties. In simplistic terms, the Conservatives, represent centre-right wing, free market, capitalist, and traditional ideologies, while the Labour Party, embody centre-left wing, social democratic, and progressive values. Design and technology has evolved and been shaped during prolonged periods of both Conservative and Labour governments.

As previously indicated, the subject emerged out of Conservative reforms, during a period which stretched from 1979–1997 and which saw 'design' and 'technology' emerge as a single entity as a national curriculum subject. In doing so, the seeds of conceptual conflation in the relationship between 'design' and 'technology' were sown. This then translated into confusion over how the subject would be taught and assessed as a single entity. Nevertheless, the structural and financial support for the subject from the government, and the unabated enthusiasm of the Secretary of State for education, can be considered to have accelerated the status of design and technology beyond its capacity for coherence.

Under the ensuing period of Labour governance (1997–2010), design and technology maintained its previous momentum, with continued growth in GCSE examination (age 16) entries peaking in 2003. In that year 414,000 students studied the subject, making it the fourth most studied subject after mathematics, English and science. However, a pivotal change occurred in 2004 when Labour 'disapplied' the subject, making it non-compulsory for students aged 14–16, although design and technology remained an 'entitlement'. This decision marked the end of sustained growth and initiated a rapid decline in GCSE entries resulting in them being halved by 2009 to 280,670. This decline in examination entries continued, moreover, as a new Conservative government took power from 2010. This heralded the promotion of a neo-traditional, knowledge-rich curriculum and led to the further marginalisation of design and technology, leaving it on the periphery of the curriculum, and experiencing the most rapid implosion of any curriculum area. Across this decade, examination entries were halved again, falling first to 136,150 and then falling further to a mere 79,000 by 2023.

Over a period of approximately 30 years, design and technology examination entries in England have fallen by approximately 85%. Correspondingly, teacher education has suffered its own crisis (Ellis, 2023) as the number of new design and technology teachers being 'trained' has dramatically fallen by two thirds, from approximately 1200 in 2010–11 to less than 400 a year by the end of the decade. This sustained collapse can largely be attributed to both the initial, artificially inflated status afforded the subject and the impact of the, often 'unintended,' consequences of government reforms. Rather than these being deliberate attempts to destabilise the subject, the disregard shown by successive governments toward design and technology has, however, revealed weaknesses in its foundation and its susceptibility to external changes.

Nevertheless, successive governments' failure to act to protect the subject and reduce disruption can be construed as a 'wilful neglect' (Payne-James et al., 2024). It also privileges the structural significance of hierarchical notions of knowledge and understanding considered necessary for a successful society, thus inferring that design and technology has little to offer. This negligence from successive governments is thus multifaceted and the embodiment of disregard, abandonment and disrespect. Such ultimately reckless and deliberate systemic neglect indicates a lack of understanding and interest in what the subject offers students. In turn, the failure of successive governments to prioritize the subject has perpetuated a cycle of disinterest and apathy that has undermined the subject's sustainability. This has resulted in a diminishment of its profile and presence on the curriculum.

Deliberate, wilful neglect, involving contempt, abandonment, and disrespect, has thus diminished the status of design and technology within England's curriculum. Over the course of around three decades, the subject has experienced a sharp increase and decrease in examination entries, political interest and funding, alongside a substantial decline in the recruitment of new teachers (Tuckett, 2022). Additionally, the breakdown of governance systems (Spendlove, 2022), such as local education authority guidance and government oversight, has compounded the subject's deterioration. Whilst this may not have been overtly intended to undermine the subject, it demonstrates a reckless indifference towards its importance in educational discourse. Furthermore, such wilful neglect extends to the broader ecosystem of education communities (ibid.), exacerbating the demise and damage to the subject. Policies such as academisation, the shift away from local education authorities and the crisis in teacher education have led to a reduction in university provision, stripping the subject of crucial professional development and diminishing the subject's research and intellectual base.

Government ministers such as Nick Gibb, who prioritised a neo-traditional curriculum, further contributed to the wilful neglect of design and technology. Gibb's focus on 'academic' subjects and his disdain for skills and creativity contributed greatly to the subject's decline. Gibb's emphasis on a neo-traditional, 'knowledge rich', cultural literacy mantra, based on the work of E.D. Hirsch (1988), manifested itself in radical changes to both the national curriculum and GCSE. Yet this wasn't a panacea for greater social mobility as was claimed at the time. The refocusing on a narrow view of 'education' was, conversely, merely a means of preserving and reproducing the class system rather than a genuine attempt to increase accessibility and equity It was a reaffirmation that knowledge is neither universal nor neutral but, rather, is always social and ideological (Berger & Luckmann, 1966).

Reform in England therefore, instead of widening the curriculum to increase the breadth of experiences or a broadening of culture as claimed, has achieved the opposite by limiting access to subjects, individuals, and institutions the government chose to privilege. In contrast, figures like Lord Kenneth Baker, who were instrumental in design and technology's early growth, exemplified a misplaced enthusiasm for a narrow, ill-informed view of the subject's potential. Baker's commitment to vocational and technical education during his tenure (Passmore, 1997), whilst contrasted with Gibb's policies, can also be seen to have negative consequences through promoting a particularly oppressive, economic, and enterprise vision of the subject.

Gibb and Baker therefore represent 30 years of extremes of Conservative policy which have ultimately shaped the subject's trajectory, with Gibb overseeing design and technology's demise through neglect, whilst Baker advocated for its artificial status and growth during his tenure. Thus, whilst Gibb and Baker embody opposing views within Conservative policy, they also exemplify the fickleness of political will and ideology. This has resulted in the failure of design and technology to establish itself within the curriculum, based upon sound principles. The subject's volatile history is also illustrated by its dependency on government support and its vulnerability to political influence. While some policymakers championed design and technology's growth, others facilitated its decline through neglect and alternative policy priorities.

Ultimately then, the decline of design and technology in England is a consequence of political 'wilful' neglect and misguided policy decisions. Reversing this trend would require sustained advocacy, policy reform, and a commitment to supporting the development of the subject in the curriculum. Design and technology's decline reflects a failure of successive recent governments to acknowledge its potential significance in offering unique opportunities. Yet it is also incumbent on the community to concede that there has been a failure to coherently and consistently communicate the values of the subject. This has resulted in its susceptibility to political whims. Consequently, as design and technology navigates uncertain terrain, its future hinges on concerted efforts to revitalise and reimagine its role within the curriculum.

5 Activism and Values Driven Pedagogy

I previously called for an exploration of an alternative version of the subject in the form design and/or technology 2.0. This represents a conceptual upgrade of the subject, drawing on the metaphor of software and product upgrades (Spendlove, 2017). In positing that a significant conceptual overhaul is required in moving towards version 2.0, it is also acknowledged that intellectual and reflective space is needed to reconceive a future version of the subject.

Within version 2.0, there are two crucial dimensions central to the conceptual upgrade. First, there is the fundamental issue that the current version 1.0 lacks a clear sustainable rationale for whether 'design and technology' should be perceived as a combined entity (design AND technology) or as two separate, related but optional domains (design OR technology). Addressing the semantic, philosophical, and pedagogical implications of this issue are deemed essential for the future existence of the subject. Second, there remains a recognized weakness in the development of authentic design capability in students within 1.0.

Consequently, I have proposed the concept of 'design thinking' to be considered as a potential catalyst for design and/or technology 2.0. This acknowledges the historical challenges in establishing traditional notions of design within current practice, whilst advocating that 'design thinking' could provide an opportunity to reconceptualise the subject by emphasising cognitive and reflective aspects over creative and aesthetic elements. Accordingly, 'design' within an education context requires rethinking and redefining by challenging the dated concept of design education as comprising predominantly aesthetic, creative, material, and product-oriented activities.

Calling for design and/or technology 2.0 therefore, represents a conceptual upgrade, one that also requires a re-evaluation of teacher development, governance structures, and prevailing political attitudes. This transition from the existing subject (1.0) to 2.0 also emphasises the importance of 'activism' (Spendlove, 2022a) and its incorporation into any future iteration of the subject. At the core of 'activist' pedagogies is the need for a thoughtful and reflective approach to 'design thinking' within 2.0. This requires an understanding of the diverse values that influence design decision-making. It is, however, acknowledged that simply increasing awareness of cognitive limitations, as a feature of 'design thinking', is insufficient unless accompanied by action. Thus, a fundamental aspect of 2.0 within an educational context should be its 'activist' orientation for both students and teachers.

Activism, as conceptualized by Jordan (2002), is identified as a moral endeavour aimed at envisioning a 'better' society. Within this context, the teaching profession itself needs to be 'educated and politically astute' (Sachs, 2003), as teachers are uniquely placed to 'see first-hand' the inequity and economics of poverty played out in their classrooms and local communities. More specifically, teachers, in the context of 1.0, need to be aware of their presumably 'unconscious' reproduction of the broader neoliberal (Giroux, 2004) project that constrains rather than liberates their students. Building upon 1.0, then, and defining a vision for 2.0 (Spendlove, 2020), the emphasis is on exploring the nature, scope, direction, and expression of activism, rather than questioning its necessity. This argument arises from the recognition that design and technology education 1.0 has often exhibited tokenism and replication, resulting in a benign educational experience that fails to genuinely address values and causes.

Whilst acknowledging the initial historical successes of establishing design and technology within the curriculum, the rapid demise of 1.0 nevertheless provides a unique opportunity to reconsider the subject's purpose within evolving economic, political, and environmental contexts. It allows for the reconceptualisation of an activist-oriented learning experience within the domains of design and/or technology, presenting it as an essential opportunity rather than a threat. The call for activism in 2.0 thus underscores the need for both teachers and students to possess agency and awareness of how their endeavours, through designing and/or technologizing, can 'activate' potential processes and outcomes. This implies agential realism and emphasises the importance of acknowledging the invisible 'habitus' or ecosystem that influences decision-making. It argues that, without a proactive stance, students studying design and/or technology may inadvertently contribute to a failing capitalistic mind-set with its reproduction of social inequities and environmental damage.

Accordingly, teachers need to be aware that the pedagogical choices they make are neither neutral nor passive. Teachers are 'producing identifications for students and are implicated in the pedagogical formation of identity'. Consequently, they are 'responsible for what meanings they make in the classroom' (Saltman, 2018, p. 4). Teachers are thus reproducing or contesting ethical and political discourses through their everyday actions and choices which in design and technology become all the more profound when seeking to 'improve the world' around them.

Embedded within 1.0 is a neoliberal legacy of Thatcherism which therefore requires additional scrutiny, rather than the passive reproduction of products without a consideration of their broader implications. Consequently, within 2.0, the contemporary challenges of global warming, structural inequalities, and wealth disparities become central to the relevance of a curriculum that fosters an activist orientation and aligns with values that give priority to environmental responsibility and equity.

2.0 is ultimately presented as an opportunity to transform the existing capitalist model of design and technology education. It advocates a shift away from conventional 'human-centred' approaches, towards a value-driven paradigm that prioritises environmental sustainability, prevents individual exploitation, and adheres to the principle of 'doing no harm' (Spendlove, 2023). In contrast to the assumptions inherent in 1.0, 2.0 challenges the legitimacy of capitalist consumption embedded in the creation of 'innovative, functional, appealing products' based solely on perceived needs and wants. 2.0 aims, moreover, to embody proactive activism, to foster a commitment to social justice within and beyond the classroom. Teachers and students are therefore encouraged to actively pursue transformative and sustainable change by addressing economic disparities such as injustice, inequality, and to question the production of superficial goods. Within this context, "dialogue, debate, dissent, and habits of engaged curiosity become the lifeblood of schooling as they are the lifeblood of democracy" (Saltman, 2018, p.8). Thus, activism within 2.0 acknowledges the complexities inherent in evolving educational dynamics and underscores the essential need for teachers to comprehend the broader political landscape.

The concept of activism within 2.0 is recognised as being potentially contentious, particularly in the current education system which prioritises compliance and which resists alignment with activist ideals. In addressing the challenge of maintaining political neutrality in schools, activism emphasises the significance of nurturing a democratic educational environment that strives for equity. This positions 2.0 as an alternative discourse for the future of design and technology education. It promotes activism as a means of combating the anti-democratic, exploitative, and environmentally harmful ideologies that are surreptitiously embedded within 1.0. As a catalyst for change, 2.0 also calls upon educators to adopt an activist stance that emphasises equity, ethics, and values and which envisions a sustainable and just future. Within the context of the decline of design and technology education 1.0, a shift towards a values-driven 2.0. is advocated. It challenges neoliberal assumptions inherent within 1.0, interrogates foundational beliefs, and promotes a pedagogy that empowers both teachers and students to actively pursue social justice, environmental stewardship, and equity. In essence, 2.0 represents a departure from the past and an essential step towards a more purposeful, authentic, and meaningful future for design and/or technology education 2.0.

6 Conclusion – From Political Influences to Pedagogical Evolution

This chapter essentially embodies a call to action. It urges individuals to dedicate themselves to further work, effort, and energy in the pursuit of improving various aspects of learning. While the scope of this chapter is centred on England's education system, it also serves as both a cautionary tale and an opportunity for other jurisdictions to critically examine and evaluate the political foundations and sustainability of their own design and technology programmes.

Within the current turbulent, political climate which is characterised by polarisation, post-truth rhetoric, and populism, I have endeavoured to examine the ways in which design and technology education may have faltered or failed, particularly by examining the previously underexplored political foundations of the subject. In particular, the impact of the neoliberal underpinnings of design and technology education, labelled as 'version 1.0', which aligns with capitalist ideologies of consumption and production, has been explored. Design and technology currently exists in an impoverished state, surreptitiously attempting to further 'our dopamine economy' (Lembke, 2021) based on a form of 'limbic capitalism' (Courtwright, 2019). Accordingly, a blind spot remains where the means to notional 'successes' within the existing subject configuration is demonstrated through adherence to, or giving the impression of, the reproduction of capitalist, consumerist and colonialist views of the world.

At the same time the subject remains exposed and vulnerable to manipulation and further wilful neglect from the same political parties who initially accelerated the subject's presence in the curriculum. Consequently, the worst outcome for the long-term viability and credibility of the subject, is that it be rebuilt on the same principles under which it previously thrived. This would not be sustainable, desirable nor morally or intellectually justifiable. Our existing culture of production and consumption can no longer be legitimised and reinforced through the cultural norms of schooling or accorded status through the rewards of assessment. Yet ironically, within the current model of design and technology in England, the means to 'success' is also the means to long-term failure, as 'performative success' is rewarded through the reproduction of the social class system, the acceleration of capitalism, and by contribution to the climate crisis. It is, therefore, only through the disruption of the existing mode of design and technology that an opportunity exists to confront and critique our relationship with society, inequity and environmentalism. By these means we can challenge preconceptions of power and influence.

As a relatively young discipline, design and technology has too often been subject to manipulation by political influences. These have accelerated and elevated its status in order to secure its place in the curriculum without acknowledgement of the significant negative consequences attached: these include the reproduction of economic and social inequalities and further contribution to climate emergencies. The subsequent constraining of the subject, within a narrow and distorted content and pedagogical framework, along with corresponding structural changes, have ultimately deprived design and technology of the opportunity to authentically shape its own identity and moral imperative. Thus, design and technology education has manifested into a state of indecisiveness, leaving it susceptible to the deliberate wilful neglect of both major political parties in England.

The discourse surrounding the intersection of politics and design and technology education remains surprisingly sparse, yet understanding the political dynamics that have shaped, and continue to shape, the subject is crucial in order to protect and sustain the authentic knowledge, skills, understanding, and values integral to a high-quality educational experience. Evidently, design and technology has succumbed to political ideology and has uncritically followed a consumerist, commodity-oriented agenda, whilst recklessly pursuing short-sighted economic and environmentally damaging pedagogies. Yet, despite design and technology having previously been the beneficiary of political gaming and favouritism, simply waiting for a return swing of the political pendulum or continuing along a path aligned with capitalist reproduction of inequalities and environmental damage, is not a sufficient, nor indeed an ethical, strategy for the future.

2.0 therefore emerges as an opportunity to challenge the self-fulfilling prophecy of 1.0 and offers a platform from which to explore alternative activist pedagogies for the future benefit of both the subject and society. By exposing the implicit, detrimental, capitalist orientation of design and technology education, we can direct future iterations towards environmental sustainability, cultural inclusivity, and earth-centred perspectives. The subject community should therefore take charge of defining its sphere of influence and reject political attempts to distort its future contribution. Likewise traditional hierarchies of subjects and competencies, mandated through political policy, finance, and rhetoric, that have perpetuated and sustained societal inequalities ultimately need to be challenged. It is, for example, imperative to challenge the subservient position of design and technology education in to enable the subject to evolve authentically and on merit.

If design and technology education is to evolve, it must acknowledge the necessity, dynamics and vagaries of political support. In particular, there is a need to acknowledge the significant danger of repeating past mistakes by seeking to gain government endorsement without critically evaluating the feasibility, sustainability, or desirability of such actions. Ultimately, design and technology education must be recognised for its potential to positively impact upon society rather than the consequences of its negative impact. Therefore, this chapter has argued for a conceptual shift towards 'version 2.0'. This advocates a values-driven approach that challenges capitalist narratives and promotes sustainability, social justice, and critical 'design thinking'. It calls for activism within the subject by emphasising the role of teachers and students in actively pursuing transformative change and addressing societal challenges of inequality and environmental damage.

Ultimately, the path to societal transformation and liberation lies in education and in particular on the need for a reimagined approach as outlined in the form of design and/or technology education 2.0. Version 2.0 calls for a transformation to a more equitable, ethical, and sustainable educational model. It envisions a future where students are empowered to effect positive change in society. This chapter is, therefore, unequivocal in its promotion of societal transformation, liberation, and environmental safeguarding. This will involve the creation of a shift in educational paradigms. The result will be a more purposeful and meaningful approach to design and/or technology education 2.0. This in turn will lead to the development of a society that can successfully navigate the inevitable, and increasingly complex, challenges that lie ahead.

References

Beckett, A. (2017). PPE: The Oxford degree that runs Britain. *The Guardian*. https://www.theguardian.com/education/2017/feb/23/ppe-oxford-university-degree-that-rules-britain (Accessed:1st November 2023)

Berger, P., & Luckmann, T. (1966). *The social construction of reality: A treatise in the sociology of knowledge*. Doubleday.

Coffey, C., Espinoza Revollo, P., Harvey, R., Lawson, M., Parvez Butt, A., Piaget, K., Sarosi, D., & Thekkudan, J. (2020). *Time to care: Unpaid and underpaid care work and the global inequality crisis*. Oxfam.

Courtwright, D. T. (2019). *The age of addiction: How bad habits became big business*. Harvard University Press.

Department for Education. (2013). *Design and technology programmes of study: Key stage 3*. Department for Education.

Ellis, V. (Ed.). (2023). *Teacher education in crisis: The state, the market and the universities in England*. Bloomsbury Publishing.

Giroux, H. (2004). *Terror of neoliberalism: Authoritarianism and the eclipse of democracy*. Garamond Press.

Hirsch Jr., E. D., Kett, J. F., & Trefil, J. S. (1988). *Cultural literacy: What every American needs to know*. Vintage.

Jordan, T. (2002). *Activism!: direct action, hacktivism and the future of society*. Reaktion books.

Lembke, A. (2021). *Dopamine nation: Finding balance in the age of indulgence*. Penguin.

Letcher, T.M. (2021). Global warming—a complex situation. In *Climate change* (pp. 3–17). Elsevier.

Naím, M. (2022). *The revenge of power: how autocrats are reinventing politics for the 21st Century*. St. Martin's Press.

Parliament, U. K. (1988). *Education Reform Act 1988*. HM Stationary Office.

Passmore, B. (1997, February 21) Kenneth Baker's dream celebrated 10 years on ... *Times Education Supplement*. Accessed February 16, 2023, from https://www.tes.com/magazine/archive/kenneth-bakers-dream-celebrated-10-years

Payne-James, J., Perera, C., & Aggrawal, A. (2024). 'Neglect,' in *Forensic and Legal Medicine* (pp. 464–472). CRC Press.

Sachs, J. (2003). *The activist teaching profession*. Open University Press.

Saltman, K. J. (2018). *The politics of education: A critical introduction*. Routledge.

Spendlove, D. (2017). Design thinking: What is it and where might it reside? In *Design epistemology and curriculum planning* (p. 39). Loughborough Design Press Ltd.

Spendlove, D. (2020). Design and/or Technology 2.0: Is this the way forward? In A. Hardy & E. Norman (Eds.), *Talking and thinking about design and/or technology 2.0*. Loughborough Design Press.

Spendlove, D. (2022). Why did design and technology education fail, and what might replace it? In *Debates in design and technology education* (pp. 65–76). Routledge.

Spendlove, D. (2022a). Activism for 2.0: 11th DATTArc-ICTE-TENZ-ITEEA 2022. In *11th DATTArc-ICTE-TENZ-ITEEA 2022*: Southern Cross University, Gold Coast Campus, QLD, Australia.

Spendlove, D. (2023, October). Do no harm 2.0. In *The 40th International Pupils' Attitudes Towards Technology Conference Proceedings 2023* (Vol. 1).

Tomlinson, J. (2021). Deindustrialisation and 'Thatcherism': Moral economy and unintended consequences. *Contemporary British History, 35*(4), 620–642.

Tuckett, S. (2022). *A spotlight on Design and Technology study in England: Trends in subject take up and the teacher workforce*. Education Policy Institute. Accessed February 9, 2023, from https://www.data.org.uk/media/4402/spotlighton-dt-report-epi-march-2022.pdf

CHAPTER 6

Back to the Future

(Re)imagining a History-based Technology Curriculum

Jonas Hallström

1 Introduction

A recurring challenge in technology education is the notion of technological change, that is, the fact that technology continually proliferates and evolves as society changes (e.g., Arthur, 2009). This is of course not something that is exclusive to technology education – a changing society affects all school subjects, more or less – but technology is particularly noticeable due to the speed of innovation and range of stakeholder influence on the subject (Layton, 1994). The most obvious consequence is that technology curricula, syllabuses, and standards may more easily become obsolete compared to those of many other school subjects. The rapid development of especially certain technology – for example, biotechnology, Internet-based social media, artificial intelligence (AI), computer systems, and more – is often slow to be included in curricula as governments and schools cannot – and perhaps should not – change the subject knowledge components and requirements at the same pace.

In response, this chapter will outline a history-based technology curriculum. This (re)imagined curriculum will be based on historical examples of technological design, innovation, and use, from different regions and cultural contexts across the globe. Furthermore, it will outline certain common principles and patterns that do in fact recur in the history of technology (Arthur, 2009; Hallström, 2023). At the same time as it is mostly old and established technologies that form the backbone of society (Edgerton, 2008), students also need to tackle current technological change, and be prepared to both use, model, and design new, hitherto unknown, technologies (Norström & Hallström, 2023). Thus, in the proposed curriculum, such historical principles will also be possible to project into the immediate future.

The object of the proposed curriculum is to put the evolutionary character of technology in its context and have students learn not only about the nature of technology as such but also the societal preconditions, driving forces, and actors (users, designers, engineers, politicians, etc.) that together propel technological development. In this way, students can learn to understand, use, design, and

© JONAS HALLSTRÖM, 2025 | DOI:10.1163/9789004710948_006

critique technology regardless of historical, geographical, and societal context (e.g., Williams, 2017), and will be better prepared to handle future technological artefacts and systems. They will also be able to appreciate the relationship of technology to STEM subjects – science, technology, engineering, and mathematics – more broadly (de Vries, 2001, 2005), as well as assess new technology in relation to citizen integrity, cybersecurity, systemic risks, sustainable development, and more. Such a technology curriculum will be instrumental in achieving a timely and relevant technology education for the benefit of students, and in fact all citizens, in an even more technology-saturated future society.

2 History of Technology as Curriculum Component in Technology Education

That the history of technology and technological change should be part of technology education is not self-evident. School subjects look different in different national and regional, political, and cultural contexts, as actors in and around the school negotiate the position in the curriculum and the actual content of school subjects (Goodson, 2014). However, in quite many countries with a broader view of technology education, there is also an appreciation of the importance of the historical perspective as a prerequisite for understanding present-day society and looking to the future. These countries often have an approach to technology education that includes a broader societal context (Nordlöf et al., 2022). In the United States, for instance, the history of technology has been a natural ingredient of technology education for decades. The Committee on Technological Literacy carried out work in the early 2000s that resulted in the report *Technically Speaking: Why All Americans Need to Know More About Technology* (Pearson & Young, 2002). It was clear to the Committee that technology education also includes knowledge of technological change, and therefore a technologically literate person – a person with a basic and comprehensive understanding of technology – should know something about:

> The ways technology shapes human history and people shape technology [...] society shapes technology as much as technology shapes society. There is nothing inevitable about the changes influenced by technology – they are the result of human decisions and not of impersonal historical forces. (Pearson & Young, 2002, pp. 4, 18)

History of technology is now also a curriculum objective in the American *Standards for technological and engineering literacy* (ITEEA, 2021). Another similarly comprehensive curricular example is from the New Zealand curriculum of 2017, which, incidentally, was the same in the 2007 iteration and describes historical elements in the subject as follows:

> [Students] learn to critique the impact of technology on societies and the environment and to explore how developments and outcomes are valued by different peoples in different times. As they do so, they come to appreciate the socially embedded nature of technology and become increasingly able to engage with current and historical issues and to explore future scenarios. (Ministry of Education, 2017, p. 3)

Technology in the New Zealand Curriculum thus emphasises the importance of students' understanding that technology develops in a societal context, in a similar way to the Committee on Technological Literacy, by using historical examples.

In Sweden, technology curriculum documents dating back all the way to the 1960s, before the subject became mandatory, have included some history of technology (Hallström et al., 2013). The Swedish Government-appointed Technology Delegation (*Teknikdelegationen*) echoed this tradition in its 2010 report on the then future of STEM education:

> Technology education, like history education, for example, should give us an understanding of the society we live in, how we got here, and where we might be going. It should also, by analogy with school science education, provide a foundation and open up opportunities for students who later want to specialise in technology. (Teknikdelegationen, p. 113)

The current Swedish curriculum (Skolverket, 2022) includes a historical perspective in the theme "Technology, people, society, and the environment" in the core content. It is also within this theme that most of the history of technology content is to be found, which falls back on the formulations of aims and abilities in the introduction to the syllabus:

> Teaching in the subject of technology should provide students with the preconditions to develop:
>
> - the ability to reflect on different choices of technical solutions, their consequences for the individual, society, and the environment, and how technology has changed over time. (Skolverket, 2022, p. 1)

Compared to previous curricula this aim is now the first one in the syllabus, something which indicates its increased importance. It is thus in technology education that explicit components of the history of technology and technological change are included, and which primarily assumes responsibility for this content in Swedish compulsory schooling.

A notable contrast to the above reports, curriculum documents, and standards is the Australian Curriculum, which has the same underlying philosophy with a focus on broad technological knowledge and technological literacy. This concerns the subject Technologies, which is divided into Design and Technologies and Digital Technologies. The Australian Curriculum is broad and future oriented, and it focuses a great deal on students' contributions to the sustainability of future technological solutions. However, this technology curriculum does not have a historical perspective, which is shown succinctly in the following extract from Technologies, sequence of achievement, years 9 and 10:

> [The students] identify the changes necessary to designed solutions to realise preferred futures they have described. When producing designed solutions for identified needs or opportunities, students evaluate the features of technologies and their appropriateness for purpose for one or more of the technologies contexts. Students create designed solutions for one or more of the technologies contexts based on a critical evaluation of needs or opportunities [...]. (ACARA, 2023)

Thus, curricula can be broad, comprehensive, and forward-looking without the historical perspective. However, it seems to be very common in countries with technology curricula and standards that aim for a broader technological literacy – which may even stretch into engineering and the whole STEM field – to emphasise the centrality of the historical perspective on technology as a way of clarifying what technology is, how it evolves, and how it may be successfully developed in the future (de Vries, 2018; Hallström, 2023).

3 Central Knowledge Components of a History-based Technology Curriculum

History of technology can therefore be seen as a pivotal component when looking to the future, which is in itself an argument for the history-based technology curriculum which I will outline in the following sections. A related issue is exactly what knowledge components should be included in a history-based technology curriculum because the curriculum should be based on how historical knowledge could support the learning of technology. This is by no means an easy endeavour given the plethora of views on technology education, and the diversity of technology curricula around the world (Nordlöf et al., 2022). Furthermore, such knowledge components must also reflect research in the history of technology to a reasonable extent, at least major theoretical and empirical contributions, and historiographical trends (e.g., Edgerton, 2010; Schatzberg, 2018; van der Vleuten et al., 2017). The knowledge components proposed here are by no means exhaustive, so there is room for a teacher to include other artefacts and systems as historical examples or address other problematics. The list that follows is based mainly on my own and other researchers' historical and technology education research and is in this sense well-founded.

- First of all, teachers as well as students who follow the history-based technology curriculum will need to acquire "empirical" knowledge, that is, knowledge of concrete technologies, innovations, and systems and what happened to, in and through them at various points in history and in different places, cultures, and societies:
- Key technological innovations and innovators: These could be technological products and methods, when and by whom they were invented or innovated, and their functions and uses. Furthermore, key actors and actor groups that have driven the development of technology, for instance, engineers, scientific researchers, designers, inventors, research labs, artisans, politicians, company executives, administrators, etc. need to be addressed (de Vries, 2005; Hallström & Kaijser, 2022; van der Vleuten et al., 2017). Not all technology was a product or gadget. Equally important, and in the 19th century and onwards perhaps more important, was the proliferation of technological systems of various kinds such as water systems, gas networks, electric grids, industrial production systems, computer systems, and food systems (Hughes, 1987; Hallström & Kaijser, 2022).

  Importantly, although they historically made up the majority, not all innovators were men. Women made up a substantial, but unfortunately forgotten or suppressed, portion of inventors, not only in recent years (Marçal, 2021). For example, patent applications as early as the 19th century show that there were indeed women inventors. They designed products that benefited women, such as the brassiere, but also techniques for obtaining freshwater from seawater as well as many other technical solutions that solved societal problems (van Dulken, 2001).
- Use and modification of key innovations: Innovations were not only designed and produced but, equally importantly, they were also used. Users often used products in ways that designers had not foreseen, and even invented new

uses, thus blurring the boundary between innovation and use, production and consumption. Users became increasingly important as consumers and recipients of marketing initiatives for new inventions, but amateur users and enthusiasts were also crucial drivers for invention, for example, in the early 20th century development of rocket technology. At other times, users modified and developed existing technologies to be more suitable, and environmentally friendly, such as the Danish windmill movement of the 1970s (Bowler, 2017; Edgerton, 2008, 2010; van der Vleuten et al., 2017). Present-day examples of such groups are maker culture and hacker culture enthusiasts, to whom tinkering and modifying instead of innovating from scratch are central (e.g., McGregor Jacobides & Winterbottom, 2020). When studying technological use, there is also a potential gender issue or bias to be addressed, namely the notion of men being primarily producers and women consumers. This idea may hold in particular contexts, such as late 19th and early 20th century middle-class households with a clear gender distinction between public and private, but not in others (cf., Cowan, 1983; Hagberg, 1986).

– Societal change influencing and/or influenced by technological change: Technological artefacts and systems were central in relation to societal and cultural changes, economic developments, power structures and class issues, gender relations, environmental changes, etc. Thus, technological change concerns the relationships between humans and technology in different contexts, and how society influences technology as well as how technology influences society. Furthermore, from a curricular point of view it is particularly important what can be learnt from different historical examples, for the promotion of current and future gender-inclusive innovation and technological development, sustainable society, and global development (Oldenziel, 1999; Pritchard & Zimring, 2020).

Secondly, teachers and students of the history-based technology curriculum would need theoretical knowledge and tools to orientate themselves in and analyse the historical development of technology. Such theoretical tools could be concepts, principles, patterns, or models:

– *Recurring principles or patterns in the history of technology*: The primary historical principle is in a sense change, although it is also a broad and elusive notion if it is not concretised. To begin with, one can look at the very concept of technology which has changed a great deal etymologically over the centuries, not to mention other concepts that have been used to convey roughly what we mean by technology today; industrial arts, technics, applied science, and many other terms. As regards applied science, it is quite a recent phenomenon and for any serious history or technology scholar a very simplified notion because although science and technology have partly a shared history – and indeed some technology is in some way applied science – they are different areas of knowledge with different purposes. Science seeks knowledge about the natural world while technology is about changing the same, which then becomes a designed, human-made world (de Vries, 2016; Hultén, 2013; Nye, 2007; Schatzberg, 2018). Continuity is another principle and as such technology has been responsible for societies staying the same as much as changing them (Edgerton, 2008). Think of the very robust Roman road network that contributed to continuity of trade for centuries after the fall of the Roman Empire, at the same time as European societies also changed significantly.

There are also theoretical patterns applicable to the evolution of several different technologies, for example, how technological innovations often "mimic" nature, such as airplanes mimicking large sea birds or birds of prey in gliding flight. Another example of the same pattern is a fishing net which might be construed as a humanly designed version of a spider's web. Today we call this pattern of technology emulating biology as biomimetics or biomimicry. Another pattern is technology transfer, that is, successful innovations in one technological field often transfer to new fields (Hughes, 1987). The steam engine, for instance, started out by pumping mines and subsequently transferred to propelling trains and ships. Other examples of successful transfers are the integrated circuit and various computer programming languages, which are used and applied in many, many different technological solutions (cf., Arthur, 2009; Bowler, 2017).

Another recurring pattern is what Edgerton (2010) calls novelty mongering, that is, the attribution of disproportional attention and significance to something only because it is new, which also implies that there are no alternatives to the new technology – but there almost always are. It may also be focusing on the latest innovations instead of the older technologies and systems that actually play a more central role to humanity, such as food production systems, or vital infrastructure like water. It may also be to focus more on innovation and production, and less on use and consumption, thereby creating a skewed history of technology.

– *Application of social science models*: Other important theoretical tools are social science theories, models, and concepts such as actor-structure; needs and drivers; power, in relation to class, ethnicity, and gender. Because these are important for understanding societal change, they will also be crucial for an understanding of technological change (e.g., Burke, 2005). Using such concepts and models will also lead to a discussion of the extent of human agency in relation to technology, and vice versa, the extent of technological agency in relation to humans. By extension, issues of technological determinism and autonomy will have to be dealt with, although these are hard to study empirically. Nevertheless, the idea of technological determinism has been and continues to be powerful which necessitates its inclusion for discussion (Edgerton, 2010; Hallström, 2022).

– *Historical contextualisation*: Contextualisation is perhaps the primary theoretical tool of the historian, which helps sorting out the evolution of technology and ultimately manifests in an empirical narrative about things past. In technology education, context can be "reconstructed" around an innovation to promote understanding of the same. For example, the current rapid proliferation of different AI applications cannot be reduced only to technological advances but has to be viewed in the current context of a few dominating international digital media companies, convergence of more and more of people's everyday communication and routine into social media and other internet-based applications, polarisation and democratic challenges, digital divides between generations, between North and South, etc. (e.g., Briggs & Burke, 2009; Vaidhyanathan, 2018).

Contextualisation could utilise internalist or externalist perspectives, originating within the history of science (Shapin, 1992). The former perspective long dominated research in the history of science and even historical studies of technology. From the internalist perspective, technological innovation is seen as driven mainly by individual scientists and engineers and/or technical innovation/improvement. The centre of interest is the artefact, but very seldom the system or social context. From the externalist perspective, society and its social, cultural, economic, and political forces dominate and are seen as the main driving forces for technology. Staudenmaier (1985) adds a third perspective – the contextualist – which constitutes a recognition that the internalist and externalist perspectives need to complement one another, that is, one needs to look for the driving forces of technology both internally and externally.

The contextual perspective acknowledges technological as well as non-technological actors and factors in shaping technology (Shapin, 1992; Staudenmaier, 1985). Edgerton argues that the existing historical work used to build up context often has an implicit view of technology and technological development that may lead the historian to wrong or contradictory conclusions about technology development. Consequently, he proposes a "post-contextualist" historiography that writes the history of content and context together (Edgerton, 2010).

4 A model for a Progression of Knowledge Components in a History-based Technology Curriculum

Empirical knowledge and knowledge of concepts, principles, and models in the history of technology also need to be linked to a thoughtful progression, so that students' learning can be facilitated by following a logical development. Historically, there are different ways of thinking about progression in schools in general. By far the oldest and most obvious progression is from simple to complex knowledge. This may mean starting with simple knowledge for younger students, or in general for absolute beginners or novices, and continuing with more complex knowledge as they get older or become more experienced in a particular area of knowledge (e.g., Herbart & De Garmo, 2013).

Another form of progression that is conceivable in technology education is a geographical progression, from local to global (Schooner et al., 2023), similar to the one that existed in the Swedish lower primary school for many years, where the first years include a great deal of knowledge components related to the students' local environment. In fact, from 1919 until the early 1980s there was a subject called Home region instruction for lower primary education (also translated as local geography but was broader than that and included both history and technology: *Hembygdskunskap*). After that, more and more knowledge of national and global geographies and issues was included as the students became older. The current Swedish technology curriculum includes both the progression from simple to complex, and from local to global, in terms of how knowledge of technology develops throughout the years of compulsory school (primary and lower secondary education, ages seven to 16). For instance, technological systems begin with simple aspects of local systems, often found in the home, such as knowing about what input and output you can get from a personal computer, for early primary education. At lower secondary level, on the other hand, students are expected to have knowledge of global technological systems such as the internet, and sustainability issues (Skolverket, 2022).

For technological history knowledge, one can also think of the progression from then to now, and even the other way round, from now to then. However, it is not so easy to determine which of these is best. Since the development of technology is in many cases evolutionary (Arthur, 2009), that is, newer technology builds upon older technology, one should presumably start in the far past and move towards the present. However, in such a case one would also have to start far from the well-known, everyday life of the students which is also challenge in history education regarding the historical thinking of students (e.g. Ludvigsson & Booth, 2015).

Assuming we employ the progression from then to now, then there will still be knowledge components that are quite simple such as memorising 1066 as the year of the Battle of Hastings. There will also be more complex processes to learn such as the causes of the Viking conquests (including technological advancements), why there was a Viking colony in Normandy, and what propelled the Norman conquest. There are, theoretically, two different ways of looking at the progression from simple to complex. I explain these with inspiration from the English sociologist of education Basil Bernstein and his concepts of vertical discourse and horizontal discourse (Bernstein, 1990). If one sees the development of knowledge in students as hierarchical or vertical, students work their way "up" towards higher and more advanced levels of knowledge, where the higher, more complex knowledge builds on the lower and simpler knowledge (Biggs & Collis, 1989). In various technology curricula we can see examples of this vertical type of progression from simple to complex. The more complex knowledge about technological change thus consists, for example, of students coming up with developed instead of simple explanations to the Battle of Hastings, and in the practical elements of carrying out design work in a more elaborate way, as opposed to a less developed or advanced solution.

Horizontal progression, instead, assumes that different competences or areas of knowledge are complementary. Knowledge is, from this perspective, a "building block" or "piece of the puzzle" for students that can be added to expand their knowledge, but at the same time, new knowledge does not necessarily build on previous knowledge in the same way as in the vertical discourse or view of progression. Within the technology curriculum there are also examples of this horizontal type of progression, for example in the use of concepts. It could be that early practical craft knowledge expands as new areas and concepts from electronics, mechanics, and materials science are added, as students get older and study technology in higher grades. Regarding technological change, an example of the horizontal progression could be to study history from the point of view of a certain product, system, or technological area, and successively add artefacts and areas as the students progress (Bernstein, 1990).

5 Conclusion

In conclusion, a history-based technology curriculum needs to involve central knowledge components concerning technological change, which should be empirical and include key technological innovations and innovators, use and modification of key innovations, and society as influencing and/or being influenced by technology. Furthermore, this curriculum should include theoretical components such as recurring principles or patterns in the history of technology, social science models, and ideas or models of historical contextualisation. A model for a progression of technological change should entail a progression from then to now, and from simple to complex, in both Bernstein's (1990) vertical and horizontal discourse. A further progression is also conceivable, namely from local to global. A historical example of such a curriculum is that of the subject Home region instruction in Sweden, and parts of it which went into later curricula, which to varying degrees included both history and technology.

Finally, in following the history-based technology curriculum, it is important to avoid the pitfalls of novelty mongering and merely focusing on what has been most visible or obvious, but perhaps not most decisive, in the evolution of technology (Edgerton, 2018). To be able to do that, students need to adhere to the words of Edgerton:

> Thinking about the technological past can give us insights into "the question of technology" – what is it, where does it come from, what does it do? But this book has attempted to do much more than take historical examples to address this perennially interesting question. It has been concerned primarily with asking questions not about technology, but about technology in history – asking questions about the place of technology within wider historical processes. This important distinction is not obvious, but it is central to a proper historical understanding of technology. (Edgerton, 2008, p. 211)

References

ACARA (2023). The Australian Curriculum Version 9.0, Technologies: Sequence of achievement. https://www.australiancurriculum.edu.au/f-10-curriculum/technologies/pdf-documents/

Arthur, W. B. (2009). *The nature of technology: What it is and how it evolves*. Free Press.

Bernstein, B. (1990). Vertical and horizontal discourse: An essay. *British Journal of Sociology of Education, 20*(2), 157–173. https://doi.org/10.1080/01425699995380

Biggs, J., & Collis, K. (1989). Towards a model of school-based curriculum development and assessment using the SOLO taxonomy. *Australian Journal of Education, 33*(2), 151–163. https://doi.org/10.1177/168781408903300205

Bowler, P. J. (2017). *A history of the future: Prophets of progress from H. G. Wells to Isaac Asimov*. Cambridge University Press.

Briggs, A., & Burke, P. (2009). *A social history of the media: From Gutenberg to the Internet* (3rd ed.). Polity Press.

Burke, P. (2005). *History and social theory* (2nd ed.). Polity Press.

Cowan, R. S. (1983). *More work for mother: The ironies of household technology from the open hearth to the microwave*. Basic books.

de Vries, M. J. (2001). The history of industrial research laboratories as a resource for teaching about science-technology relationships. *Research in Science Education, 31*, 15–28. https://doi.org/10.1023/A:1012610328351

de Vries, M. J. (2005). *80 years of research at the Philips Natuurkundig Laboratorium (1914–1994): The role of the Nat. Lab. at Philips*. Pallas Publications.

de Vries, M. J. (2016). *Teaching about technology: An introduction to the philosophy of technology for non-philosophers*. Springer.

de Vries, M. J. (2018). Technology education: An international history. In M. J. de Vries (Ed.), *Handbook of technology education* (pp. 73–84). Springer.

Edgerton, D. (2008). *The Shock of the old: Technology and global history since 1900*. Profile Books.

Edgerton, D. (2010). Innovation, technology, or history: What is the historiography of technology about? *Technology and Culture, 51*(3), 680–697.

Edgerton, D. (2018). *The rise and fall of the British nation: A twentieth-century history*. Penguin.

Goodson, I. (2014). Context, curriculum and professional knowledge. *History of Education, 43*(6), 768–776.

Hagberg, J.-E. (1986). *Tekniken i kvinnornas händer: hushållsarbete och hushållsteknik under tjugo- och trettiotalen* (Diss.). Liber.

Hallström, J. (2022). Embodying the past, designing the future: technological determinism reconsidered in technology education. *International Journal of Technology and Design Education, 32*, 17–31. https://doi.org/10.1007/s10798-020-09600-2

Hallström, J. (2023). Teknisk förändring som innehåll i teknikundervisning: Kunskap, begrepp, progression. *ATENA Didaktik, 5*(1). https://doi.org/10.3384/atena.2023.4481

Hallström, J., Hultén, M., & Lövheim, D. (Eds.). (2013). *Teknik som kunskapsinnehåll i svensk skola 1842–2010*. Gidlunds förlag.

Herbart, J. F., & De Garmo, C. (2013). *Outlines of educational doctrine*. Macmillan.

Hughes, T. P. (1987). The evolution of large technological systems. In W. E. Bijker, T. P. Hughes, & T. J. Pinch (Eds.), *The social construction of technological systems: New directions in the sociology and history of technology*. The MIT Press.

Hultén, M. (2013). Technology for all: turning a keyword into a school subject in post-war Sweden. *History of Education, 42*(5), 622–637.

International Technology and Engineering Educators Association (ITEEA). (2021). *Standards for technological and engineering literacy: The role of technology and engineering in STEM education*. ITEEA. https://www.iteea.org/STEL.aspx

Layton, D. (1994). A school subject in the making? The Search for Fundamentals. In D. Layton (Ed.), *Innovations in Science and Technology Education* (Vol. V). Unesco Publishing.

Ludvigsson, D., & Booth, A. (2015). Introduction: Building Knowledge, Building Connections. In D. Ludvigsson & A. Booth (Eds.), *Enriching history teaching and learning: Challenges, possibilities, practice. Proceedings of the Linköping conference on history teaching and learning in higher education*. ISAK, Linköping University.

Marçal, K. (2021). *Mother of invention: How good ideas get ignored in an economy built for men*. Abrams.

McGregor Jacobides, E., & Winterbottom, M. (2020). Tinkering *with* technology education. In de Vries, M. J., Fletcher, S., Kruse, S., Labudde, P., Lang, M., Mammes, I., Max, C., Münk, D., Nicholl, B., Strobel, J., & Winterbottom, M. (Eds.), *The impact of technology education: International insights*. Waxmann.

Ministry of Education. (2017). *Technology in the New Zealand Curriculum*. http://nzcurriculum.tki.org.nz/The-New-Zealand-Curriculum/Technology

Nye, D. E. (2006). *Technology matters: Questions to live with*. The MIT Press.

Nordlöf, C., Norström, P., Höst, G., & Hallström, J. (2022). Towards a three-part heuristic framework for technology education. *International Journal of Technology and Design Education, 32*, 1583–1604. https://doi.org/10.1007/s10798-021-09664-8

Norström, P., & Hallström, J. (2023). Models and modelling in secondary technology and engineering education. *International Journal of Technology and Design Education, 33*, 1797–1817. https://doi.org/10.1007/s10798-023-09808-y

Oldenziel, R. (1999). *Making technology masculine: Men, women and modern machines in America, 1870–1945*. Amsterdam University Press.

Pearson, G., & Young, A. T. (2002). *Technically speaking: Why all Americans need to know more about technology*. National Academies Press.

Pritchard, S. B., & Zimring, C. A. (2020). *Technology and the Environment in History*. The Johns Hopkins University Press.

Schatzberg, E. (2018). *Technology: Critical history of a concept*. The University of Chicago Press.

Schooner, P., Höst, G., Klasander, C., & Hallström, J. (2023). Teachers' cognitive beliefs about their assessment and use of tools when evaluating students' learning of technological systems: a questionnaire study. *International Journal of Technology and Design Education, 33*, 937–956. https://doi.org/10.1007/s10798-022-09763-0

Shapin, S. (1992). Discipline and bounding: The history and sociology of science as seen through the externalism–internalism debate. *History of Science, 30*, 333–369.

Skolverket. (2022). *Läroplan (Lgr22) för grundskolan samt för förskoleklassen och fritidshemmet, Kursplan i teknik*. https://www.skolverket.se/undervisning/grundskolan/laroplan-och-kursplaner-for-grundskolan/laroplan-lgr22-for-grundskolan-samt-for-forskoleklassen-och-fritidshemmet?url=907561864%2Fcompulsorycw%2Fjsp%2Fcurriculum.htm%3Ftos%3Dgr%26cur%3DLGR22&sv.url=12.5dfee44715d35a5cdfa219f#anchor_5

Staudenmaier, J. M. (1985). *Technology's storytellers: Reweaving the human fabric*. Society for the History of Technology and the MIT Press.
van der Vleuten, E., Oldenziel, R., & Davids, M. (2017). *Engineering the Future, Understanding the Past: A Social History of Technology*. Amsterdam University Press.
van Dulken, S. (2001). *Inventing the 19th Century: 100 Inventions that Shaped the Victorian Age: From Aspirin to the Zeppelin*. New York University Press.
Vaidhyanathan, S. (2018). *Anti-social media: How Facebook disconnects us and undermines democracy*. Oxford University Press.
Williams, P. J. (2017). Critique as a disposition. In P. J. Williams & K. Stables (Eds.), *Critique in design and technology education* (pp. 135–152). Springer.

CHAPTER 7

Empowering Students to Bridge the Gap into a Knowledge-Driven Technological Realm

The Case of Technology Education in Israel

Moshe Barak

1 Introduction

Israel is considered one of the leading countries in various fields of science and technology, and many of the world's leading high-tech companies such as Microsoft, Google, Intel, and Hewlett-Packard (HP) have established research centres in Israel. However, a question arises regarding the role of technology education in empowering high school graduates and preparing them for success in a knowledge-driven technological society. An additional question is to what extent does technology studies in high school prepare graduates for higher studies in science and technology or encourage students to integrate as professional workers into a technological field. Although several articles have been published in recent decades on technological education in Israel (Barak, 2006, 2009, 2020a; 2020b; Doppelt, 2005, 2009), there is room to examine these questions in more depth.

2 Literature Review

2.1 *Technology Education and Vocational Education*

The terms 'technology education' and 'vocational education' sound very similar, and it is sometimes difficult to distinguish between them. This often reflects the tension between training for specific skills competencies and educating for more generic core competencies. These terms share similarities but differ in their focus and objectives, as will be described below.

2.2 *Focus*

Technological education focuses on teaching theoretical principles and practical applications in specific technological fields such as engineering, computer science, information technology, etc. It often involves a deeper understanding of scientific principles alongside technical skills.

© MOSHE BARAK, 2025 | DOI:10.1163/9789004710948_007

Vocational education emphasises hands-on skills and practical training for specific trades or occupations such as plumbing, carpentry, automotive repair, culinary arts, etc. It is geared towards immediate application in a particular job or industry.

2.3 *Breadth of Knowledge*

Technological education provides a broader understanding of underlying scientific and technical principles, allowing for adaptability across related fields within the technological domain.

Vocational education focuses on specific skills required for a particular trade or occupation, with less emphasis on broader theoretical understanding beyond what is directly applicable to the job.

2.4 *Duration and Depth of Study*

Technological education often involves lengthier academic programmes, such as bachelor's or master's degrees, offering a more comprehensive and in-depth study of the chosen technological field.

Vocational education usually involves shorter-term training programmes or certifications that focus on practical skills needed for immediate entry into a specific occupation or trade.

2.5 *Adaptability and Versatility*

Technological education allows for greater adaptability across various technological sectors, providing a foundation that could be applied to different industries or fields within the same domain.

Vocational education offers immediate job skills but may require additional training or education to transition to different trades or industries.

The following is a concise comparison between technological education and technological education (Pavlova, 2008; Stevenson, 2003):

- General knowledge vs. Specific knowledge
- Theoretical knowledge vs. Practical/functional knowledge
- Conceptual understanding vs. Proficiency in skills
- Creative abilities vs. Reproductive abilities
- Intellectual skills vs. Physical skills
- Preparation for life vs. Preparation for work

3 Engineering Education

Engineering is a subfield of technology. It refers to the application of scientific principles, mathematical methods and empirical evidence to design, create, maintain and improve structures, machines, systems, processes and materials. Engineers use their knowledge to solve real-world problems, innovate new technologies, and develop solutions that benefit society across various fields such as civil, mechanical, electrical, chemical engineering, and more. A key concept in engineering is optimisation, meaning solving a problem or developing a new product using a minimum of resources such as materials or energy while meeting the product's requirements. Engineering involves critical thinking, creativity, problem-solving and collaboration to tackle complex challenges and bring ideas from concept to reality.

Both technological and vocational education play crucial roles in meeting the demands of the workforce. Technological education tends to focus on higher-level technical roles and broader adaptability, while vocational education offers direct, hands-on skills training for immediate entry into specific trades or professions (Williams et al., 2015). This review of technology education vs. vocational education could help in understanding the framework of technology education in Israel, as will be described in the following section.

4 Technology and Engineering Education in Israel

4.1 *Study Tracks in High Schools*

The Israeli education system comprises the following age levels: kindergarten (age 5), primary school (grades 1–6), middle school (grades 7, 8, 9), and high school (grades 10, 11, 12). Higher education takes place in colleges and universities. The current paper focuses on teaching and learning technology in high school.

Until the 1980s, senior secondary education comprised two separate tracks: general high schools were intended for high-achieving students, and vocational schools for low-achieving students. This system has gradually changed in recent decades, and most high schools today are comprehensive schools meant to provide a home for students from an entire spectrum of achievements. All students study compulsory subjects such as Mathematics, English, History & Citizenship, and Hebrew. In addition, each student majors in at least one scientific subject such as physics, biology or chemistry, or a technological subject in one of the following study tracks:

- *Engineering track* (38%): engineering-oriented, for example, computer engineering, electrical engineering, mechanical engineering, or biotechnology engineering.
- *Technological track* (54%): quasi-academic track, for example, electricity, computer aided design and manufacturing (CAD-CAM) or industrial management.

– *Occupational track* (8%): occupation-oriented, preparation for a specific workplace such as fashion, hair design, cosmetics, tourism, and the art of cooking and baking.

Students in all study tracks take at least two matriculation exams, called the *Bagrut*, at different levels (1, 3 or 5 units). Many engineering-track graduates pursue engineering studies at university, and technological-track graduates often continue to study in technological colleges.

5 Student Projects in the Final Year of High School

Within the *Bagrut* exams framework, students learning in the engineering and technological tracks prepare a final project during their final year in high school. These projects often involve the design and construction of a technological system under the teacher's supervision. The project's final evaluation and grade are determined by an external examiner sent to the school by the Ministry of Education, together with the teacher. The following sections present examples of students' final projects in the engineering and technological tracks.

5.1 *Example 1: Engineering-Track Students' Final Project – A Beach-Cleaning Robot*

As previously mentioned, engineering refers to the application of scientific principles, mathematical methods, and empirical evidence to design, create, maintain, and improve structures, machines, systems, processes and materials. In this paper, we present the case of two high school students who developed a robot model to clean sea sand. The students worked on the project for about eight months and performed the following steps, as is customary in the world of engineering:
– Defining the problem.
– Carrying out a literature review on beach pollution. For example, it was found that Israel is in third place regarding coastal pollution among the Mediterranean countries.
– Identifying the pollution sources such as local waste and waste that reaches the country's shores from neighbouring countries through sea waves.
– Examining several solutions for cleaning the beach, for example, using a powerful vacuum cleaner or a vehicle that blows the sand to the sides and filters the clean sand.
– Choosing the most suitable solution according to a comparison table.
– Constructing their machine by using industrial components, parts manufactured at school, LEGO, Fishertechnik, or Tetrics modules.

In the current example, the students arrived at a solution to the problem according to steps accepted in the world of engineering, using mathematics and physics to explain the structure and operation of the components they used, such as electromagnetism, direct current and alternating current motors, and an ultrasonic sensor. To enable the machine to move in a sandy environment, the students chose a drive system in the form of an isosceles trapezoid and a drive chain, as illustrated in Figure 7.1. The students built a model of the system and tested its operation on a sandy surface in the school yard to identify failures and problems.

Figure 7.2 shows the structure of the completed cleaning robot, including two main parts, the drive system and the waste collector.

The students prepared a booklet documenting the work process on the development of the system, including presenting the problem and the solutions considered, choosing the solution with the help of a 'rating table' having the following categories: quality of beach cleaning; safety; resistance to environmental conditions such as strong sun, sand, and corrosion; keeping a distance from the seawater. They presented their project to an examiner from the Ministry of Education and received high scores for the project work and their oral presentation.

In working on the project, the students acquired a variety of general skills related to engineering problem-solving and design, for example, defining a

FIGURE 7.1 A drive system in the form of an isosceles trapezoid based on a chain

FIGURE 7.2 Example of a student project – a beach-cleaning robot

problem, testing several solutions and making informed decisions about a particular solution, creativity in building a computerized technological system, troubleshooting, evaluating the final product, teamwork, documenting the work process, and oral presentation of their work (Dopplet, 2009; Doppelt & Barak, 2002; Doppelt & Barak, 2021).

5.2 *Example 2: Technological-Track Students' Final Project – Design of the Electrical System in a Bakery*

As previously mentioned, technological-track studies are often intended for students of average achievements, for example, those studying mathematics or English at the level of 3–4 (out of 5) 'study units' for the *Bagrut* exams. These students are also candidates for post-secondary studies in a technological college for the degree of Certified Technician.

Planning the electrical system for a bakery is quite complex. It involves designing sockets for electrical appliances in the kitchen, such as an oven, refrigerator, and air conditioner, and designing the bakery's main electrical panel. An important component of the electrical system is planning the lighting in different areas of the bakery such as the guests seating area and the kitchen, as shown in Figure 7.3. The designer must choose the type and number of lamps, their location, and the light intensity of each lamp. Lighting

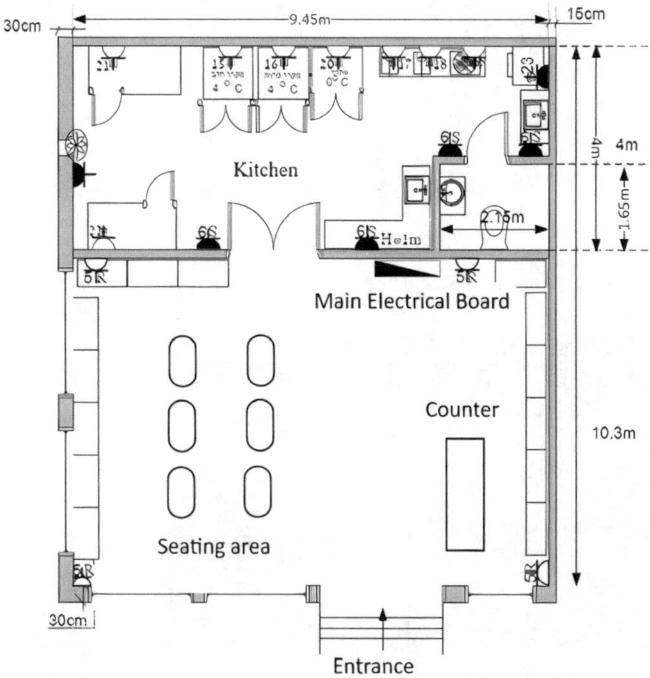

FIGURE 7.3 Electrical system in a bakery

design is a challenge; for example, lighting above tables in a luxury restaurant differs from lighting in a fast-food restaurant. In addition, the electrical system must be designed to prevent overload that could cause a fire and include protection against electrocution. The planning is carried out according to mandatory standards and the electrical system is tested by an electricity expert.

The students in the class designed the main electrical panel for their project, which includes a main circuit breaker for each of the electrical lines in the bakery, as illustrated in Figure 7.5.

Figure 7.6 shows an example of circuit breakers on the main panel.

	Single-phase socket
	Water-protected single-phase socket
	Three-phase socket
	Water-protected three-phase socket

FIGURE 7.4 Electricity components of the electrical system in a bakery

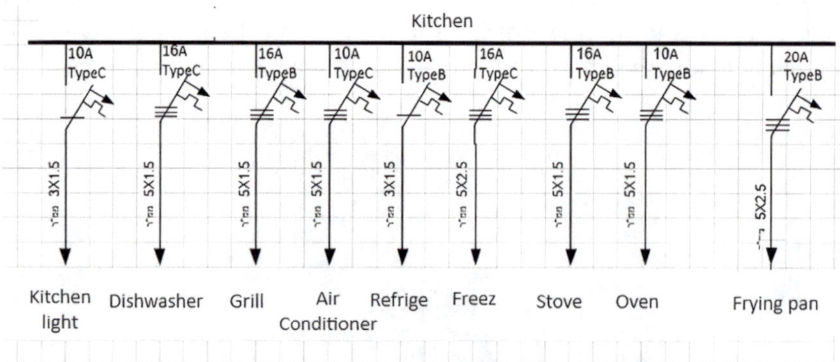

FIGURE 7.5 Part of the bakery's main electrical panel

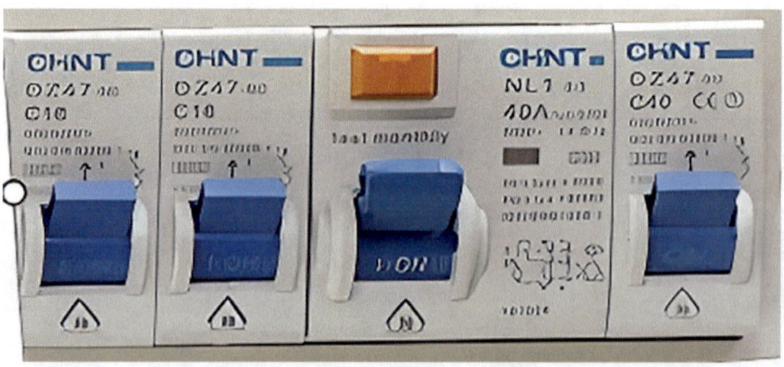

FIGURE 7.6 Example of circuit breakers on the main electrical panel

Students completing the study according to this programme receive a certificate of "Assistant Electrician". This opens the door for them in the working world and encourages them to continue their studies to obtain a certified electrician's license. Let us look more closely at the skills and knowledge the students gain from participating in this programme.

6 Skills and Knowledge that Students Acquired in the Project Work

6.1 *Specific vs. General Skills and Students' Projects*

We have seen that while the beach-cleaning robot project was developed by outstanding students, the project of designing the electrical system for a bakery was carried out by students of average achievements. The students in both groups gained different skills during their project work in terms of specific versus general skills, as will be discussed below (Christenko, Martinaitis, & Gaušas, 2020).

- *Specific skills* are those that are specialized and are often directly applicable to a particular job, task, or field. They are concrete abilities that are acquired through training, education, or experience, for instance, building or testing electrical circuits, programming languages, data analysis, graphic design, and accounting.
- *General skills*, on the other hand, are broader and are transferable across various roles, industries, and situations. They are often referred to as soft skills or transferable skills. Examples include communication, problem solving, teamwork, adaptability, leadership, and time management. While specific skills are essential for performing tasks or roles effectively, general skills are valuable across different domains and are crucial for overall personal and professional development.

In the current study, the students who developed the robot design project obtained general skills, and they perceived their future in further academic studies and work in the world of engineering and the high-tech industry. The students who tackled the project of designing an electrical system for a bakery, on the other hand, acquired specific knowledge and skills for integrating into work in this field. They could continue studying in a technological academic college and receive a certified electrician's license.

6.2 *Technological Knowledge and Students' Projects*

According to de Vries (2017), technology is generally seen as the development and use of an enormous variety of artifacts that we find around us. The impact of this on our lives is so important that we speak of "technological literacy" as a requirement that should be learned by every citizen in school. In the Handbook of Technology Education (de Vries, 2017), the author mentions the problems one runs into when trying to identify distinct characteristics of technology. Based on the literature, de Vries lists four characteristics of technology: (1) the context-dependent nature of technological knowledge; (2) the often-non-propositional nature of technological knowledge; (3) agreement as an origin for technological knowledge (e.g., agreements on technical standards); and (4) normativity in technological knowledge. Technological knowledge encompasses the understanding, expertise and skills required to develop, use, and manage technology. It involves problem solving and innovation, interdisciplinary knowledge, ethical and social consideration, and system thinking. According to de Vries (2005), the normative component for technology education is important. Students must learn to make judgements about effectiveness, as this is a prominent characteristic of technological knowledge, which makes it distinct from scientific knowledge. Students must also learn to deal with ethical and other values when doing technological project work.

6.3 *Engineering Knowledge and Students' Projects*

As previously mentioned, a key concept in engineering is optimisation (Rao, 2019), meaning solving a problem or developing a new system using a minimum of resources such as materials or energy while meeting the product's requirements. Engineering knowledge has to do with expertise (Litzinger et al., 2011). For example, planning an electrical system is done according to an official standard to avoid overloading the power lines and to protect the users from electrocution. In planning the lighting in a bakery there is room for creativity to a certain extent, for example, in choosing the type of lamps, but even in this area there are binding standards for the light intensity required in different areas of the bakery. In the current study, the term technological knowledge is broader than engineering knowledge and to a certain extent includes engineering knowledge.

In summary, the two examples from high school students' projects that we have seen in the current study, a robot for cleaning sea sand and the design of an electrical system for a bakery, demonstrate to some extent the definition above of technological knowledge and engineering knowledge. These examples also illustrate the distinction between general and specific skills mentioned in the previous section. Our two examples can be summarised as follows:

- The students with average achievements who developed an electrical system for a bakery used distinct engineering knowledge and acquired mainly specific skills.

– The high-achieving students who developed the beach-cleaning robot used technological knowledge and gained mainly general skills.

Although these distinctions between engineering and technological knowledge or between specific and general skills are not precise or definite, the above analysis could help in the construction or evaluation of technology and engineering curricula.

7 Concluding Remarks

In the Introduction, we raised two questions: first, what is the role of technology education in empowering high school graduates and preparing them for success in a knowledge-driven technological society; and second, to what extent do technology studies in high school prepare graduates for higher studies in science and technology or encourage students to integrate as professional workers in a technological field.

Regarding the first question, we presented two examples in this study of preparing students to integrate into the knowledge-rich technological world. One group of outstanding students faced an open and challenging task in the field of robotics – to design, construct and evaluate a robot model for cleaning sea sand. The second group of students of average achievements were engaged in a semi-structured project of designing the electrical system for a bakery. The students in both groups acquired the fundamental skills related to their project work, a requirement frequently encountered in a knowledge-driven technological society. These include, for example, defining goals of a task, gathering information, self-learning a new subject, experiencing teamwork, meeting a schedule for completing a task, preparing a report on the project, and presenting the work to others. The knowledge that the students acquired may be useful and essential for any person in modern society, for example a doctor or a businessman, and not only for those planning to engage in engineering.

Regarding the second question of choosing the field of occupation for a high school graduate, technology studies in high school offer two options to the graduate: continuing to higher studies or integrating into the working world. Many of the students in classes such as the robot group further their academic studies in the field of engineering or science in university faculties. Many of the students in classes such as the electrical system design group receive a good foundation and often continue studying for a certified technician and senior technician diploma in a technological college. In summary, graduates of the first group are suitable for further academic studies in university, and graduates of the second group are suitable for studies in a technological college that prepares them for integration into the working world.

It is important to note that in Israel's education system, technological studies in high school constitute a significant contribution to graduates, but do not limit them to a certain field. As mentioned in this paper, technological studies in most cases take place in a comprehensive high school where all the students study compulsory and elective subjects such as Mathematics, English, History & Citizenship, Hebrew, science, and technology, and receive a matriculation certificate, the *Bagrut*, depending on the skills and preferences of the student. This opens the door for academic studies in many fields, not necessarily technology.

Finally, it is worth noting that technology today offers a rich variety of fields that may interest high school students for a final project, for example, technology related to media, communication, smart house, agriculture, food, transportation, environmental protection, solar energy, and wind energy. These areas are undergoing rapid development in Israel and in many other countries and could be a fascinating challenge as a project for high school students. The integration of innovative subjects into technological education depends primarily on the skills of the teacher and the school's resources.

References

Barak, M. (2006). Engineering and excellence: An old-new agenda for technology education in Israeli high schools. In *International Handbook of Technology Education* (pp. 477–486). Brill.

Barak, M. (2009). Robotics projects and learning concepts in science, technology and problem-solving. *International Journal of Technology and Design Education, 19*(3), 289–307.

Barak, M. (2020a). Problem-, project- and design-based learning: Their relationship to teaching science, technology, and engineering in school, *Journal of Problem-based Learning*, *72*(2), 94–97.

Barak, M. (2020b). Renovating project-based learning in Israel to foster learning STEM, computational thinking, and design arts. *Australasian Journal of Technology Education, 6*. https://doi.org/10.15663/ajte.v0i0.69.

Christenko, A., Martinaitis, Ž., & Gaušas, S. (2020). Specific and general skills: concepts, dimensions, and measurements. *Competition & Change*, *24*(1), 44–69.

de Vries, M.J. (2005). The nature of technological knowledge: philosophical reflections and educational consequences. *International Journal of Technology and Design Education*, *15*, 149–154.

de Vries, M.J. (2017). Philosophy of technology: themes and topics. In: M.J. de Vries (Ed.), *Handbook of Technology Education*, pp. 7–16, Springer International Handbooks of Education.

Doppelt, Y., & Barak, M. (2002). Pupils identify key aspects and outcomes of a technological learning environment. *Journal of Technology Studies*, *28*(1), 22–28.

Doppelt, Y., & Barak, M. (2021). Design-based learning in electronics and mechatronics: Exploring the application in schools. In *Design-based concept learning in science and technology education* (pp. 101–134). Brill Sense.

Doppelt, Y. (2005). Assessment of project-based learning in a mechatronics context. *Journal of Technology Education*, *16*(2), 7–24.

Doppelt, Y. (2009). Assessing creative thinking in design-based learning. *International Journal of Technology and Design Education*, *19*, 55–65.

Litzinger, T., Lattuca, L. R., Hadgraft, R., & Newstetter, W. (2011). Engineering education and the development of expertise. *Journal of Engineering Education*, *100*(1), 123–150.

Pavlova, M. (2008). *Technology and vocational education for sustainable development: Empowering individuals for the future* (Vol. 10). Springer Science & Business Media.

Rao, S. S. (2019). *Engineering optimization: Theory and practice*. John Wiley & Sons.

Stevenson, J. (2003). Examining cognitive bases for differentiating technology education and vocational education. In *Initiatives in technology education–Comparative perspectives: Proceeds of the American Forum* (pp. 194–206).

Williams, P. J., Jones, A., & Buntting, C. (Eds.). (2015). *The future of technology education*. Springer Singapore.

CHAPTER 8

Technological Concepts and Technical Contexts for Technology Education

Andrew Doyle

1 How Do You Describe the Content of Technology Education?

I have struggled with this question for about 10 years. As an undergraduate student in the Irish educational context, I studied two *subjects* from the national curriculum. The terminology around the subjects have shifted in recent years, but in 2011, I was studying to become a Materials Technology (Wood) and Technical Graphics teacher. The origins and trajectory of technology education in different national contexts has been well charted, and the Irish context mirrors the vocational past found in many countries around the world (Banks & Williams, 2022). While an in-depth discussion on this history is beyond the scope of this chapter, it is important to note. The relevance here stems from the clarity through which the *content* of these previous subject may be represented. As the primary focus was on the preparation of learners for the world or work, the skills, knowledge and competencies associated with wood- and graphical communication-oriented professions formed the primary content of subjects. With the development from vocationally oriented technical education, technology education is instead presented as a learning area of importance for all learners. While the goals for technology education are well articulated, and have been for some time, the perennial debate in technology education is the stymied transition from technical education. It is my contention that the technical contexts that defined vocational education, which are still present in curricular representations of technology education, contribute to these tensions.

In this chapter, I seek to problematise representations of the *content* of technology education. While this approach may at first appear like a step backwards, towards a more archaic conception of curriculum or indeed teaching, the nuance lies in how *content* as a concept may serve as delineator between technical and technology education. Based on my PhD research (Doyle et al., 2023) with educators in Ireland, Sweden and New Zealand, three different approaches to conceptualising technology education will be presented. The implicit representations of *content* within each conceptualisation of technology education will then be problematised and discussed relative to

© ANDREW DOYLE, 2025 | DOI:10.1163/9789004710948_008

technological concepts and technical contexts (Rossouw et al., 2011). The chapter will conclude with some questions intended to prompt discourse on how *content*, and in turn learning, may be articulated in technology education.

2 Difference: The Role of Technical Context

Technical education (see Keirl [2017] for more detail on the distinction between technical education and technology education) was developed with an industry-oriented agenda. Dakers (2022) notes how this approach ultimately manifested in a two tier education system based on meritocracy. The transition from primary education to secondary education was also the point at which learners were separated. High achievers were placed in an academic pathway, and other learners were placed on a vocational pathway. Whereas the academic route focused on the development of the learner, the technical education approach held a vocational agenda whereby the service of industry was the primary motivation. More detail is not required here as this history and implications are well charted (Buckley, 2023). However, if we return to our opening question, *how do you describe the content of technology education?* the distinct heritage of a technology education that's evolved from a technical predecessor is worthwhile exploring. The challenge here stems from the clarity through which the *content* of these technical predecessors to technology education may be represented.

The vocational agenda inherent to technical education, manifested as a very specific curriculum, and in turn, *content*, that could be articulated. As the purpose of such subjects was to prepare learners for the world of work, in a specific trade, then the knowledge, skills or competencies associated with this trade can be organised and articulated as content. In a similar way to how the pedagogical approach of early attempts at technology education followed that of the master apprentice model of the medieval guild (Banks & Barlex, 1999), the *content* followed a specific series of skills development (Doyle, 2023). In place of the broader goals of technology education, the content of specific vocations dictated the content of curriculum. If the technical context, in this case defined by vocation is changed, then the *content* changes also. Wood as a material is changes to food, and a discussion around appropriate angles at which to cut dovetails in hardwood versus softwoods is shifted to a discussion around the effects of humidity on proving times. As noted earlier, this historical context and the implications on technology education today are well charted, and so I will not devote further time to this here. Understanding the heritage of different technology educations is important if we want to problematise today's representations. Where this discussion becomes particularly interesting, is if the role of technical context is considered relative to multiple technology educations. As such a discussion requires one the ask questions about the different, but more importantly the commonality between Food and Wood as *contexts* for technology education.

It's worthwhile pausing at this stage to consider the use of the word 'content' to discuss technology education. If you have read this far and are slightly uncomfortable with its use I understand. I know that I am. Having worked in Initial Technology Teacher Education for several years now I am always apprehensive to use the word. I have been conditioned or conditioned myself away from its use. I suspect this apprehension stems from the difficulty in demarcating the limits of well understood content associated with technical education, and concisely articulating the content of technology education.

Despite my discomfort with the use of *content* to adequately describe technology education, I found myself asking a participant about the content of technology education in an interview as part of my doctoral research. The question arose towards the end of an interview in which the participant had, in a very articulate manner, described the nature of teaching and learning in their technology subject. Following this, the participant was reflecting on a peer teachers experience of covering a technology class and their discomfort with the apparent mayhem that was occurring. The covering teacher, in this case a mathematics teacher, found it difficult to reconcile that all students could be engaging with very different activities in the technology lesson. This was not an issue for the participant, as I suspect it is not an issue (or at all surprising) to any technology education professional. In any case, the covering teacher in this instance could not reconcile the apparent sporadic nature of student activity. When comparing this to maths, which they described as having a well-structured content basis that informed the structure of lessons, technology didn't have a coherent structure. How could it? My participant reflected on how this is not problematic in technology. Stating how this was common practice, and something which out to be celebrated about the learning area. Despite my internal agreement, I decided to pick at the anecdote. I asked them.

3 How Would You Describe the Content of Technology?

On face value this is a very simplistic question. It is worthwhile reemphasising that this participant had discussed the nature of teaching and learning in technology education for almost an hour. Learning intentions, and the evolution of practice from year to year, different rationales for teaching technology, appropriate pedagogical approaches and broader subject philosophies had all be discussed, despite this, when posed with the question, the participant exhibited dismay.

I decided to use the anecdote above at the end of future interviews, and to conclude with this question, *how do you describe the content of technology education?* With each use, the question was met with similar levels of discomfort (and in a few instances, what appeared to be distain). where it was met with similar levels of discomfort (or distain). A selection of responses is presented below.

> "I would find it very difficult to, and you'd be waiting a couple of hours for a reply, if I'm honest".
>
> "I think it's so broad. Like, it's so broad and it's so dynamic and so evolving all the time, because I guess like the word "technology"... even technology when you look at it on a wider platform, like internationally, like, it is so utterly changing".
>
> "Okay. Yeah, technology ... it's huge. Yeah, I suppose ... coming up with solutions to problems I suppose, isn't it?"; "It's life skills. It's, it's setting students up to the, um, you know, worthwhile citizens".
>
> "Our expectation is that they are going to become self-managing and that they are going to be independent, uh, risk takers. And like I'd also be inquiring, so they need to go and find out information about things".
>
> "The content is fluid. And it has to be".

Bordering on the apologetic in asking this question, the more it was asked, the more it became apparent that we do not, as technology educators, think about content as a technical educator would have. This is not a critique of the participants in the study, quite the opposite, their expertise had been exhibited in their responses to prior questions. Instead, the question appeared to probe at an approach to thinking about teaching and learning that technology education was beyond. There was no table of contents for technology education.

The tension for participants appeared to focus on the overlap between technical education and technology education. Irrespective of the evolutions in policy and curriculum, the philosophical shift from vocational to holistic or general education, and the development of pedagogical approaches to facilitate such change, the remnants of technical education remain. By this, I mean the vocational contexts. Associated with each vocational context is a body of explicit content that, by comparison to the nature of technology education, is easy to explicate, organise, and easy to write as a table of contents. While the degree to which technical contexts influence policy and practices in different national contexts, their mere existence was sufficient to derail a discussion around the content of the technology education. As I asked this question over and over again to participants in different technical and national contexts for technology education I was reminded of Dakers' subtitle for his 2014 book, *breaking with the past*, and the rather bleak view of attempts to incorporate technological literacy into extant technology education practices, stating "it cannot be done, it has not been done, it will not be done" (2014, p. 1). In existing efforts to breaking with the past, what role do the technical contexts play?

4 Is the Content of Technology Education Defined by a Technical Context?

In discussing the epistemological basis of technology education, the relevant content knowledge is often defined as that which is needed to progress learners thinking and action towards the resolution of a problem. The associated fluidity through which technological knowledge is conceived (de Vries, 2005; Williams & Lockley, 2012), is presented as indicative of authentic technological activity. This suggests that the content of technology education is not defined by the technical contents, but rather elements of the technical context may be deemed relevant on an ad lib basis. If the knowledge associated with technical context is useful in navigating a problem, then this knowledge may be termed technological in nature. While this distinction has utility in demarcating the boundary between technological knowledge and other forms of knowledge, a challenge arises when this circumstantial knowledge is represented as the full extent of technological knowledge. The knowledge associated with navigating the problem at hand, whether prescribed or defined by the learner, should also be considered as relevant. As arguably, this is the *content* of technology that is common across technical contexts.

5 Commonality: The Holistic Nature of Technology

When asked to write this chapter, I took some time to reflect on the Marc's extensive contributions to my understanding of technology education. Several discussions, presentations, and publications come to mind, however, one stood out. Published in 2011 with Ammeret Rossouw and Michael Hacker, *Concepts and Contexts in Engineering and Technology Education: An International and Interdisciplinary Delphi Study*, presents an apparently simplistic model of technology education that is proposed as an instigator for curriculum development. The nuance of the model lies in the identification of technological concepts that transcend national and technical contexts. Within the 2011 article, technical contexts are presented as "relevant and meaningful contexts through which these concepts can be taught and learnt" (Rossouw et al., 2011, p. 409). In an effort to assist learners in understanding the relationship between technical contexts, unifying concepts were identified by participants in this Delphi study, and it was envisioned that the presentation of such concepts with give insight into the holistic nature of technology.

In the same way that the convergence of different technical contexts appears to have resulted in challenges in articulating the content of technology education, vocational education was just one origin for technology education. There are multiple different approach found in practice today, for example, a craft-oriented approach; an industry-oriented approach; a science-oriented approach; a 'high-tech' approach; an engineering concepts approach; a key competencies approach; a design-oriented approach; or; a social issues approach (de Vries, 2000). In problematising these categorisations, de Vries (2000) noted that with increasing contact between academics internationally, the difference

TABLE 8.1 Concept list for technology education (Rossouw et al. 2011)

Main concept	Sub-concepts
Designing ('design as a verb')	Optimising
	Trade-offs
	Specifications
	Invention
	Product lifecycle
Systems	Artefacts ('design as a noun')
	Structure
	Function
Modelling	
Resources	Materials
	Energy
	Information
Values	Sustainability
	Innovation
	Risk/failure
	Social interaction
	Technology assessment

among the approaches found in various countries has gradually becomes less apparent than the categorisation above suggests. It is from this perspective, to support the synthesis of various aspects of technology education, that de Vries and colleagues later proposed a concepts-contexts model for curriculum development (de Vries, 2011, 2014; Rossouw et al., 2011).

In proposing a concept-context approach, de Vries (2014) highlights the intention to confront learners with the same concepts in different contexts "so that they get to know the concept in a more in-depth and versatile way, and in order to embed the concept in a network of concepts that also provides insights into relations between concepts" (p. 335). The metaphor of a chameleon is used to emphasise the potential of the concept-context approach to facilitate conceptual learning. If a concept is only perceived in one context, in this case a chameleon in a grassy field, that would give the impression that chameleons are green. It is only once one has seen a green chameleon in a grass field, a blue chameleon near water and a grey chameleon one the pavement that the conceptual understanding of what a chameleon is can be reached. If a person does not grasp the concept of a chameleon "it is difficult to separate those characteristics that define a chameleon and those that are not essential for chameleons but context-bound" (de Vries, 2014, p. 331).

6 Can the Content of Technology Education Stand Independent of Technical Contexts?

My doctoral research (Doyle, 2020) investigated teachers' conceptions of the purpose of teaching technology, through reflection on the enacted practice. Inspired by and congruent with the concept-context approach, both technical contexts and national contexts did not supersede technology education, as the intention was to understand teachers' conceptions of technology education, not materials, food, textiles, digital, etc. In the analysis, the role of technical context, stemming from anecdote described previously, and representations of content, referred to as subject matter, were lenses through which the goals of technology and progression within technology were investigated. Ultimately this analysis resulted in the presentation of three different conceptions of the purpose of teaching technology; (1) Obtain knowledge and skills for application, (2) Ability to act in a technological way, and (3) Ability to think in a technological way (Doyle et al., 2023). In the following sections, I will briefly discuss these different approaches to technology education relative to the concepts-context framework.

The first approach to technology education termed *obtaining knowledge and skills for application* adopted a sequential approach to skills development. The technical skills and knowledge that were held as fundamentally important form the context and in turn content of the subject. Relative to the concept-context curriculum model, a singular context approach was adopted. This context provided the structure of technology education, and the concepts identified by Rossouw et al. (2011), while acknowledged, were not a significant focus.

The second approach identified held the *ability to act in a technological way* as the purpose for engaging with technology education. Within this approach, the applied concepts from Rossouw et al. (2011), designing, modelling and values, formed the primary content of technology education. The approach to technical contexts differed here, as in some instances a variety of technical contexts was sought, and in other, consistency of technical contexts was adopted. Importantly however, the technical contexts themselves were not represented as content, rather means to an end. The technological actions of designing, modelling, and making value-laden decisions was instead presented as the focus.

The third and final approach focused on learners developing the *ability to think in a technological way*. The primary distinction between the previous approach is that engagement with manual technological activity was no longer necessitated. Again, a variability of technical contexts was advocated within this approach, and interesting, the rationale for shifting technical context is to ultimately refocus on technological concepts (i.e. the content). The specific concepts from Rossouw et al. (2011) that appear central here were studying design, understanding systems, investigating resources and value-laden decision making.

In addition to the concepts identified by Roussow et al. (2011), a notable example of technological content that appears to stand independent of technical contexts is the increasing focus placed on the philosophy of technology. Presented as a focus on Nature of Technology (NoT), and drawing from relevant philosophical, historical and sociological literature, the intent of studying NoT is to provide a robust description of what technology is, how and why technology is developed, how technology and science impact each other, and how individuals and society direct, react to, and are at times unknowingly changed by technology (Clough et al., 2013; Pleasants et al., 2019). Recently, NoT has received attention within social studies education (Krutka et al., 2022). However, it can also be found in technology education curriculum, notably New Zealand (Ministry of Education, 2017), where NoT is presented as one of three stands of the curriculum, alongside technological knowledge and technological practice.

7 Indifference: Fluidity of Technical Context for Technological Concepts

In returning to the idea of content for technology education, the variability of technical contexts and what are deemed relevant technological concepts present an expansive, and ever-growing body of content for technology education. The approach of teaching concepts in multiple contexts is described by Rossouw et al. (2011, p.423) as a "middle road between total unbelief in the possibility of teaching and learning abstract concepts … and a naive belief in the possibility to teach such concepts directly at a high level of abstraction only to apply them later in various contexts". In tandem with a focus on specific concepts, this approach aligns with the everchanging needs of the 21st century. The adaptability association with familiarisation with and navigation of novel contexts has the potential to foster a mindset conducive to change.

Implicit within the concept-context approach is a shift away from a singular technical context for technology education. Again, this may be an uncomfortable thought for many – requiring a fundamental change in curriculum for many contexts. Despite this, the fragmented purposes for technology education discussed in the previous section suggest the need for conceptual coherence within technology education. I began this chapter by describing my background studying to be a Materials Technology (Wood) and Technical Graphics teacher. The strength that traditionally came from this position is that the subjects, and their content was well understood. While it is acknowledged that they may present a more reductive view of technology education, their place and status within curriculum holds. Similarly, while the weight of content associated with technical contexts most likely precludes engagement with broader concepts of technology education, their place and status within curriculum holds. By comparison, the decline of Design and Technology as a subject area in the English national context can be seen as a cautionary example. It illustrates how moving away from a technical focus, without a clear conceptual framework, can lead to the creation of a subject fraught with challenges. Although a multitude of factors have influenced the demise of technology education in this context, Spendlove (2022) highlights the lack of a clear understanding of what the subject is as a significant contributor.

And so, given the thirteen years since publication of the initial concepts-contexts paper, and the increasing collaborations, and understanding of technology education internationally, I would like to ask Marc, how do you describe the content of technology education?

References

Banks, F., & Barlex, D. (1999). 'No one forgets a good teacher!': What do "good" technology teachers know? *Journal of Design & Technology Education, 4*(3), 223–229.

Banks, F., & Williams, P. J. (2022). International perspectives on technology education. In A. Hardy (Ed.), *Debates in Design and Technology Education* (2nd ed., pp. 26–44). Routledge. http://dx.doi.org/10.4324/9781003166689-4

Clough, M. P., Olson, J. K., & Niederhauser, D. S. (Eds.). (2013). *The nature of technology: Implications for learning and teaching*. Sense Publishers.

Dakers, J. R. (2014). *New frontiers in technological literacy: Breaking with the past*. Palgrave Macmillan.

Dakers, J. R. (2022). *A Nomadic Pedagogy about Technology: Teaching the Ongoing Process of Becoming Ethnotechnologically Literate*. BRILL.

de Vries, M. J. (2000). Technology education: Towards a new school subject. In M. Ben-Peretz, S. Brown, & B. Moon (Eds.), *Routledge International Companion to Education*. Routledge.

de Vries, M. J. (2005). Technological knowledge. In *Teaching about technology: An introduction to the philosophy of technology for non-philosophers* (pp. 29–48). Springer.

de Vries, M. J. (2011). A concept-context framework for engineering and technology education. In M. Barak & M. Hacker (Eds.), *Fostering human development through engineering and technology education* (pp. 75–87). SensePublishers. https://doi.org/10.1007/978-94-6091-549-9_5

de Vries, M. J. (2014). Chapter 23—The concept–context approach to learning material properties in design(-related) education. In E. Karana, O. Pedgley, & V. Rognoli (Eds.), *Materials experience* (pp. 329–336). Butterworth-Heinemann. https://doi.org/10.1016/B978-0-08-099359-1.00023-0

Doyle, A. (2020). *Consolidating concepts of technology education: From rhetoric towards a potential reality* [PhD dissertation, KTH Royal Institute of Technology]. http://urn.kb.se/resolve?urn=urn:nbn:se:kth:diva-272837

Doyle, A. (2023). Task-based learning: An opportunity for focused learning in technology education. In D. Gill, D. Irving-Bell, M. McLain, & D. Wooff, *The Bloomsbury handbook of technology education* (pp. 240–254). Bloomsbury Academic. http://www.bloomsburycollections.com.ezproxy.waikato.ac.nz/collections/monograph-detail

Doyle, A., Seery, N., Gumaelius, L., Canty, D., & Hartell, E. (2023). Subject(s) matter: A grounded theory of technology teachers' conceptions of the purpose of teaching technology. *International Journal of Technology and Design Education*. https://doi.org/10.1007/s10798-023-09859-1

Keirl, S. (2017). Critiquing as design and technology curriculum journey: history, theory, politics and potential. In P. J. Williams & K. Stables (Eds.), *Critique in design and technology education* (pp. 109–133). Springer. https://doi.org/10.1007/978-981-10-3106-9_7

Krutka, D. G., Metzger, S. A., & Seitz, R. Z. (2022). "Technology inevitably involves trade-offs": The framing of technology in social studies standards. *Theory & Research in Social Education, 50*(2), 226–254. https://doi.org/10.1080/00933104.2022.2042444

Ministry of Education. (2017). *Technology in the New Zealand Curriculum.* https://nzcurriculum.tki.org.nz/content/download/168478/1244184/file/NZC-Technology%20in%20the%20New%20Zealand%20Curriculum-Insert%20Web.pdf

Pleasants, J., Clough, M. P., Olson, J. K., & Miller, G. (2019). Fundamental Issues Regarding the Nature of Technology. *Science & Education, 28*(3), 561–597. https://doi.org/10.1007/s11191-019-00056-y

Rossouw, A., Hacker, M., & de Vries, M. J. (2011). Concepts and contexts in engineering and technology education: An international and interdisciplinary Delphi study. *International Journal of Technology and Design Education, 21*(4), 409–424.

Spendlove, D. (2022). Why did design and technology education fail, and what might replace it? In *Debates in Design and Technology Education* (2nd ed.). Routledge.

Williams, J., & Lockley, J. (2012). Using CoRes to Develop the Pedagogical Content Knowledge (PCK) of Early Career Science and Technology Teachers. *Journal of Technology Education, 24*(1), 34–53.

CHAPTER 9

The Wisdom, Power and Promise of Marc J. de Vries' Thoughts

Kendall N. Starkweather

> When you're good at something, you'll tell everyone. When you're great at something, they'll tell you.
>
> WALTER PEYTON

Once in a generation, a person comes along with the wisdom, power and promise to make an extraordinary contribution to their profession that is felt around the world. Marie Curie was that person for science with discoveries in radiation research. Wernher von Braun was the space architect known for rocketry. Jonas Salk was the medical research scientist who developed polio vaccines. Marc J. de Vries is that person for the field of technology education and the study of technological literacy.

When announcing that he was retiring to a group of colleagues, the very accomplished Dutch educator, Dr. Jan Raat, announced that he had a young colleague with extraordinary thinking skills who would become a significant technology education leader. That person was Dr. Marc J. de Vries. Dr. Raat's prediction was correct. Dr. de Vries has become known throughout the world of technology educators and technological literacy researchers as one of the premier thinkers in his field.

Dr. de Vries has a personality focus that is not on himself, but upon the contributions of others as it relates to teaching about technology. He is known to be extremely humble, kind and a very likeable person who is not in the field to be notable, but for his enthusiasm and love for teaching and research in his chosen profession.

His United States of America (USA) impact has been with the articulation of his perspectives through writing, research, publications and presentations for over four decades. USA educators have followed his work through the Pupils Attitudes Toward Technology (PATT) initiative that is synonymous with

© KENDALL N. STARKWEATHER, 2025 | DOI:10.1163/9789004710948_009

his name. Dr. de Vries has been the coordinator and leader of PATT presentations at the International Technology and Engineering Educators Association's (ITEEA) Annual Conference for over 20 years. PATT, through Dr. de Vries, has become a major component of ITEEA's International Conference and provided featured speaker's at the International Luncheon where delegates from around the world gather to share work, enjoy friends and promote the advancement of technological literacy.

PATT held in conjunction with ITEEA Conferences became a constant venue for featuring the profession's leaders from around the world. These meetings were in addition to numerous other PATT conferences held in many other countries. A professional educator wanting to enter and interact with fellow educators from other countries would become a part of the PATT/ITEEA initiatives.

Dr. de Vries has been a frequent speaker throughout the USA at many of the annual conferences. He has been a very active consultant to many research projects centered in the USA sponsored by foundations, state associations and universities. His participation was invited and so frequent that it was not unusual and almost expected at every major research project being conducted pertaining to technology education.

His advice has been used in projects and initiatives funded by the National Aeronautics and Space Administration, National Science Foundation, National Academy of Sciences and National Academy of Engineering. These agencies are the major USA government agencies that provide funding, research venues and assist in guiding the direction of science, technology, engineering and mathematics (STEM) education.

The one major project that has shaped the teaching of technology in the USA has been the Standards for Technological Literacy. This project was guided by ITEEA with the majority of funding coming from the National Science Foundation and the National Aeronautics and Space Administration with significant guidance from the National Academy of Engineering. Dr. de Vries was one of the early consultants and frequently was an advisor for these standards that helped to determine what kindergarten through grade twelve students should know and be able to do pertaining to technological literacy. They have been adopted by the majority of the USA states. The standards were translated into six different languages and have been a widely used reference by many countries. Dr. de Vries was one of the initial architects helping to create a strong foundation for the standards.

The standards provided a content for the study of technology literacy that never before existed. They were acknowledged to be companion documents for the STEM movement that was becoming more recognised in the USA school curriculum. Dr. de Vries' knowledge of how design, technology and engineering were being addressed by educational systems around the world was of much value to the writers. His wide interests in the ethics of technology, philosophy and pedagogy of technology and engineering, and reformational philosophy became an asset to the project and provided knowledge not available without the involvement of Dr. de Vries' expertise.

Dr. de Vries has been awarded the highest recognition that ITEEA can bestow upon one of its members - The Academy of Fellows Distinction. He was the first person outside the USA recognised for his scholarship, leadership and prominence in the field of technology and engineering education.

People often believe that every person in all of history has been placed in the time and place to help advance a cause or belief. Dr. de Vries' place, time and work for the advancement of technology and engineering education has provided educators with his wisdom, power and promise to further the development of education. His success has come from his heart. He has empowered us to know that we can change the world.

CHAPTER 10

A Technology Curriculum for 2030 and Beyond

Hilda Ruth Beaumont

1 Being True to the Nature of Technology

Writing in the 1970's Jacob Bronowski captured the uniqueness of humankind as technological beings (Bronowski, 1973[1]).

> Humans are singular creatures. They have a set of gifts which make them unique among the animals: so that, unlike them, they are not figures in the landscape – They are shapers of the landscape. (p. 19)

> Among the multitude of animals which scamper, fly, burrow and swim around us, humans are the only animals who are not locked into their environment. Their imagination, reason, emotional subtlety and toughness, make it possible for them not to accept the environment but to change it. (p. 20)

> They have what no other animal possesses, a saw of faculties which alone, over three thousand million years of life make them creative. Every animal leaves traces of what it was; humans alone leave traces of what they created. (pp. 36, 37)

This elegant description of humans' unique ability to envisage what might exist in the future and intervene in the world to bring this vision into reality provides a powerful justification for the inclusion of technology in the school curriculum. In England the task of turning this vision into a curriculum reality for the new National Curriculum fell to *the Working Group for Design and Technology* (DES/WO 1988) A major contributor to this was David Layton and one can see his intellectual sophistication in the following extracts from the Working Group's Interim Report.

> What is it that pupils learn from design and technological activities which can be learnt in no other way? In its most general form, the answer to this question is in terms of capability to operate effectively and creatively in

© HILDA RUTH BEAUMONT, 2025 | DOI:10.1163/9789004710948_010

> the made world. The goal is increased 'competence in the indeterminate zones of practice'. (p. 3)

> As opposed to scientists, who are concerned to explore and understand what is, designers and technologists are concerned with what might be, the conception and realisation of 'the form of things unknown'. (p. 4)

> Understanding of technological change and of the ways in which it is restructuring the workplace and influencing lifestyles is a crucial aspect of an education in design and technology. The consequences of technological change are profound and pervasive. Furthermore, technological revolutions are irreversible; no technological change can be uninvented after it has taken place. We need to understand design and technology, so that we can acquire a sense of its enormous transformatory power. By the end of the period of compulsory education pupils should have some understanding of the value options and decisions that have empowered the technological process in the past and which are doing so today. (p. 6)

Regrettably it has proved difficult for schools and teachers to respond effectively to this brave vision of technology education which aimed to embrace both two intertwined aspects of technology – capability and perspective. HildaRuth Beaumont and Torben Steeg defined these as follows. Technological capability is designer-maker capability, capturing the essence of technological activity as intervention in the made and natural worlds. Technological perspective is an appreciation of 'how technology works' which informs critique and enables young people to consider how technology might be used to provide products and systems that help create the sort of society in which they wish to live (Beaumont & Steeg, 2024).

Responding has been hampered by government interventions which revised the National Curriculum requirements (Department for Education, 2013), privileged traditional academic subjects through the introduction of the English Baccalaureate (EBacc) (Department for Education, 2019), introduced new assessment arrangements (Ofqual, 2016), and by the rising costs of materials, tools and equipment. And this has been compounded by the disagreement between scholars as to technology's intrinsic nature and the way it influences our lives. Kevin Kelly and Jacques Ellul are in agreement that we have little control over technology and the way it plays out in the world. Kelly is a technological optimist and views new and emerging technologies as providing ever-increasing opportunities for humans to 'self-realise' in ways not available to those in previous generations (Kelly, 2010). The thinking of Jacques Ellul (1964) is in direct contrast to that of Kelly in that he is most definitely a technological pessimist. He views the intrinsic nature of technology with its emphasis on efficiency and standardisation as inimitable to the human spirit and the wide-ranging influence that technology is having, and continuing to have, on society as both dehumanising and outside human control. Thomas Hughes (1994) offers an intermediate path between Ellul's pessimism and Kelly's optimism. He has developed the idea of technological momentum, particularly with regard to the large technological systems that underpin much of modern life. He argues that as such systems come into being it is relatively easy for society have an influence on their development but once they become established this is much more difficult requiring significant disruption and at that stage in their development these technologies become autonomous and outside human control. And Brian Arthur (2009) suggests that our relationship with technology is one in which there is a conflict between our trust for the natural compared to our suspicion of the artificial.

> These two views, that technology is a thing directing our lives, and simultaneously a thing blessedly serving our lives are simultaneously valid. But together they cause unease, an ongoing tension, that plays out in our attitudes to technology and in the politics that surrounds it. (p. 214)

> We trust nature, not technology. And yet we look to technology to take care of our future – we hope in technology. So we hope in something we do not quite trust. (p. 215)

Having explored the broad intentions of technology education and highlighted the contested nature of technology and society's ambivalent nature towards technology it is time to consider ideas about the future and education.

2 Education and the Future

Writing for UNESCO, Keri Facer has written a paper which argues that ideas of the future play a central role in educational thinking, policy and practice (Facer, 2021). Facer presents five features and associated questions that considers education's relationship to the future:

1. Education in the future – what might education look like in the future?
2. Education as preparation for the future – how do we prepare young people for the future?

3. Education about futures – how can students learn to read learn to think reflexively about futures?
4. Liberating education from the future – how do we liberate education from the future?
5. Education healing futures – how might educational future repair past injustice?

This section will focus on the first three of these features. Regarding *Education in the future* Facer raises the issue that education is more and more being influenced by actors outside education whose goal is financial profit rather than educational development. She warns that education predicated on financial profit will not necessarily provide an education that people want but rather on what can be supplied by available technology. This has implication for the future of technology education. Regarding *Education as preparation for the future* Facer draws a distinction between students seen as adaptive and agentic. The adaptive student is educated so that they are given the knowledge, understanding and skills to be able to thrive in a future as envisaged by politicians and industrialists. Students are not encouraged to critique this envisaged future. The agentic student is educated to critique and challenge the future proposed by those in power and create new possibilities and trajectories that resist the futures that are currently envisaged. The role of technology features large in politicians' envisaged futures and technology education must surely consider the extent to which the curriculum and teaching is geared to adaptive or agentic education. Regarding *Education about fu*tures Facer cites two approaches; futures literacy and the capacity to aspire. She notes that developing futures literacy engages learners with challenging prevailing assumptions about the future in ways that allow the discovery of new possibilities in the present or new routes to create desired futures to be uncovered. She also notes that the capacity to aspire is underpinned by general ideas as to what constitutes the good life and that this can be used in futures literacy as a driver for developing futures possibilities.

3 Teaching the Future as a Concept in Technology Education

The thinking of Facer as outlined above may be seen as taking place mainly in higher education but the seeds of thinking about the future may be planted in secondary school technology education. Technological capability requires the conception of that which does not yet exist to be clarified to the point where it can be brought into existence in the near future. Technological perspective requires critique of present situations with the intention of identifying how improvements might be made in the future. Hence the idea of 'future' is intrinsic to technology education. Facer's thinking provides three challenges to educators concerned with the school technology curriculum:

1. Beware of those players whose motivation is profit.
2. Encourage learners to become agentic as opposed to adaptive.
3. To explicitly engage learners with considering what constitutes a good life; a life worth living.

4 Essential Components?

In this section Nick Givens and James Pitt both highly successful technology educators of many years' experience identify and justify one feature that each consider an essential component of a future technology curriculum. Nick argues for the significance of evaluation and James for collaborative community-based projects.

Nick's priority is that students should learn to:

- Evaluate existing products in terms of their potential environmental and social impacts, positive and negative.
- Evaluate their own provisional design briefs to ascertain whether they can be justified in terms of potential environmental and social impacts.
- Evaluate their own designs, models and completed products against design specifications that they themselves have developed.

He argues that implicit in this is that students should learn how to identify assessable and increasingly rigorous criteria for the above evaluations, throughout their progression in technology education.

In the first two parts of his 'essential' Nick is moving away from the idea that evaluation is concerned primarily with whether an outcome meets its specification but should also consider environmental and social impacts. To some extent this is moving into the area of critique. In the third part he requires evaluation against specification. To be consistent with the first two parts it will be important that such specification is written to respond to environmental and social impacts. He makes that final point that the students should take responsibility for developing evaluation criteria that become increasingly more rigorous as they move through their technology education. He suggests that such progression could be aided if, when evaluating existing products, students learned to interrogate those products so as to infer the original designers' brief and specification. It is important to note that this focus on evaluation cannot take part as a stand-alone feature of a technology curriculum but must be embedded within a broad technology curriculum which would include other topics and a range of related learning activities.

For James, education is all about empowerment; providing an environment for people to become creative, social learners who collaborate in the creation of a more just and sustainable society at all levels. James argues that propositional knowledge in a curriculum for the future needs to be kept to a minimum and he thinks a basic competence in systems thinking is more than enough. He believes that all technology education should be built around group projects, chosen by the learners, and facilitated by the teachers, based on meeting real-life needs, and assessed by learners using criteria they have chosen. James' 'essential' will embrace Nick's 'essential' in that teaching about evaluation as envisaged by Nick will enable learners to develop the criteria for assessing their group projects. It is worth noting that James is not alone in advocating team-based design activity. Emily Pilloton (2009) writes compellingly about the way groups of citizens have been empowered by becoming involved in design activities aimed at helping their communities. And on a much smaller scale it is pleasing to report that a paper presented at the PATT 2024 conference describes how learners aged 12/13 years were working in small teams to identify and solve practical problems in their school (Jones, 2023).

It is instructive to consider James' and Nick's essentials in the light of Keri Facer's challenges. Nick's concern with evaluation linked to critique and James' focus on collaborative projects in response to learner chosen problems are unlikely to be considered as fertile ground for financial gain. Nick's view that students themselves identify criteria for evaluation and James's insistence on student autonomy in the collaborative projects they choose both lead towards learners becoming agentic as opposed to adaptive. Both their essentials are to some extent underpinned by learners' views on how technological activity can contribute to a good life, to read to a good life, a life in their future.

5 The Role of Virtue

Shanon Vallor (2016) explores the role that virtue might play in the way society uses technology. Vallor grounds this exploration in the philosophies of Aristotle, Buddha and Confucius and develops the following taxonomy of technomoral virtues:

- Honesty: respecting truth
- Self-control: becoming the author of our desires
- Humility: Knowing what we do not know
- Justice: upholding rightness

- Courage: Intelligent fear and hope
- Empathy: compassionate concern for others
- Care: loving service to others
- Civility: making common cause
- Flexibility: skilful adaptation to change
- Perspective: holding onto the moral whole
- Magnanimity: moral leadership and nobility of spirit

She argues that a combination of these virtues will amount to technomoral wisdom and considers how society might meet with the future using this wisdom enabling us to live well with new and emerging technologies. She considers in some detail how this wisdom and the associated virtues should influence the use of social media, surveillance technologies, robots for both civil and military purposes and the enhancement of humans. As I write all these uses are under active discussion in the UK.

It may seem a tall order to embed virtue and virtuous behaviour into a school technology curriculum, but I think it is worth considering as part of teaching technological perspective. A key driver of much if not all technological innovation and development has been the pursuit of financial gain. And some argue that without this incentive such development is stifled. Rarely are the following question asked by those seeking venture capital funding: "We can, but should we"? And "If we do what are the likely consequences, intended and unintended"? To be fair, some such as Tim O'Reilly (2017), have been critics of the way technology is used; in particular the use of algorithms to maximise shareholder profits in the short term by treating humans as a cost to be eliminated paying little if any attention to the impact of such decisions on opportunities for workers and the lives of actual people. Engaging learners with virtue and virtuous behaviour could be achieved through the judicious use of science fiction. Not the dystopian tales which focus on the consequences of non-virtuous behaviour but stories that explore what it means to be human in the context of a life worth wanting. The writing of Becky Chambers (2021 and 2022) in her *Monk & Robot* series explores just such issues in a highly accessible and thought-provoking way.

6 Key Questions from the Royal Society

Peter Finegold Head of Policy Education and Skills at the Royal Society recently posed the five following questions as the basis for a fundamental review of the education system (Finegold, 2023) on the grounds that whichever party forms the next government, it will need to face up to the fact that technology and society are moving quickly and leaving education to play catch-up.

- What is education for?
- What do young people need to know and to do?
- What is the future role for teachers and how can the status of teaching be improved?
- How to nurture the skills essential to thrive in the face of a complex and uncertain world?
- What should schools be measuring, for what purpose and for whom?

For the purposes of this chapter, I will revise and consider three of these questions such that they are particularly relevant to technology education as follows:

1. What is technology education for?
2. What do young people need to know, understand and be able to do as a result of their technology education?
3. How can technology education nurture the knowledge, understanding, skills and values essential for young people to thrive in the face of a complex and uncertain world?

With regard to question 1 Beaumont and Steeg (2024) have argued that there are four complimentary justifications for technology education: preparation for work, personal empowerment, cultural transmission, and preparation for citizenship. The precise wording of these justifications will need revising in the light of Facer's advocacy of developing young people who adopt an agentic as opposed to adaptive stance.

With regard to question 2 Beaumont and Steeg (2024) have identified 6 Big Ideas for design & technology that could be used as the basis for identifying what young people need to know understand and be able to do through their technology education. These are: an appreciation of the fundamental nature of technology, a knowledge and understanding of materials, manufacture and functionality, and an understanding of and ability in both design and critique. A matter of concern here is how learners are enabled to not only acquire this knowledge, understanding and skill but also deploy it as an essential part of their technology education. I return here to the essential two aspects that can be used to describe technology education at a high level of generality: technological capability and technological perspective. In pursuing these goals, a curriculum fit for the future will have to give both equal priority and achieve a balance in both acquisition and deployment. This chapter is not the place to delve into the detail of exactly what will be taught nor how it will be taught but it is important to highlight the following considerations. It will be essential to teach the intellectual tools for critique so that you people are equipped to consider their position in society from both adaptive and agentic stances. It will be essential to equip young people with the digital tools available for making design decisions in the way the outcomes of their designing and making will look, function and be manufactured. It must be noted that not only are these tools changing rapidly but that new tools are regularly becoming available. And there is a tension between the roots of technology education in handicraft and modern digital interpretations. Ed Charlwood (Beaumont & Steeg, 2024) makes the point that a mixed economy will be required in which 'hands on' and 'bits on' work in harmony.

With regard to question 3 a to read in complex and uncertain world brings us back to young people engaging with the idea of future. Whatever the future holds it is certain that technology will play a significant part and young people's understanding of technology, along with their ability to use various technologies and have reasoned positions on how technologies of various sorts should be deployed in society will be important. Underpinning this will be the idea of **their** future and the extent to which they feel they have any control over what this might be. Hence in any technology curriculum for 2030 and beyond it will be important that young people are enabled to be future literate and develop the capacity to aspire in envisioning what might be in terms of a life worth wanting. Utilising technomoral wisdom as envisaged by Shanon Vallor will be important here in helping young people work out for themselves how they want technology to be deployed in their lives. And it is important that the curriculum for 2030 and beyond can be interpreted by teachers and learners in terms of their situation. What learners produce in terms of their technological capability either as individuals or in groups, as decided by their teacher or themselves will, and should be, context dependent. Similarly, in terms of their technological perspective, their thinking will be located in the situation in which they find themselves. But it is important that it is not limited to that, and they are able to 'look over the horizon' and consider the ramifications of technology on the world and other peoples in different parts of that world.

7 Conclusion

There should be two main thrusts in a school technology curriculum for 2030 and beyond. These are technological capability and technological perspective. Technological capability engages young people in 'doing' technology; technological perspective engages young people in understanding the nature of technology and how it influences society and the wider world. In acquiring technological capability, it will be important that learners have the opportunity to work as individuals and teams in responding to technological challenges. It is essential that they are suitably prepared in learning the Big Ideas which form the epistemological basis for the subject and can deploy these in developing and demonstrating their technological capability. In acquiring technological perspective, it will be important that young people are engaged with futures thinking and enabled to adopt both adaptive and agentic responses in deciding what sort of world they wish to live in and how this might be achieved both for themselves and the wider global community.

Note

1 Writing in the 70's Bronowski used the term 'man' to describe humans; this text has been adapted to use the term humans.

References

Arthur, W. B. (2009). *The nature of technology*. Allen Lane.

Beaumont, HR, and Steeg, T. (2024). *Design and Technology in Your School*. Routledge.

Bronowski, J. (1973). *The Ascent of Man*. London BBC.

Chambers, B. (2021). *A psalm for the wild-built*. Thomas Docherty Associates.

Chambers, B. (2022). A prayer for the crown shy. Thomas Docherty Associates.

Department for Education. (2019). *English Baccalaureate (EBacc)*. Accessed January 31, 2024, from https://www.gov.uk/government/publications/english-baccalaureate-ebacc/english-baccalaureate-ebacc

Department for Education. (2013). *National curriculum in England: design and technology programmes of study The statutory programmes of study and attainment targets for design and technology at key stages 1 to 3*. Accessed January 31, 2024, from https://www.gov.uk/government/publications/national-curriculum-in-england-design-and-technology-programmes-of-study

Department of Education and Science/Welsh Office. (1988). *Interim Report of the Working Group for Design and Technology*. Department of Education and Science/Welsh Office.

Ellul, J. (1964). *The technological society*. Vintage.

Facer, K. (2021). *Futures in education: Towards an ethical practice UNESCO*. Accessed October 15, 2023, from https://unesdoc.unesco.org/ark:/48223/pf0000375792

Finegold, P. (2023). *Why the next government should commit to a fundamental review of the education system*. Accessed October 15, 2023, from https://royalsociety.org/blog/2023/08/government-review-education-system/

Hughes, T. (1994). *Technological momentum.* In M. R. Smith & L. Marx (Eds.), *Does technology drive history*? MIT.

Jones, P. (2023). Preliminary study of how 21st-century skills are developed during a participatory user-centred curriculum intervention at Key Stage 3 in Design and Technology. In S. Davies, M. Mclain, A. Hardy, & D. Morrison-Love (Eds.), *The 40th International Pupils' Attitudes to Technology Conference Proceedings 2023* (pp. 474–484), 31 October–3 November, Liverpool John Moores University, Liverpool, UK. https://doi.org/10.24377/PATT40.2023

Kelly, K. (2010). *What technology wants.* Penguin.

Ofqual .(2016). *GCSE Subject Level Conditions and Requirements for Design and Technology Crown copyright.*

O'Reilly, T. (2017). *WTF what's the future and why it's up to us.* Penguin.

Pilloton, E. (2009). *Design Revolution.* Thames and Hudson.

Vallor, S. (2018). *Technology and the Virtues: A philosophical guide to a future worth wanting.* Oxford University Press.

CHAPTER 11

The Design Concept/Methodology – A Personal Professional Voyage

Osnat Dagan

1 Introduction

Technology educators agree that technology is human activity that changes the natural environment to adapt it to human needs, using knowledge, materials, energy, and cultural resources. They interpreted technology as a manifestation of human intelligence.

The term 'technology education' has several interpretations. Some interpret it as vocational education whose graduates are qualified engravers, millers, and mechanics, mainly people without academic education; others interpret it as scientific and technological education for excellent students in which they study biotechnology, electronics, robotics etc. These graduates choose their professional development, but they must continue studying at university to become professionals (engineers). Others interpret the term as technology literacy for all students in grades K-12, to prepare them for citizenship in a technological society. The graduates of this view acquire 21st-century skills: problem-solving, design, system thinking, critical thinking, teamwork and more. Therefore, it is important to distinguish between these concepts as each of them has different implications for study content, skills, the required teachers, and the ways to train them.

For some time, the technology educators' community has dealt with the question of what constitutes technological knowledge; how to define the relationships between science and technology, and whether technology is applied science, considering that science has a long history while technology is a young discipline. As science is a difficult and abstract subject, teachers gave examples from objects in which science laws are implanted and this is 'technology as applied science'. de Vries (1996) stated that technology is more than applied science, giving a few examples in which he showed there are three types of technology, depending on the level of use of scientific knowledge and contribution to product design: a) experience-based technology; b) macro-technologies; and c) micro-technologies. He further claimed that the 'design process is the heart of technology' and recommended we limit ourselves to

© OSNAT DAGAN, 2025 | DOI:10.1163/9789004710948_011

experience-based and macro-technologies when teaching it in schools and use micro-technologies only in high schools (as it is too abstract).

Technology education began in the early 18th century. In this era of Craft (known as Sloyd in some countries), it focused on hands-on learning and practical skills development in traditional crafts and trades. The pedagogical approach emphasised apprenticeship and experimental learning, where students learned through direct engagement with materials and tools. Craft as part of technology education lasted till the beginning of the 1990s, when technology literacy developed worldwide, enabling technology education for everyone over the world. It was influenced by the philosophy of technology such as Mitcham's (1994) definition, which divided it into four categories: a) technology as objects (artefacts), when the artefact has a dual physical and functional nature; b(technology as knowledge distinguished from other types of knowledge (which include: skills, laws, rules and theories); c) technology as activity (action of making – design and process of using); and d) technology as volition (conceptual problem concerning technology) and as volition concerning technology.

A conceptual framework was developed that contains two main axes: a) problem-solving design process; and b) system thinking. Mioduser (1998) defines the learning space of solving technological problems according to the following categories: a) the functional aim: knowledgeable user, problem solver, professional, craftsman and expert, and b) the cognitive aim – the knowledge and the mental models. For example: when the functional aim is an expert (higher level), the cognitive aim is designing, and the required knowledge is systematic methodology of design, formal representations, interdisciplinary formal knowledge, and heuristics. These categories enable us to analyse and understand technology education in various layers.

In various countries, different emphasis has been placed on technology teaching; for example, the Craft that became Craft, Design & Technology (CDT) evolved into the National Curriculum in 1990 as Design and Technology in the UK, and into Technology for All in the USA, and spread across the world at the same time. In this transformation from Craft to technology literacy, technology gained a broader perspective as a disciplinary and cultural phenomenon (de Vries, 1997). Technology education for All refers to various thinking and making skills such as: problem-solving, critical thinking, system thinking, creativity, teamwork etc. These skills are based on expert thinking and more complex communication than routine cognition (Levi & Murnane, 2004) advocating integration of the teaching of skills in areas of knowledge and is suitable for today's job market requirements. This approach emphasises the development of higher-order thinking skills such as synthesis, explanation, creation, and evaluation, as well as communication and collaborative skills and the understanding that those who could understand "big ideas" would become valuable citizens (Jones et al., 2013). In all programs, design was a central component that began as a linear model but was found to be iterative (de Vries, 1997).

The notion of Science Technology and Society (STS) followed technology literacy during the 1990s and spread from the USA to other countries as a method of integrating science and technology, including both social and cultural aspects. Israel was influenced by this trend and curricula for the combined subject of STS were developed (for middle school in 1995, and elementary school in 1999).

During the last 30 years, technology education has suggested another direction that integrates cross-curricular activities with Science, Technology, Engineering and Mathematics (STEM). The role of technology varies within this multidisciplinary subject depending on the aims of different countries. The overall aim of STEM is to respond to economic and social needs (Banks & Barlex, 2020; Dagan, 2020). STEM literacies are needed for the maintenance and development of the individual and society in the 21st century. These literacies integrate STEM literacies (which are common to all disciplines) with those that are unique for each S.T.E.M. discipline (Tang & Williams, 2019). The differences lie between the types of knowledge (knowing how and knowing that) in sciences, which is about the "simplified version of reality" in the design in technology and engineering, which is part of the reality and scientific knowledge that is needed in problem-solving (de Vries, 2021).

There are various methods of STEM implementation, as Dugger (2010) mentioned: a) teach each of the four STEM disciplines individually as an independent subject with little or no integration with the other subjects; b) teach each of the four STEM disciplines with more emphasis on one or two of the four (science and math); and c) integrate one of the STEM disciplines into the other three (engineering to science, technology, and math). In response, various curricula and initiatives have emerged (Bybee, 2013; Cagle et al., 2018; Sanders & Wells, 2006; Dagan et al., 2023).

Technology education research claims that this integration of these subjects helps teachers and students develop an understanding about the nature of technology without it being blurred by the integration process. The nature of technology, when taught side by side with the nature of science, will enhance technological and scientific literacy (Jones et al., 2013; Barlex, 2009; Williams, 2009; de Vries, 2021). A solid understanding of the essence of technology is needed to retain its uniqueness and its place among the other three STEM subjects (Jones et al., 2013). When teaching STEM, design could be a challenging project in which students use knowledge from all four disciplines while solving a problem (de Vries, 2021).

2 Technology Education in Israel

The changes in technology education in Israel over the years follows the international trends and added some innovative trends such as the science and engineering track.

Since the establishment of the state of Israel (1948) until the 1990s, technology education in elementary, middle and high schools were Craft, usually held separately for boys and girls, e.g., carpentry and sewing, respectively. Vocational education was part of the technology education in high schools and mainly for students lacking sufficiently high academic achievements. Following the 1993 Harari Report, things changed. The report claimed that "Science and Technology are part of the general education that is needed today and, in the future, even more for every person" (p. 9 in Hebrew). The report's main recommendations were: scientific-technological education for all students from kindergarten to the end of high school; development of a new compulsory subject known as "Science and Technology" to be studied in sequence starting from kindergarten, through elementary school and up to middle and high school; Integrating ICT in teaching and learning processes and the development of science and technology-rich learning environments.

The main aim of this STS program was to develop integrations within the subjects by creating social and cultural links. These concepts still exist in the revised curriculum till now. Over the years, 21st-century skills were integrated in all subjects throughout the K-12 curriculum.

Subsequently, STEM began as local initiatives in private programs, and in some matriculation tracks, mostly in high schools (Dagan, 2020).

Below is a description of the current situation regarding technology education for the different age groups (Figure 11.1).

Early age (kindergarten) curriculum for integrating science and technology in kindergarten culture was published in 2005 as the result of understanding its importance. Early childhood is the period when the basis for effective learning is embedded, and constant and continuous learning is built. The children have an internal ability to imagine, invent and create, which leads to the potential to develop as technologists from the beginning of their lives (Stables, 1997). Therefore, technological education was added as a central part of the

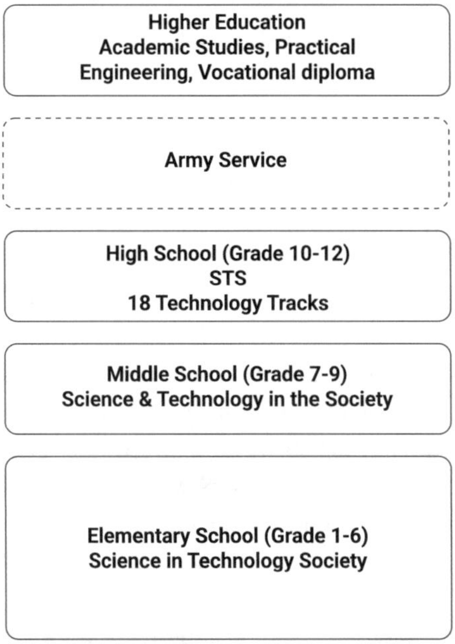

FIGURE 11.1
An overview of the education system in Israel

curriculum for this age group in Israel, as in many other countries. This curriculum was never implemented in the kindergarten system and there is no demand to teach science and technology. However, there are local initiatives, some of which form the basis for technology literacy that focuses on problem-solving and design (Mioduser, 2009; Mioduser et al., 2008).

Elementary school: The curriculum for STS has existed since 1999–2000. The science and technology syllabus for elementary and middle schools aspires to develop citizens with the following skills: scientific and technological literacy, curiosity, motivation to learn, independent learning, the ability to use and implement scientific and technological knowledge, thinking and making skills to understand phenomena in the world, solve problems, designing, decision making and taking a stand about daily life dilemmas. Students should be able to ask questions, find and evaluate information, distinguish between opinions and facts, be open to new information, and understand that science and technology are developed by humans.

In the elementary school STS curriculum, one of the main seven subjects is 'Man-made world', with four main topics: objects, design, systems, and the impact of technology on society and the environment. Curriculum materials and textbooks were developed for the Science and Technology content knowledge (Material, Life, and Earth and Space Sciences and Technology). Although textbooks and teacher guides were developed and teacher training took place, most teachers skipped the technology chapter in the textbook. There are various explanations, but it is mainly because the teachers of this subject had a scientific background and neither the Content Knowledge nor the Pedagogical Content Knowledge needed for teaching the technology topics in the curriculum.

Middle School: In 1996, the Science and Technology curriculum that integrates social aspects was published. In the middle school syllabus, seven study topics were identified. One of these is 'technology systems and objects' which include two sub-topics: 'technology systems' and 'from need to product' (design process).

In that period, many academic groups developed curriculum materials and teachers were invited to participate in massive teacher training. After this tremendous effort, the Ministry of Education cut the study hours for this subject and as a result also the study content. The science teachers who did not feel confident in teaching technology taught only the science subjects and those needed to succeed in the international tests; design and systems were not among them.

High School: With the founding of the State of Israel, most high schools were academic and students who were not accepted went to work or were integrated into vocational training in schools under the supervision of the Ministry of Labor. Till the 1960s, vocational training took place within the framework of a four-year program without eligibility for a matriculation certificate. In the 1970s, a major reform was carried out in the education system that included a transition from vocational training to technological education. In this framework, comprehensive schools were established. During this period, technology education had three tracks: a) practical vocational training for students at a low level; b) a regular vocational track; and c) a vocational high school track for high-level students who can also obtain eligibility for matriculation in technological subjects. Students could not move from one track to another.

Another reform occurred in the 1980s, aiming to cancel the tracks and develop a modular method of study and examination, the development of innovative programs based on broad scientific knowledge as well as the creation of technology-rich learning environments. This is how programs developed that distinguish between introductory subjects, core subjects and specialised subjects.

In the 2000s, following the State Comptroller's report on the implementation of the reform in technological education, an advisory committee was established, and a structure for technology education was formulated. Its guiding principle was the integration of sciences with innovative technologies while updating the study topics to be more relevant. As part of the student's specialisation, a final project is required. In this structure, the method of modularity was integrated and still exists. Today it is possible to study STS – science and technology literacy, or to join one of the 18 technological tracks. These tracks are divided into three: engineering sciences tracks, technological tracks, and vocational tracks. Some of them, mainly in the engineering tracks, have STEM components. These tracks aim to motivate the students to be part of the science and technology community as researchers, engineers, and practical engineers.

3 My Path via Technology Education – My Design Voyage

In the early 1990s, after 20 years of teaching mainly sciences in elementary, and middle school and doing my MA. Degree in ICT, for our middle school we developed an "Introduction to Technology" lab in which we exposed pupils to technology such as robotics, Leg-Logo, the lathe, the milling machine, the 3D printer and a computerised warehouse. We wanted these technology experiences to encourage them to choose technology tracks in high school. This was my first encounter with technology. The lab was unique, and it was organised in stations for active learning. For each station, we developed activity instructions. Students worked in the lab 90 minutes per week for three years in groups of 2–3 and we, the teachers, acted as facilitators. In their third year, they created a design project via Project Based Learning (PBL). During the six years of teaching and learning in this lab we added technologies and improved the learning materials. Pupils explored and experienced technology while working in the lab and even enjoyed themselves. Some of them later joined the technology track in high school, but we were sceptical about how much they learned about technology.

This experience brought me to the ORT Israel R&D centre. It was just in time for the publication of the "new syllabus for science and technology" (the STS syllabus). My first project there was to find ways to make connections between the seven topics, between science and technology issues. This was the beginning of my path in developing curriculum materials for systems, control systems and mainly, in design subjects.

The situation in technology education in middle and high schools at that time was that there were only Craft teachers, this is what they knew to teach. Efforts in various directions were needed to change it into an STS subject: development of curriculum material for the seven topics and the connections between them, development of teacher guides and massive teacher training. I remember myself working with the craft teachers (in-service courses) helping them with CK as well as PCK. ORT Israel's R&D centre was given an important role as the developer of the technology subjects (design, systems, and communication systems) and the ways to integrate sciences and technology and be part of the teacher training. At that time, I was part of this challenging effort.

We began to develop the concept for the new subject which led to textbooks (for design and systems) and design activities such as videos and a kit of cards with design briefs, a design website and a CD about how the system works.

4 Design Models – from Linear Model to Iterative, Design Tools and Design Skills

The design process as part of the curriculum for middle school was new for us as developers and for the teachers. We studied this subject and were aware of the dilemma of how to teach design and decided that to make it easier for the teacher, we would adapt the cyclic but linear model as presented in Figure 11.2 and developed 20 Tik-Tech cards with several design briefs in four areas: robotics, models, Lego, and challenges, with the support of a teacher's guide and teacher training. While moderating teachers in their classrooms they were very happy with the process and the students were very enthusiastic. One of them told us at the end of a design project: "Now I know how to solve every problem. I have the path in my mind".

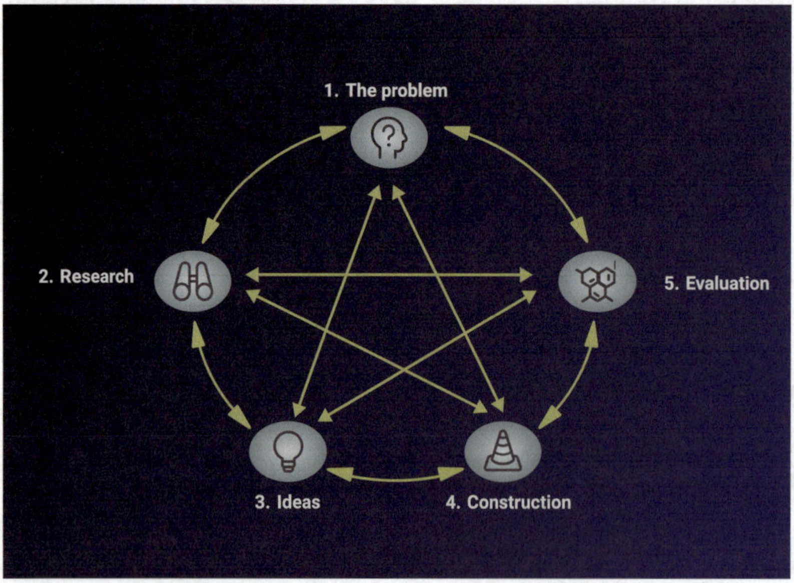

FIGURE 11.2 Cyclical but linear model (ORT R&D centre, 1996)

At the same time, Prof. David Mioduser developed a textbook for teaching design for 7th grade called "Everything can be designed". The design model was linear with many options for repetition (as shown in Figure 11.3).

In those days, we knew that many textbooks used linear models (Wright & Smith, 1998; Norman et al., 1995; Hutchinson, 1993) but we were aware of the criticism that experts do not perform a linear process; that this step-by-step method does not take into consideration students' alternative mental models; that there is more than one way to solve a problem; that there are different processes for different problems and situations; and that students perform each step separately without seeing the overall process in order to meet the teacher's requirements (de Vries, 1996; McCormick, et al. 1994; Kimbell et al., 1991). de Vries, (1996) even emphasised that in real-world industrial processes, design is constructed according to a specific product for specific needs. We also knew about other models that presented various pedagogical methods for design (see, e.g., Lawson, 2006; Johnsey,1998) and Kimbel et al. (1991) "Bridging between mind and hand", and Mioduser (1988) "Design Tools model" (Figure 11.4). These models drove my PhD research (under Prof. Mioduser's supervision). Its research questions examined the connections between learning design in either of the two instructional approaches (structural – step-by-step and functional – by design tools) and: 1) the students' mental models of technological problem solving; 2) students' patterns of use of the various design functions when designing a solution, in the following aspects: the quality and the scope of use of these functions, and the students' conceptions of the

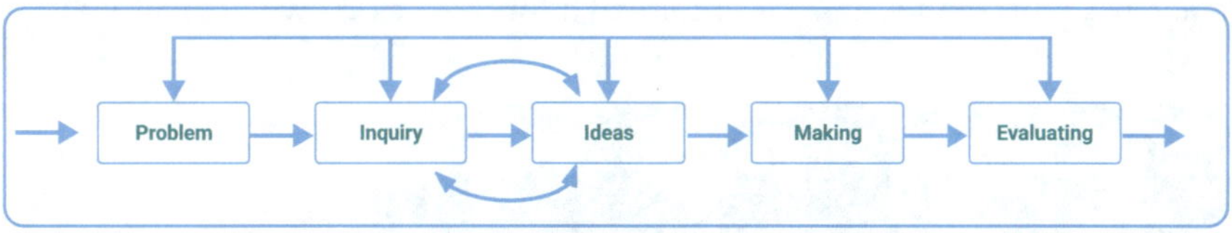

FIGURE 11.3 "Everything can be designed" design model

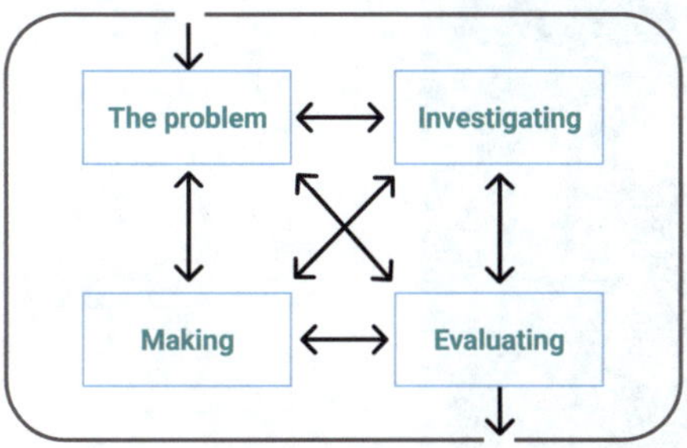

FIGURE 11.4
Design Tools model (Mioduser, 1988)

problem-solving process and 3) the quality of the solutions generated by the students for different problems (Mioduser & Dagan, 2007).

Even though the researchers agreed that it is necessary to teach using the functional method, no curriculum materials were found for this. In this research, a curricular plan, and materials for teaching the design process in the functional approach was designed and examined by comparing it to the structural approach. The functional approach was found more appropriate as an instructional approach for teaching/learning design, in the various aspects examined (construction of mental models, quality of the process and quality of the solutions). This method enables the students to develop problem-solving capabilities according to their conceptions, learning styles and creativity. The constructivist principles of this method enable the students, who are involved and take ownership of their learning, to construct their knowledge and conceptions during the learning process. While searching for the appropriate path to solve a problem, they reflect on the process and gradually acquire technological problem-solving abilities at both the cognitive and metacognitive levels.

We used this understanding in developing an internet site for middle and high school, where students can solve a problem by deciding on their path through the design process by moving the pentagon to the needed function, receiving instructions on what to do, and applying it to their design activity, and document it in their e-portfolio that is built as they work (see Figure 11.5). While developing textbooks for elementary school, we integrated the design tools model (Figure 11.4) in the updated Science and Technology textbook for grades 4, 5 and 6. We found that it was too complicated for the teachers to teach with this model, and we simplified it for them as shown in Figure 11.6.

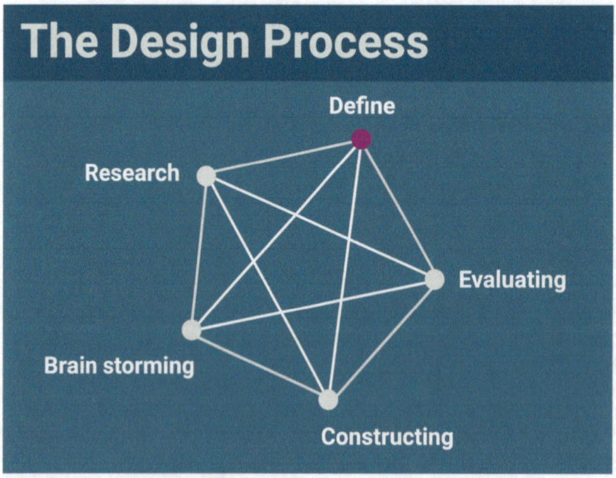

FIGURE 11.5 Design tools Internet site

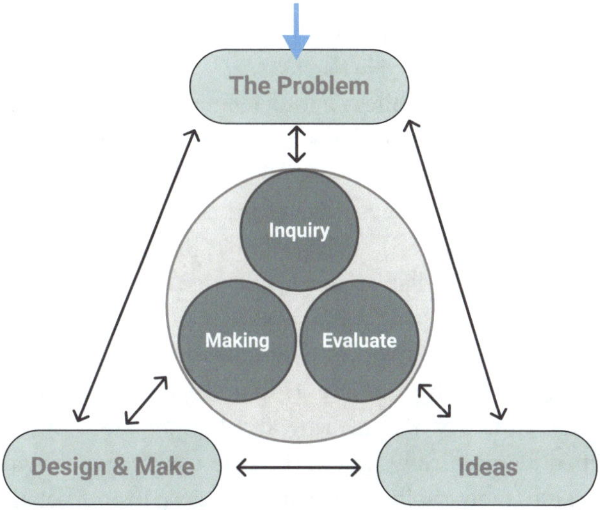

FIGURE 11.6 Design model for 4th grade

In 2012 I began to teach technology thinking at Beit Berl College to students training to be kindergarten and elementary school teachers, and I used the design tool model for teaching design. During our inquiries while developing the new program for M.Ed. in 'Integrative STEM Education' we travelled to Delft University, the Netherlands where I learned from Remke Klapwijk (2019) how to integrate the design skills – Make Design Learning Visible (MDLV) (Klapwijk, 2018; Stables, 2020).

While it was difficult to teach a hands-on course during the COVID-19 period, I prepared a design area in the virtual world (Dagan, 2022) in which I combined the design tools model (Mioduser, 1988) with the MDLV model. I feel that we need to add a place (a new hexagon) in the MDLV model for inquiry, investigating and collecting information that could happen many times during the design process when knowledge acquisition is required, and later I was introduced to a new model (Bloom & Canty, 2022), that cover this aspect. On the other hand, I am not sure that this model presents clearly and precisely the iterative quality of the design process, but I will try to work with it next semester.

5 Teacher Training

From the beginning of integrating technology with sciences, it was not clear who would teach it in schools - science teachers or craft teachers. Neither had the necessary Content Knowledge and Pedagogical Content Knowledge. Many hours of teacher training were delivered which taught the knowing-how and knowing-what that is important to design, and new skills such as cooperative teaching (de Vries, 2021) mostly for in-service teachers.

While teaching a course at Beit Berl College about 'technology thinking' for kindergarten and elementary school to students as part of their pre-service preparation I taught them the CK (design, systems, robotics) and the PCK they needed. Over the years, at the declarative level- there has been a subject known as science and technology in schools (K-9), with textbooks for every year that cover all the topics required in the syllabus, but in practice, the technology subjects are rarely taught in schools.

The uniqueness of our new 'Integrative STEM Education' M.Ed. program that we are now delivering for the third time is 1) the design integrates the four disciplines: Science, Technology, Engineering and Mathematics; 2) the core courses are developed hierarchically; and 3) the core courses are taught by a team of faculty using PBL methodology (Dagan et al., 2023; Ragonis et al., 2023).

6 Technology Education – My Vision

Technology is around us everywhere and the children need to understand how it works and acquire the skills and values they need to become productive citizens: that is the reason for technology for all at the literacy level from K-12. The literacy level will focus on the topics of design, systems, artefacts, and knowledge needed for problem-solving; skills such as creativity, teamwork, critical thinking, decision making and problem-solving; and values such as the impact of technology on the environment and society, as well as ethics in using technologies.

I believe that we need to begin teaching technology literacy from kindergarten with simple issues that they can understand. I would adapt Mioduser et al.'s (2008) six strands: s1) the designed world; s2) problem-solving; s3) design; s4) notations; s5) smart artefacts; and S6) whole kindergarten project and continue through these strands and taking into consideration the curricula for all ages, work in a spiral fashion, expanding and deepening the work at each stage. In high school, it is right to suggest that students study technology in various tracks towards a profession of their choice, as is done now. I believe that we need to allow the students to experience active learning via PBL, inquiry-based learning and even better design-based learning and develop their ability to learn rather than teaching them, especially in this period with the opportunities that open for us with Generative Artificial Intelligence.

Technology education must be part of every child's toolkit, and it could be done through technology as a subject, Design and Technology, STS, or STEM.

Teacher training is the key issue for teaching technology in all school grades, and it could be done during pre-and in-service training – teacher training that will focus on CK and PCK, that will expose teachers to the necessary content, skills and values for teaching this subject. This will allow them to experience collaborative teaching and enable the students to be responsible for their learning.

For that to happen, we need a Minister of Education who believes in the necessity of technology education for all students in all age groups and a group of educators who will follow the Minister's lead.

Acknowledgement

I would like to thank Prof. Marc de Vries from the bottom of my heart for his contribution to my professional development by reading his proceedings, by sharing his knowledge with me during his visits to Israel (ORT Israel) and my visit at Delft, and by creating PATT family and including me in this community. Thank you for giving me food for thought when reading your proceedings, and during presentations and even mingling, I learnt a lot from you. I enjoyed and learnt a lot by being your teacher assistant during "The philosophy of technology" course, looking forward to the next one. I thank you deeply for your friendship, which is very dear to me.

References

Banks, F. (1994). Learning through design and technology. In F. Banks (Ed.), *Teaching technology* (pp. 59–67). The Open University.

Barlex, D. (2009). The STEM programme in England–help or hindrance for design & technology education? In *Proceedings of the Pupils Attitude Towards Technology PATT-22 conference* (pp. 42–53).

Bloom, N., & Canty, D. (2022). Exploring pre-service technology teacher education students developing conceptions of design pedagogy. In D. Gil, J. Tuff, T. Kennedy, S. Pendergast, & S. Jamil (Eds.), *PATT39 conference PATT on the Edge. Canada* (p. 283–290). https://research.edgehill.ac.uk/ws/portalfiles/portal/50821651/patt39_proceedings_june_17_2022.pdf

Bybee, R. W. (2013). *The case for STEM education: Challenges and opportunities.* NSTA Press.

Cagle, N. L., Caldwell, L., & Garcia, R. (2018). K-12 diversity pathway programs in the E-STEM fields: A review of existing programs and summary of unmet needs. *Journal of STEM Education, 19*(4), 12–18.

Dagan, O., (2020). STEM education in the secondary school in Israel. In D. Barlex & F. Banks (Eds.), *Teaching STEM in the secondary school.* Routledge.

Dagan, O. (2022). Constructing design skills in HE via virtual worlds environment – A case study. In. D. Gil, J. Tuff, T. Kennedy, S. Pendergast, & S. Jamil (Eds.). *PATT39 conference PATT on the Edge*. pp. 274–282.

Dagan, O., Ragonis, N., Goldman, D., & Wagner, T. (2019). *Integrative STEM education—A new M.Ed. program: Development, objectives, and challenges.* Paper presented at the *PATT 37 Developing a knowledge economy through technology and engineering education, University of Malta, Msida, Malta.*

Dagan, O., Ragonis, N., & Goldman, D., (2023). Insights from the implementation of the course 'Development of an interdisciplinary STEM project via PBL approach' in an Integrative STEM Education M.Ed. program. *PATT40. Liverpool. UK.*

de Vries, M. J. (1996). Technology education beyond the 'technology is applied science' paradigm. *Journal of technology Education, 8*(1), 7–15.

de Vries, M., & Tamir, A., (1997). Shaping concepts of technology: What concepts and how to shape them. *International Journal of Technology and Design Education, 7*, 3–10, Kluver.

de Vries, M. J. (2021). Design-based learning in science and technology as integrated STEM. In *Design-based concept learning in science and technology education* (pp. 14–24). Brill. https://doi.org/10.1163/9789004450004_002

Dugger, E. W. (2010). *Evolution of STEM in the United States. Paper presented at the 6th Biennial International Conference on Technology Education Research in Australia.* http://www.iteea.org/Resources/PressRoom/AustraliaPaper.pdf

Hutchinson, J. (1993). *Design and problem solving in technology.* McGraw-Hill School Pub.

Jones, A., Buntting, C., & de Vries, M. J. (2013). The developing field of technology education: A review to look forward. *International Journal of Technology and Design Education, 23*, 191–212. https://doi.org/10.1007/s10798-011-9174-4

Johnsey, R. (1998). *Exploring primary design and technology.* Cassell.

Kimbell, R., Stables, K., Wheeler, T., Wosniak, A., & Kelly, V. (1991). *The assessment of performance in design and technology.* London Schools Examinations and Assessments Council (SEAC).

Klapwijk, R. M. (2018). Formative assessment of creativity. In M. J. de Vries (Ed.), *Handbook of technology education* (pp. 765–784). Springer International Publishing.

Klapwijk, R. M. & Stables, K. (2019). *Make design learning visible – toolbook. Formative assessment tools for design thinking.* TUDelft. Goldsmiths.

Lawson, B. (2006). *How designers think: The design process demystified.* Routledge.

Levy, F., & Murnane, R. J. (2004). *The new division of labour: How computers are creating the next job market.* Princeton University and the Russell Sage Foundation.

Mawson, B. (2003). Beyond the design process: An alternative pedagogy for technology education. *International Journal of Technology and Design Education, 13*(2), 117–128.

McCormick, R., Murphy, P., & Hennessy, S. (1994). Problem-solving processes in technology education: A pilot study. *International Journal of Technology and Design Education, 4*, 5–34.

Mioduser, D. (1998). Framework for the study of cognitive and curricular issues of technological problem solving. *International Journal of Technology and Design Education, 8*, 167–184.

Mioduser, D., & Dagan, O. (2007). The effect of alternative approaches to design instruction (structural or functional) on students' mental models of technological design processes. *International Journal of Technology and Design Education, 17*, 135–148.

Mioduser, D., Tal-Levi, S., Kuperman, A., & Rudolph, L. (2008). 'Design & Learning' (D&L) in the kindergarten. In D. Kipperman, O. Dagan & M. de Vries (Eds.) *Critical issues in technology education.* (Eds.), *PATT20, Israel.* (pp. 262–274).

Mitcham, C. (1994). *Thinking through technology: The path between engineering and philosophy.* University of Chicago Press.

Norman, E., Cubitt, J., Urry, S., & Whittaker, M. (1995). *Advance design and technology.* Wesley Longman.

Ragonis, N., Goldman, D., & Dagan O. (2023). Educating the educators: An innovative M.Ed. program in Integrative STEM education incorporating Open Schooling principles. In *Proceedings of the STEM & Open Schooling for sustainability education conference, 11–12 May, Leiden.*

Sanders, M., & Wells, J. G. (2006). *Integrative STEM education.* http://www.soe.vt.edu/istemed/

Stables, K. (2020). Signature pedagogies for designing: A speculative framework for supporting learning and teaching in design and technology education. In J. P. Williams & D. Barlex (Eds.), Pedagogy for technology education in secondary schools: Research informed perspectives for classroom teachers (pp. 99–120). Springer. https://doi.org/10.1007/978-3-030-41548-8_6

Stables, K. (1997). Critical issues to consider when introducing technology education into the curriculum of young learners. *Journal of Technology Education, 8*(2), 50–65.

Tang, K. S., & Williams, P. J. (2019). STEM literacy or literacies? Examining the empirical basis of these constructs, *Review of Education*, doi: 10.1002/rev3.3162

Williams, P. J. (2009). Technological literacy: A multiliteracies approach for democracy. *International Journal of Technology and Design Education, 19*, 237–254.

Write, R. T., & Smith, H. B. (1998). *Understanding technology.* Goodheart-Willcox.

CHAPTER 12

A New Future for Technology Education in Wales

A Case Study of Curriculum Ambition

Frank Banks

1 Background and Context – Back to the Future?

In 2006 Marc de Vries in reviewing contributions to the PATT-15 conference was pondering the collection of articles that considered, in retrospect, the two decades of developments in technology education since the first PATT (Pupils Attitudes Towards Technology) conference was held in 1985. He wondered:

> Why look back if you need to move forward? […] In the development of a school subject or learning area, a period of 20 years is relatively short. Most school subjects have developed in the course of a process that took perhaps several ages (depending on what one counts as the true starting point of development). For technology education, it has been a turbulent period. (de Vries, 2006:2)

Most PATT conferences over the years have had contributions describing the development of Technology Education in a delegate's home country. This chapter takes a similar path in outlining how the new curriculum in Wales is liberating schools to engage in exciting developments in the new Area of Learning and Experience (AoLE) of Science and Technology. In 1990, England and Wales were praised for the bold adoption of a compulsory Technology curriculum for all pupils from the ages 5 to 16, but what was its subsequent 'turbulent period' and how did that first novel 'National Curriculum Technology' document come about?

In discussing the long journey to the new Curriculum 2022 for Wales which gives schools the freedom to innovate, I would like to start a decade earlier than that first PATT conference. On 18th October 1976 James Callaghan, the then Prime Minister of the UK and the Member of Parliament for the Welsh constituency of Cardiff South-East, gave a speech at Ruskin College, Oxford where, stung by criticism of state schools by the opposition Conservative party, he initiated a 'great debate' in education. In 1992, looking back on that speech, he reflected how he was advised by civil servants to 'keep off the grass' as he

© FRANK BANKS, 2025 | DOI:10.1163/9789004710948_012

would be criticised for making such an intervention as it would be interpreted as political indoctrination:

> By some of the educational elite it was thought an unseemly intrusion of the Prime Minister to poke his nose into educational matters and stir up trouble on matters best left to those who know most. [...] In 1976 I asked, [...] Why did some school-leavers lack the basic tools to do a job when they left school? Why did so many girls abandon science even before leaving school? On the other hand, was the curriculum sufficiently demanding? [...] I suggested the introduction of a basic curriculum (not, let me emphasise, a straitjacket) with national standards of performance to ensure basic literacy and numeracy. The goal, I said, should be to equip children to the best of their ability for a lively, constructive place in society, and at the same time to fit them to do a job of work. 'Not one or the other', I emphasised, 'but both'. (Callaghan, 1992:9)

I started my teaching career in 1976, one month before Callaghan's original Ruskin speech. Then, teachers were free to design the curriculum for their particular school, specific to the needs of the students and their community. Local government, responsible for funding state schools, had educational advisers to support the teaching of each subject. There was then, as now, an externally set examination taken at the age of 16. The 'back wash' of needing to teach the topics for that examination's syllabus was the only constraint on the curriculum design. In 1976, Technology was already being taught in some schools long before any legal requirements of a National Curriculum. For example, I taught pre-university students Engineering Science, and a colleague taught Control Technology to Years 10 and 11 (ages 14 to 16), part of the course package *Project Technology* (Schools Council, 1975 and 1981) which focused on project work-based on mechanisms, structures, electronics, and pneumatics.

Margaret Thatcher became Conservative Prime Minister in 1979. Thatcher, a chemistry graduate, tacitly agreed with Callaghan that '[t]he goal should be to equip children to the best of their ability for a lively, constructive place in society, and at the same time to fit them to do a job of work'. Using generous funding from the Department of Employment, she proposed TVEI (the Technical and Vocational Education Initiative). This was good news for those teaching Technology as it legitimised projects with real-world outcomes and funded new equipment, as well as in-service support using a 'classroom on wheels' which enabled a craft teacher and a science teacher from each school in the county to attend local 4x1-week courses in the new subject. Then in 1988, the Department of Education and Science and the Welsh Office introduced

'Science in the National Curriculum' and in 1990 'Technology in the National Curriculum.'

In a clear case of the Department of Education not being aware of what the Department of Employment had done, or perhaps to thwart its upstart involvement in the school curriculum, the phased introduction of first Science and then two years later Technology broke apart all the successful cooperation arrangements that TVEI had facilitated. Although some thought that making Technology a compulsory subject for all pupils from the ages of 5 to 16 was a victory suggesting that the then education secretary, Keneth Baker, did not fully appreciate what he was signing off, closer inspection shows that he was well aware of the content of the new compulsory subject – but it was like the age-old problem at education conferences – what is the definition of 'Technology'? (de Vries, 2009:1) In 1990, Technology in the National Curriculum had 5 Attainment Targets:

- AT1 Identifying needs and opportunities.
- AT2 Generating a design.
- AT3 Planning and making.
- AT4 Evaluating.
- AT5 Information technology capability. (DFES & WO, 1990: III)

It was the requirement for compulsory Information Technology that Baker had focused on. But Technology as set out in AT1 to AT4 was vague, and many found it difficult to translate into meaningful workshop activities – after all, the attainment targets could equally apply to writing an essay as much as creating an an electronic system. The Engineering Council funded a review of the curriculum conducted by Alan Smithers and Pamela Robinson. It was titled 'Technology in the National Curriculum. Getting it right,' and the first sentence was, 'Technology in the national curriculum is a mess'.

> The main reason why technology in schools seems so elusive is that it embodies the aspirations of different interest groups which have been kept together only by pitching its objectives and content at such a high level of generality that it can include almost anything. (Smithers and Robinson, 1994: 37)

In 1995, Information Technology and Design and Technology (D&T) were separated into two separate subjects and for D&T there were now just two Attainment Targets – Designing and Making. In the 1994 draft proposals, John Patten, the then education secretary, set out what changes he intended to make. The last sentence is most revealing – the rather insultingly bald statement, 'I am copying this letter and the proposals to the Secretary of State for Wales' (SCAA, 1994:i)

2 Background and Context – The Rise of the Red Dragon

The 1990 and 1995 'national' curriculum, then, was a curriculum designed in London for schools in England and Wales, but the requirement to teach its content was enshrined in law. However, Wales has a distinctive culture and *Cymraeg*, the Welsh language, is the first language of many children, especially those living in north and west Wales. The National Curriculum was also described as an 'entitlement curriculum' and the learning of *Cymraeg* (either for first language or second language children) became an additional compulsory subject from ages 5 to 16 in Wales. Also, for all subjects, *Cwricwlwm Cymreig* produced by the Curriculum Council for Wales based in Cardiff, was non-statutory guidance to help in the teaching of the National Curriculum subjects specifically in the context of Wales. The extra requirement for schools to teach *Cymraeg* to all students, and the over-zealous enthusiasm of the various subject committees, led to a vastly overloaded curriculum which was most severely felt in Wales. Indeed, as Marc de Vries says, 'For technology education, it has been a turbulent period' (de Vries, 2006:2). Headteachers in Wales successfully argued that Design and Technology, an expensive subject to resource, should cease to be compulsory after Year 9 (age 14) to ease the extensive curriculum demands for ages 14–16. Wales, too, has different cultural priorities to England. Jonathan Sacks, a former Chief Rabbi said:

> To defend a country, you need an army. But to defend a civilisation you need schools. You need education as the conversation between the generations. (Sacks, 2021:207)

Wales is known as the Land of Song and cultural competition festivals in poetry, music and literature known as 'Eisteddfodau' have been held since 1176. In Welsh schools, an annual Eisteddfod often also includes art and design-and-make tasks too. The school curriculum organised by the parliament in Westminster and centred on the requirements of England was expected to be adopted in the very different cultural context of Wales. Some insisted that subjects such as Mathematics, Science, and Technology are not culturally specific, so what was the problem? Others argued that, like Art, History and Geography, *all* subjects need to build on examples from the locality and community. This was especially true for schools teaching through the medium of Welsh. Ysgol Gymraeg yr Urdd, the first such school, opened in 1939, and now 439 Welsh medium schools are teaching 23% of all pupils in Wales. Despite this different cultural context, it was only in 1967 that an Act of Parliament separated the two countries. Until then Wales was considered part of England. Given that recent history, it is not surprising that the Union Jack, ostensibly combining the flags of the constituent countries of the United Kingdom, nevertheless does not include the ancient 'Red Dragon' flag of Wales.

In May 1999, under the Labour Prime Minister Tony Blair, an Assembly was set up in Cardiff, which in 2020 became the parliament for Wales known as *Senedd Cymru*. The responsibilities devolved to Cardiff include (among others):

- Education and training
- Culture
- Welsh language
- Health and health services

Welsh politicians, now responsible for Education, initiated plans for a new curriculum for Wales:

> We face complex technological change and huge shifts in competitive pressures from within Europe and beyond. Many of our communities are vulnerable and disadvantaged. Our skills base is relatively low. A great deal remains to be done to heal the significant divisions in our society – for the health and well-being of everyone who lives here; to ensure that people can realise their ambitions; and to build a truly enterprising and creative culture, making the very most of our distinctively rich and diverse inheritance. We are at a turning point for education and lifelong learning in Wales. It is highly significant that the National Assembly's first major legislation and decisions were related directly to what is done in schools and to the arrangements for learners post-16. We now have legislation in place for a revised National Curriculum following on from a large-scale review taken forward specifically for Wales. (Davidson, 2001:1)

Despite this ambition, the Programme for International Student Assessment (PISA) outcomes for Wales since 1999 have been stubbornly resistant to improvement, and the Senedd has turned to the Organisation for Economic Co-operation and Development (OECD) for support (OECD, 2014, 2017). Changes, particularly to teacher professional development, were implemented, yet the 2022 PISA report still makes for grim reading. Although exacerbated by the 2020–2021 global pandemic which closed many schools across the world, Wales still comes out worse in the UK PISA tables:

- All UK countries have lower average scores for mathematics and reading relative to their performance in 2018. The differences between the 2018 and 2022 average scores were statistically significant in both cases.
- The average mathematics score for Wales (466) was significantly different to the average scores for Northern Ireland (475) and England (492), but not significantly different to the average scores for Scotland (471).

- The rank order of UK nations is broadly consistent across the different mathematics sub-domains, suggesting that individual nations do not have relative strengths or weaknesses (relative to each other) in terms of the different areas of mathematics that were assessed in PISA 2022.
- The average science score for Wales (473) was significantly different to the average scores for Scotland (483), Northern Ireland (488) and England (503). (Ingram *et al.* 2023:148)

These outcomes, combined with the advice received by the OECD, will no doubt re-double political pressure to secure the successful implementation of Curriculum 2022 (Senedd, 2017).

3 Curriculum for Wales 2022

The new Curriculum for Wales enables each school in Wales to develop its own curriculum – in many ways it's like travelling back to 1976 when Callaghan started his 'great debate', but there is now a framework to support each school's individual curriculum plans (see Hwb, 2022a). There are four fundamental purposes to the Welsh curriculum supporting the aspiration of all children in Wales to become:

- ambitious, capable learners, ready to learn throughout their lives.
- enterprising, creative contributors, ready to play a full part in life and work.
- ethical, informed citizens of Wales and the world.
- healthy, confident individuals, ready to lead fulfilling lives as valued members of society. (Hwb, 2022b)

Indeed, as we have seen, Thatcher and Callaghan, would agree: 'The goal should be to equip children to the best of their ability for a lively, constructive place in society, and at the same time to fit them to do a job of work. Not one or the other, but both' (Callaghan, 1992:9). Along with these purposes, schools also need to consider how to address the three cross-curricular skills of literacy, numeracy, and digital competence; each school has the freedom to answer for themselves the key questions:

- What should we teach and why?
- How should we teach it?
- How will this support our learners to realise the four purposes?

There are six Areas of Learning and Experience (AoLE): Expressive Arts; Health and Well-being; Humanities; Languages, Literacy and Communication;

Mathematics and Numeracy; Science and Technology. Within the Science and Technology AoLE, the support to teachers about 'what matters' sets out that:

1. Being curious and searching for answers is essential to understanding and predicting phenomena.
2. The world around us is full of living things which depend on each other for survival.
3. Matter and the way it behaves defines our universe and shapes our lives.
4. Forces and energy provide a foundation for understanding our universe.
5. Computation is the foundation for our digital world.
6. Design thinking and engineering offer technical and creative ways to meet society's needs and wants.

Number 6 applies to technology education:

> By applying their experiences, skills and knowledge, learners can design and shape innovative engineered solutions. Being part of a user-centred design process will encourage them to use creativity to develop ideas, manage and mitigate risks, and minimise complexities. When engineering products, services and systems, they will need to understand and control the interactions between materials, structures, components and users. The application of engineering processes allows learners to develop accuracy, precision, dexterity and craftsmanship. By designing and engineering outcomes in response to needs and wants, learners can become enterprising problem solvers. (Hwb, 2022c)

The AoLE of Science and Technology not only gives new opportunities to consider collaborations that could be considered as STEM but also enables collaboration across all the Areas of Learning and Experience in a school's new approach. In England, a new 'lightweight' National Curriculum for Design and Technology was introduced in 2013 for those up to the age of 14 (DfE, 2013) but that requirement no longer applies to Wales! However, Wakefield and Hardy suggest a set core:

- *Knowledge and understanding of materials*, their properties and uses.
- *Designing* and working with clients/users/target groups to explore and generate solutions.
- *Making* by using a range of tools and equipment via a range of different material areas.
- *Evaluation* and reflection on pupils' work and the work of others. (Wakefield and Hardy, 2013:23)

These core values can be seen in the examples from Welsh schools that follow.

4 Curriculum for Wales 2022 – Science & Technology AoLE: Case 1

In the King Henry VIII 3–19 School, Abergavenny, Jordan Curran, Subject Leader of Technology, collaborates with the Science department on a project for Year 7 (11–12 years old) based on a Theme Park: 'Monmouthshire County Council are building a theme park somewhere in Monmouthshire. This will create many jobs and attract tourists to Monmouthshire.' The project is introduced through a video (see Resources Case 1) and, in a coordinated approach (Banks & Barlex, 2021: 36), part of the project is covered in Science lessons, and part in Technology lessons, with supporting ICT lessons in spreadsheets and website design as follows:

In science students learn about:
- 'Murder most foul' a murder has happened due to a land deal for the land of the theme park – forensic science activities.
- Forces acting upon a rollercoaster (and potential and kinetic energy).
- Chemistry of fireworks.
- Heart rate and fear.

In technology, students design and make:
- an aerodynamic rollercoaster cart for a new ride based on a theme of their choice. They must apply their scientific knowledge to make their rollercoaster cart as fast as possible, such as by changing weight and wheel diameter, and there is a leader board and competition – *The Velocity Project* (see figure 12.1).
- The data gathered from the weight and speed of their cars is saved to a spreadsheet which is exploited as meaningful data in ICT.
- a functional souvenir product for a gift shop at the Theme Park. Students use both recycled HDPE (High-Density Polyethylene) and pewter to practise with, and then decide on the most appropriate material for their product. They explore the melting points and scientific principles of pewter and HDPE – *The Giftshop Project*. Also, students design realistic-looking packaging for their products.

Building on the work of the Plastics Lab is a club run by a group of dedicated students in years 7 & 8 (mostly girls) who make products from HDPE, such as coasters, bookmarks and key rings which they sell to a local plastic-free shop.

In Year 8, students (12–13-year-olds) are asked the question in science and technology lessons, 'What will the world look like in 50 Years?'. Everything they learn in the AoLE Science and Technology lessons helps them address this question. At the end of the year, students are tasked with creatively answering

FIGURE 12.1 The rollercoaster track

FIGURE 12.2 Prototype of a building for a brand-new city

this question via a newspaper article, magazine feature, a podcast and/or TV news segment, or a designed website. This year's work is also introduced by a video.

In science students learn about:

- Space,
- Disease, respiration. photosynthesis, ecology, genetics,

- Speed, acceleration, crumple zones, electricity and waves, electromagnets, light, camera, eye,
- States of matter, the periodic table, cars of the future, building an eco-home, rusting, electroplating, rates of reaction.

All links to what the world will look like in 50 years.

In technology, students design and make:

- a colour-changing, USB-powered, LED light. This is used as a stimulus to discuss how the design of this light might change in 50 years. (Wood is sustainable but acrylic is not etc.) - *USB Light Project*
- a prototype of a building for a patch of land in a brand new city. The Welsh Government are proposing to build a new city to be ready by 2073 (see figure 12.2.)

As part of Digital Competence/ICT students consider:

- AI in the household – homes of the future
- Programming.

5 Curriculum for Wales 2022 – Science & Technology AoLE: Case 2

At Lewis Girls' School in Ystrad Mynach, Matt James is Senior Leader of Learning for Design and Technology, STEM, and the Digital Competence Framework. The school has adopted a novel approach to the whole curriculum, taking advantage of the freedom across all traditional subjects by organising their curriculum through the Areas of Learning and Experience. The Science and Technology AoLE has been carefully crafted using a Collaborative approach (Banks & Barlex, 2021:38) between Design and Technology, Textiles, Food, Science, Digital Design, and a separate STEM lesson each contributing to an overall project. In Year 7 (ages 11–12-year-olds) the overarching theme is 'Sustaining Planet Earth – what does it take to sustain a planet?' (see figure 12.3). In STEM lessons, pupils learn about different types of structures and the way that forces can affect them, before designing a structure to hold hydroponics equipment. This is linked to growing larger volumes of food in a smaller footprint area and the importance of 'Sustaining Planet Earth'. In groups they design, build, and test a tower to destruction, measuring the loads applied and calculating the strength-to-weight ratio to find which team's tower has been most successful (see figure 12.4). The time allocation for the separate subjects and their coming together in the STEM contribution for a year 7 pupil is every

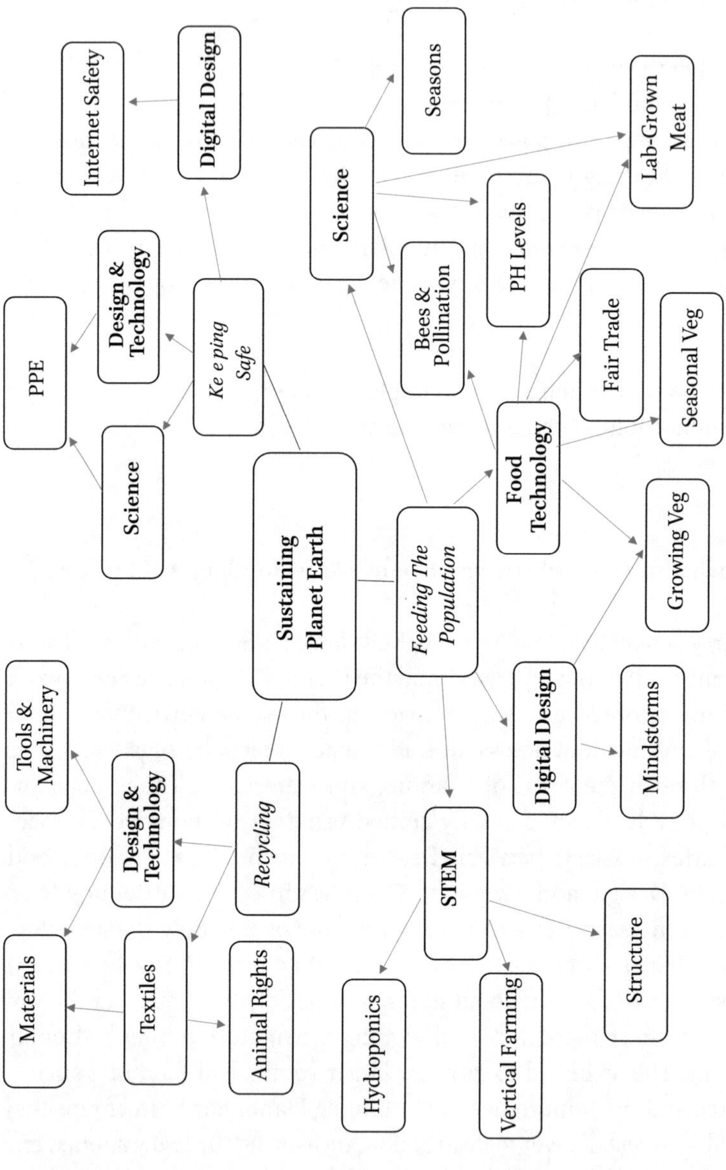

FIGURE 12.3 A whole year connection – Year 7 Lewis Girls' School

FIGURE 12.4
Testing Hydroponic Tower

fortnight: 1 hour of D&T, 1 hour of Textiles, 1 hour of Food, 5 hours of Science, 3 hours of Digital Design (ICT) and a 1-hour STEM lesson.

Similarly, Year 8 all have separate D&T, Textiles, Science, STEM, and IT lessons. As in Year 7, all lessons link to a common theme, now – 'Protecting Planet Earth'. The STEM lessons look at energy particularly fossil fuels versus renewable energy sources. Pupils use the school's custom-made equipment to conduct a range of energy experiments and analyse the effect of different insulation levels in model houses.

In Year 9, as part of the theme 'Planet Earth and Beyond', pupils build and test both water rockets and a 'Mars Rover' (see figure 12.5). For the rocket project, pupils working in pairs use a computer simulation to model their design before building their water rocket, changing variables such as the amount of water and the shape of the rocket's fins. This process is linked to the Bronze CREST award allowing pupils gain this qualification as they complete the project (see Resources Case 2).

6 Assessing Curriculum 2022

The PATT-8 conference was devoted to Assessing Technology Education (Mottier & de Vries, 1997). It set out the assessment strategies in a range of countries, tackled the thorny problem of assessing attitudes, and illustrated

FIGURE 12.5 Track for testing Mars Rovers

several assessment methods and instruments. These issues still need to be borne in mind. In January 2024, Qualifications Wales published its new suite of qualifications for students aged 14–16. It might seem strange to teachers in other countries that the UK, including Wales, spends so much time, effort, and money on testing children at the school-leaving age of 16. Qualifications Wales, a quarter of a century since Wales assumed responsibility for education, announced with great excitement new 'Made-for-Wales' GCSEs.

GCSE (General Certificate of Secondary Education), an academic qualification initiated in 1986, will sit alongside a new qualification, The Vocational Certificate of Secondary Education (VCSE). Like the new GCSEs it will be introduced in 2027 when the current cohort following the introduction of Curriculum 2022 will have reached the appropriate age. It has been decided:

- It will no longer be possible to sit separate GCSEs in physics, chemistry, and biology, only 'Double Award Science'.
- There will be one maths and numeracy GCSE combining the separate qualifications.

- In English, Welsh, maths, and the sciences pupils will have the option of taking a double GCSE or a less-demanding single-grade GCSE.
- There will be more digital and on-screen assessments. (Qualifications Wales, 2024)

VCSE, the new vocational qualifications, are intended to be more 'practical' and geared to particular jobs or careers. There is a hope that the VCSE will be considered by parents and employers to be on a par with the better recognised and academic GCSE. Undoubtedly the new qualification will rationalise the current plethora of about 300 vocational qualifications as it is intended that there will only be about 13 to 16 VCSE examinations in total. The Welsh Examination Board (WJEC) has also announced its suite of new GCSEs for examination in 2027 (WJEC, 2024) but surprisingly there is no GCSE currently proposed for Design and Technology. It might be that the subject will be included in VCSEs, or in some new qualifications yet to be elaborated for learners aged 14–16:
- Foundation qualifications.
- Skills for Life and Skills for Work qualifications.
- Personal Project qualification.

The possibility of a Personal Project qualification could be very attractive to some learners.

7 Conclusion

The opening section of this chapter 'Back to the future?' seems particularly apt. Almost 50 years ago I and my colleagues had the freedom to introduce Technology and Engineering into our school's curriculum. As predicted by de Vries (2006:3, 2009:1), over the years the top-down closely prescribed curricula have not significantly improved learning. The Westminster Government itself has recognised the flaws in over-prescription, as for each iteration of the National Curriculum it has slackened the reins. The Welsh Government has decided to return the design of the curriculum to the hands of a headteacher and their staff, supported by the local community that the school serves. To ensure progression from the foundation stage to the end of compulsory schooling there needs, of course, to be some agreed fundamental purposes and suggested themes within each Area of Learning and Experience, but as in 1976 when Callaghan opened the great debate, now each school has the freedom to address the key questions:

– What should we teach and why?
– How should we teach it?
– How will this support our learners to realise the four purposes?

References

Banks, F., & Barlex, D. (2021). *Teaching STEM in the Secondary School: Helping teachers meet the challenge* (2nd ed.). Routledge.

Callaghan, J. (1992). The Education Debate 1. In M. Williams, R. Daugherty, & F. Banks (Eds.), *Continuing the Education Debate*. Cassell.

Davison, J. (2001). *The Learning Country: A Comprehensive Education and Lifelong Learning Programme to 2010 in Wales*. National Assembly for Wales.

de Vries, M. J. (2009). The developing field of technology education: An introduction. In A. Jones & M. de Vries (Eds.), *International handbook of research and development in technology education*. Sense Publishers.

de Vries, M. J. (2006). Two decades of technology education in retrospect. In M. J. de Vries & I. Mottier (Eds.), *International handbook of technology education reviewing the past twenty years*. Sense Publishers.

DFES & WO. (1990). *Technology in the national curriculum*. HMSO.

Hwb. (2022a). Accessed February 3, 2024, from https://hwb.gov.wales/curriculum-for-wales/designing-your-curriculum/

Hwb. (2022b). Accessed February 3, 2024, from https://hwb.gov.wales/curriculum-for-wales/designing-your-curriculum/developing-a-vision-for-curriculum-design/#curriculum-design-and-the-four-purposes

Hwb. (2022c). Accessed February 3, 2024, from https://hwb.gov.wales/curriculum-for-wales/science-and-technology/statements-of-what-matters/

Ingram, J., Stiff, J., Cadwallader, S., Lee, G. & Heather Kayton, H. (2023). PISA 2022: National Report for Wales: Research report, Oxford, University of Oxford.

Mottier, I., & de Vries, M. (1997). *Assessing Technology Education*, Eindhoven, PATT Foundation.

OECD. (2014). *Improving Schools in Wales: An OECD Perspective*, Paris, OECD.

OECD. (2017). *The Welsh Education Reform Journey: A Rapid Policy Assessment*, Paris, OECD.

Qualifications Wales. (2024). Accessed February 4, 2024, from https://qualificationswales.org/news-views/qualifications-wales-announces-exciting-new-range-of-qualifications/

Sacks, J. (2021). *The Power of Ideas*. Words of Faith and Wisdom. London, Hodder.

SCAA. (1994). *Design & Technology: Draft Proposals*, London, School Curriculum and Assessment Authority.

Schools Council. (1975). *Control Technology teachers' handbook*. Hodder and Stoughton Educational.

Schools Council. (1981). *Schools Council Modular Courses in Technology: Teacher's Master Manual*. National Centre for School Technology.

Senedd. (2017). *Education in Wales: Our national mission - Action plan* 2017–21, Cardiff, Senedd.

Smithers & Robinson. (1994). *'Technology in the national curriculum. Getting it right' in Frank Banks Teaching Technology*. Routledge.

Wakefield, D., & Hardy, A. (2023). Government policies and design and technology education. In A. Hardy (Ed.), *Debates in Design and Technology Education* (2nd edn.). Routledge.

WJEC. (2024). Accessed February 2, 2024, from https://www.wjec.co.uk/articles/made-for-wales-qualification-outlines-and-consultation-report-now-available/

Resources

With thanks to Jordan Curran and colleagues, King Henry VIII 3–19 School, Abergavenny – Case 1:

Velocity Project: https://youtu.be/b2EIzlS9Nng?si=50IN-KbuUbSCmQbU
Pewter: https://www.youtube.com/watch?v=OoAKafi-yDo
HDPE/Plastics Lab: https://www.youtube.com/watch?v=W2tikauHBio&t=67s
News segment of sales of items: https://www.youtube.com/watch?v=4YobYd3yipQ
USB Light Project: https://www.youtube.com/watch?v=oZEia-_4ObQ

With thanks to Matt James and colleagues, Lewis Girls' School, Ystrad Mynach – Case 2: https://www.lgs.wales/page/?title=I+DISCOVER&pid=63

CHAPTER 13

Developing a New Technology Curriculum for an Urban Setting

Tales of Trials, Tribulations and Tiny Triumphs

Gerald van Dijk and Elwin Savelsbergh

1 Introduction

Technology curricula are evolving at a rate of perhaps no other school subject, as a result of innovations within technology itself and due to new visions about technology education, such as described in this book. Moreover, education as a whole is constantly reinventing itself in new societal contexts, such as increased cultural diversity. In this chapter, these changes intermingle. It is a story about a school where most students had a migrant background and where a new design-based technology curriculum was introduced. We use this story to arrive at four considerations for teachers, researchers and curriculum developers who are concerned with inclusive technology education for the future.

2 A New Curriculum

In general secondary education in The Netherlands, technology subjects have been scarce and few students have opted for further studies in technology, particularly in urban contexts. Therefore, a new technology subject has been introduced, under the label 'Technology and Applications (T&A)', which a school can offer students in the final two years of secondary education (vmbo-stream).

At the Herman Dooyeweerd High School (HDH), management and teachers decided to offer T&A because of excellent job opportunities in technology vocations. Many students had opted for administrative studies, which as the headmaster told us, was in line with the parents' preferences, but jobs were increasingly scarce. Many parents and grandparents had come to The Netherlands to work in factories. Technology studies were associated with low status and dirty work, and parents wished for something better for their children. T&A was regarded as an opportunity to present a richer, more attractive picture of careers in technology.

© GERALD VAN DIJK AND ELWIN SAVELSBERGH, 2025 | DOI:10.1163/9789004710948_013

3 The Curriculum Documents for T&A State That it is About

– Learning about a broad range of technologies including topics such as 'programming' and 'health'.
– 'Doing' technology by designing and making solutions for real problems in the world outside school.
– Career orientation, by means of interaction with 'problem owners' outside school.
– Project work as a means to develop general skills such as being creative, collaborating and communicating and applying knowledge from school subjects such as science.

The exam program does specify some technology content and compulsory contexts, but compared to the sciences it is rather open for interpretation. In order to promote quality and consistency, there are regional networks where teachers can voluntarily share practices and discuss implementation. So, there is quite some room to tailor the curriculum to match local needs.

4 Teachers and Their Ideas at the Start of Implementing T&A

Because of shortages of certified technology teachers, many classes in foundational education in The Netherlands are taught by teachers who are trained for different subjects than T&A. At HDH this was the case too. Two teachers were appointed to run T&A. One of them was Leo, a good-humoured arts teacher who could tell stories about his students and their families with compassion. Artistically designing things was Leo's passion, as long as it didn't involve anything digital. But then there was Nico, his vocational technology co-teacher. Nico was the tech-savvy type, always trying new technologies. According to the head of school, Leo and Nico would make a good pair to run classes together, and one can see why. The school decided to collaborate with us as researchers. Our role would be to help build T&A at the school level, to co-construct teaching materials, and to monitor the implementation process.

At the start of the project, we conducted interviews with Leo and Nico to gain more insight into the context. The beliefs they voiced, were:

a. Most students live in small and crowded rental flats, where (technical) improvements on the house are not allowed.
b. Some students have family members who work on car maintenance and repairs, but these typically aren't up-to-date technologies.

c. Creativity is an important objective within T&A. Students need to learn to think out of the box because they aren't used to it.
d. Students have difficulty with open-ended tasks, because of under-developed self-regulation.
e. Students have difficulty presenting their work up-front in class or presenting to outsiders, which is related to low academic self-esteem.
f. Students are at risk of dropping out of the school system.

So, from the onset, it was clear that there were ample opportunities to work toward a valuable curriculum for the students, but we also saw challenges:
- Curriculum: A subject with little tradition, that was very open in terms of content, pedagogy and quality control, and that aimed to increase interest in technology.
- Teachers: Little experience with the curriculum, knowledgeable about the lives of the students, somewhat negative about the students' technical competencies.
- Students: Low academic self-esteem and some 'at risk'.

This led us to the following research question:

> How can equity and interest in design and technology be promoted through an open design-based curriculum in a hyper-diverse urban context?

5 Method of Researching and Developing, and Our Position

The context of a new curriculum, in a school with a population that we knew little about, required an open research approach, rather than a method whereby a preconceived curricular design is tested. Of course, as researchers we did have our orientations towards technology education and inclusion, and those ideas did permeate the course of this study. Our position is best described as an adherence to notions of powerful knowledge, which implies that a systematic and coherent knowledge base is conditional for deep thinking and the development of higher-order skills (Young & Muller, 2013). We are wary of curricula that emphasise 'general skills', such as creativity and collaboration, and that leave students guessing about what counts as knowledge and skills that matter for success in the educational system and for life and work (Bernstein, 2003; Hirsch, 2016). However, the general literature about knowledge-rich curricula gives little guidance for implementing a subject like T&A in an urban setting.

Our research method can be characterised as Participatory Action Research, an iterative process of change in a local context, with overlapping roles of researchers, designers and teachers, often aimed at improving the situation of underprivileged groups (De Oliveira, 2022). In our case, the first author worked with two teachers for three years as co-designers, and sometimes as co-teachers to be able to get a feel of teaching in this context. A design task would typically take 7 weeks, after which analysis was done and the task was improved, including recommendations for teaching.

Data consisted of interviews with all students from three classes for three years, regular interviews and design sessions with the teachers, field notes, lesson observations, and students' products. All data were coded and thematically analysed.

6 Tales of Trialling

In this section, we present three tales that centre around design tasks.

- Preventing flooding
- Designing wooden figures for a game (2 trials)
- Designing an improvement for a sensor casing (2 trials)

6.1 *Preventing Flooding*

In many neighbourhoods, flooding risks are increasing because of climate change. This design task, commissioned by the local water management board, was meant to minimise flooding in the students' residential area. It was expected that the students would be motivated for this task because of the link with their home environment.

> Nico: "What would be really nice, thinking about T&A, to do related things in the students' neighbourhood. The water management board for instance. [...] It is a neighborhood with little greenery, and lots of stone. So, excess water is quite a thing.

The project started with two lessons where Nico elaborated on climate change, rates of rainfall, and existing solutions. Students then made 3D-models that illustrated an idea, with visual references to their living environment (see Figure 13.1)

However, we learned that for many of these students, their neighbourhood was not inspiring as a problem-solving context, at least not as we had envisaged.

FIGURE 13.1 Example of an incomplete student model, with yellow tubing that would function as water drainage pipes

This is illustrated by Mohammed's response to the suggestion that he could ask people in his neighbourhood about their living environment.

> They speak a different language, or we don't talk to them at all, because many people move houses in a short time.

Similar feelings surfaced often when we tried to relate students' work to their 'lived experiences'. When we discussed this with Leo and Nico, they asserted that students weren't proud of all aspects of their living environment and that a focus on solving neighbourhood problems, was perceived as 'rubbing it in'. What also became clear was that students hadn't used scientific knowledge about rates of rainfall in their designs.

At the end of this project, the teachers expressed disappointment about the students' achievements. One cause that they mentioned was that the design task had been too abstract and that students had become bogged down in decorating their work instead of designing conceptually and technically.

NICO: It is too abstract, we ourselves would also struggle with a task that big. [...] The steps are too large. We expect too much of our students.

This resonated with many of the student's comments, and the need for clarity about learning objectives and intermediate steps recurred during the study.

When we asked whether there was a problem that the students' designs, such as in figure 13.1, could never work in reality and would not be acceptable in any engineering study, an interesting discussion emerged. Leo felt that shooting excess water into space would be a fine solution because students would then have thought creatively. Designing realistic solutions could be learned during further studies, he argued. His disappointment mostly concerned the level of technical skills that the students had developed, rather than design performance to arrive at a realistic solution. This is an interesting observation given the T&A ambition to inform students about the nature of technical jobs.

6.2 *Designing Wooden Figures for a Game* (2 *Trials*)

Trial one of this task was organised with a game designer who developed a game for a physiotherapist who works with disabled children, with the aim is creating opportunities for those children to become more independent. Wooden figures were needed to visualise the children's living environment on a board game. The designer visited the classroom and introduced the task. The students asked questions and were positively engaged, which the teachers attributed to the fact that the students liked the designers' businesslike appearance. Abdul, a student who normally spent little time on task, was particularly interested:

ABDUL: What makes you do this job, what drives you?
DESIGNER: I think, as a designer one has a lot of impact. I feel joy when I think: this makes the world a little bit better.

The students then sketched ideas and tried to make products with the laser cutter. This was hard and most of the work on the computer had to be done by Nico, so that the students would at least have a tangible object in order to show their ideas in 3D.

A year later, three things were changed. Students now worked with booklets that guided them through the design process, they held 10-minute talks with their teachers about their progress, and a different laser cutter was used, that was easier to operate. For the 10-minute talks, we developed the following structure:

1. Ask the student to explain what design problem needed to be solved, elaborating on their answers.

2. Asking the student to show and tell using sketches and models what design process was underway.
3. Explore, dialogically, ways to improve on their ideas for the design.

Student Mourat was positive about the guidance that the booklet had provided, such as making a morphological table with solutions for each item in the program of requirements.

> We had to make a morphological chart. Yes, it was more based on your ideas, making your planning, how it would look like. I say honestly, it is better to first make a planning and only then you can make a change to it [the design]. If you want that. But a planning is actually priority and important.

Meryem, a student who had shown a lot of progress, was less enthusiastic.

> I didn't like it [the written scaffolds]. I would rather just have received the [design] task than that list of things to fill out. That is more something for subjects like math. For some students it is probably handy, but not for me. The program of requirements is useful. Whether I will use things like the morphological chart next time, depends on the task.

Although students made use of the design scaffolds to work from stage to stage, they would often forget about previous work. They would for instance make a foam model with little resemblance to the ideas they had chosen from their morphological charts, without an argumentation for such deviations. Moreover, the teachers' enactment of the design tools did not help the students to streamline and converge their ideas. There were only a few instances when the teachers mentioned these scaffolds. And in their feedback they made little reference to the students' written ideas. Instead, they offered many new ideas that weren't clearly related to the students' paperwork. This lack of alignment between written scaffolds and the teachers' guidance occurred often, but less severely a year later.

Figure 13.2 shows exemplary of work that students now produced at the intermediate and final stage of the project.

Leo and Nico asserted that the students were now proud of their products and they emphasised how important this achievement was in terms of developing creativity and motivation. We would add that being able to develop ideas through a graphic model (the image at the left) and making a product by means of digital fabrication, arriving at a realistic solution for an authentic

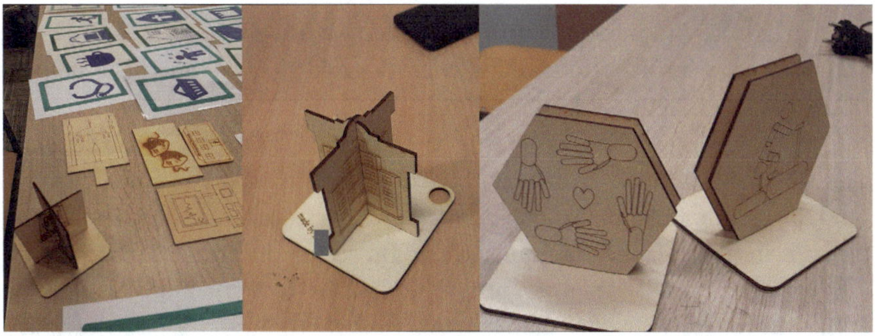

FIGURE 13.2 Wooden figures for a game, with drawings as intermediate steps in the design process

design problem, is a proper fit with a part of the T&A curriculum. Moreover, the students had incorporated buildings from their home environment, that were meaningful to them such as a mosque, without strict instructions to do so.

6.3 *Airborne Particles: Designing an Improvement for a Sensor on a Bicycle* (*2 Trials*)

This assignment focused on (re)designing a casing for a particulate matter sensor to carry on a bicycle. The standard sensor sent data to the website of the national health institute RIVM. It was thought that real-time reading of particulate matter levels would engage users more directly with their environment. The task was to design conceptually, but not to make 'real-time reading of levels' work because this was deemed too difficult. The other requirement was that the casing should look more attractive for youngsters, in order to attract new participants in this 'citizen science' project. The health institute provided a website with background information and the students were required to present their products to representatives of the agency. This was done during an online session, where students from a higher track school also presented their projects.

An interesting observation was made when students presented their designs to the representatives of the health agency. They felt honored that such important people listened to them and they enjoyed the compliments they got. However, they also expressed discontent, because they felt pupils from the other school were taken more seriously. When, after the presentation, a politician jokingly offered a job to the students of the other school, one of the HDH students said: 'This is mean because we didn't get a job offered, and that's because we are at a lower level school'. More remarks that signify indignant feelings were made, such as: 'Why is the politician [physically] present in their school and not in ours?' What transpired was that students' perceptions of their low SES status and educational level, were sometimes pervasive when the design process led them to interact with stakeholders or students outside school. Leo and Nico were sensitive to this and tried to avoid such stereotypical perceptions whenever they could.

A year later, when the task was repeated with a different class, a separate task was given in order to engage the students with sensor technology without reprogramming the particulate matter sensor: reprogramming a temperature sensor to display readings on a Micro:Bit screen. All students finished the task on time. Half of them progressed beyond the task and added code for a temperature interval rather than one temperature threshold. This success was attributed to the well-structured nature of the coding task, which also provided detailed information for intermediate steps.

An observation that wasn't directly linked to one design task, was related to the students' backgrounds. Quite a few students did help their parents with technical chores in the house. Moreover, some relatives worked in traditional and modern technical businesses, such as tailoring/fashion design and electrical engineering. Lastly, apart from being users of high-tech, some students did engage in technical hobbies, such as repairing electronics, making wallets, or designing expensive wedding clothes for real customers, a businesslike endeavour that some students took pride in. Apart from individual talks with students, they wrote a mind map of 'Me and Technology', consisting of categories such as technology activities with friends, using or repairing things, family members with technological jobs, etc. The students participated with some enthusiasm in this activity, but they wondered about its function.

Lastly, the teachers' resolve to protect their students became clear when grades had to be administered, which had an impact on enrollment for further studies. Leo and Nico insisted on being lenient because low marks would be disastrous for some students' futures. '*He would end up on the streets if he wasn't given access to further studies*', Leo remarked about Youssef's unduly high grade.

7 Reflection and Considerations for Teachers and Curriculum Developers

The tales above are now used to arrive at four considerations for foundational technology education in contexts where many students have a low-SES and migrant background, and that are concerned with learning objectives, design tasks, funds of knowledge and scaffolding.

These considerations, to a varying degree, answer our research question. *How can equity and interest in design and technology be promoted through an open design-based curriculum in a hyper-diverse urban context?*

7.1 *Consideration 1: Clarity of Learning Objectives Can Prevent Frustration*

The findings indicate that the rather open and generic T&A curriculum created uncertainty among teachers and frustration among students about learning objectives, which hampered students' interest in the activities. This could be the case in any school where T&A is offered, but it is a finding that resonates in particular with educational and sociological studies among low-SES student populations. Hirsch (2016), Bernstein (2003). Young and Muller (2013) warn that open curricula tend to be detrimental to these students because they do not get the support from home to be able to decode unclear messages about 'what counts as learning'. Students from middle-class communities do not experience that to the same extent, because their home culture is much closer to school culture (Nash, 1990). 'Academic belonging' must be promoted for these groups too and this requires some clarity about learning objectives (Loui, 2022). In subjects that emphasise designing, such as T&A, this can create tension. Whether a design solution is 'good' is often highly dependent on context. This, in itself, requires a curriculum with a degree of openness and in the case of T&A, this is a welcome addition to the more closed and teacher-centered curriculum that is prevalent in many schools. So, the question is how students can develop more clarity on what counts as learning within subjects such as T&A, with the aid of their teachers

At the level of the macro-curriculum, we suggest that recent international work on standards for technology is used to strengthen technology curricula that are too open. The Irish technology curriculum, which not only features content standards but also a national infrastructure of control, with benchmarks for teachers and external examinations, is one of many examples that would be worth looking into for the Dutch T&A curriculum makers (Buckley et al., 2023).

7.2 *Consideration 2: Proper Design Tasks Can Promote a Sense of Ownership and Agency, but Tasks May Reaffirm Feelings of Low Status If They Are Improperly Constructed*

Our findings show that students did like to design artefacts about their home environment, such as a game figure of a mosque, in particular, if the choice of the artefact was their own. However, problems that residents face in their neighborhood may not be directly suitable as design problems, because students' feelings of low status about their living environment may hamper motivation. Such vulnerabilities are the reason for careful considerations when setting design tasks. Moreover, concrete, tangible and working prototypes were favoured over conceptual solutions for high-over problems, such as excess rainwater in their home environment. As an illustration: the wooden figure for the game made them proud, because it was convincing that it could work in real life, whereas the yellow tubing to remove excess water remained an abstract idea that probably wouldn't have any value in real life, and thus didn't bolster a feeling of competency (Ryan & Deci, 2020). Lastly, tasks that involved problem owners with successful businesses, seemed to increase interest in the process of designing.

7.3 *Consideration 3: Productive Building on Students' Background Knowledge, Skills and Cultural Attitudes Requires a Teacher's Active Engagement with Those Backgrounds*

Research literature about cultural inclusion often emphasises the need to establish linkages between the students' background knowledge that they bring from home, their cultural identities and the curriculum (Lee & Buxton, 2010). Our study shows that teachers can develop insight into the students 'funds of knowledge' if they invest time in talks that link the students' home environment to the learning content that comes with the design task, even though the teachers live in a very different neighbourhood and culture (Fox-Turnbull, 2015; Lopart & Esteban-Guitart, 2018). The student who designed and made clothes for weddings, at home, was just one example of useful background knowledge, skills and cultural identities for design tasks. Apart from individual talks with students, whole class activities whereby students talk or write about their technology identities, including family tech activities, leisure tech, and using or repairing things, are worth the time and effort, provided that teachers use the outcomes to develop further interests.

7.4 *Consideration 4: Design Education Emphasises Autonomy and This Needs to be Matched with Explicit Teaching and Scaffolding Toward Mastery of Skills and Functional Use of Disciplinary Knowledge*

Accomplishment bolsters a feeling of competence, which is a vital factor for motivation (Ryan & Deci, 2020). Our study shows how this played out in an urban context with an open technology and design curriculum. It confirmed what has been noted about design education in general: Students' mastery of knowledge and skills strongly depends on the teachers' pedagogical content knowledge (PCK) to integrate explicit teaching into an otherwise 'student-centred approach' of designing (Van Breukelen, 2017). Without such PCK, students' design performance can be expected to remain at a mediocre level, with negative consequences for their motivation. The students needed explicit teaching or training with process-oriented design stages such as writing and drawing a list of requirements, a morphological chart and sketching. The same has been found for mastering technical skills such as programming, CAD drawing, using traditional tools and applying knowledge from science classes. Students benefitted from whole class guided instruction, including modelling of skills toward mastering. They also needed timely and detailed feedback on the consistency within their design process. In general, deviating from previous ideas can be functional in design processes, but derailment from previous steps needs to be prevented with help from the teacher.

Concerning academic standards, our findings indicate that a fuzzy curriculum can have a negative impact on the teachers' guidance. Firstly, if teachers believe that the development of creativity should be paramount because low-SES migrant students need this development harder than others, then such beliefs can come at a cost of limited guidance towards designs that work. This can hamper students' performance and thus motivation, and their understanding of technological practices in jobs. Secondly, teachers' caring attitudes can have a positive impact, but also a negative impact on students' performance, and hence their motivation (Ryan & Deci, 2020). We can't be judgmental about a teacher who gives a pass for poor work if that enables a vulnerable student to move on through the education system, but we won't hesitate to label such practice as 'discrimination based on low expectations' if it occurs systematically. This study highlights that risk is a result of a lack of teaching tradition in T&A, and a rather open curriculum with little quality control at the national level.

The considerations above are not only relevant for teachers but also for curriculum developers at a national level. They need awareness of the situatedness of 'good design education' in urban settings, of the importance of facilitating teachers to develop their PCK and the importance of specific qualifications for technology teachers.

The term 'knowledge', as in PCK, does not fully cover the gist of our findings. The 'tribulations and the tiny triumphs' that we noted, are related to beliefs about 'good design education in an urban setting'. de Vries (2023) questions the PCK literature in this respect: '*It must be acknowledged that knowing false beliefs of teachers is no less relevant than knowing true beliefs, but it does not help conceptually to reckon all beliefs as knowledge*'. de Vries therefore suggests looking into 'Pedagogical Content Beliefs'. Without blaming teachers for '*false beliefs*' in any way, we assert that communities of curriculum developers, researchers and teachers that work on subjects such as T&A, need to investigate deeper into their 'Pedagogical Content Beliefs' about the opportunities and challenges of design education in urban settings.

Lastly, we like to reflect on the nature of this contribution to our knowledge of education. Qualitative researchers in education often aim to contribute to 'theory'. For instance, DiSessa and Cobb (2004) argue in favour of 'ontological innovation' to denote that these research processes should lead to new ontological categories, that help us to understand what is happening, whereby parallels are drawn with 'hard sciences'. We take a more humble stance and we stay away from generalisable theoretical notions. Instead, we stay close to the nature of technological knowledge, as de Vries and others have stated, has to a large degree evolved using collaboratively made 'norms' (de Vries et al., 2013). The steam engine has evolved through tinkering, experimentation and documentation of norms. Only much later explanations in terms of thermodynamics were available, which subsequently helped to improve its efficiency. A qualitative norm we would like to suggest for design education in urban settings with students from a low-SES and often migrant background is to take into account the four considerations and the empirical background, that we have described in this chapter.

References

Bernstein, B. (2003). *Class, codes and control, volume 4: The structuring of pedagogic discourse.* Routledge.

van Breukelen, D. (2022). Teaching science through design activities. In P. J. Williams & B. von Mengersen (Eds.), *Applications of research in technology education: Helping teachers develop research-informed practice* (pp. 99–116). Springer

Buckley, J., Seery, N., Canty, D., & Dunbar, R. (2023). Features of quality and assessment standards in newly reformed Irish junior cycle technology education. In S. R. Bartholomew, M. Hoepfl, & P. J. Williams (Eds.), *Standards-based technology and engineering education: 63rd yearbook of the council on technology and engineering teacher education* (pp. 219–233). Springer.

De Oliveira, B. (2022). Participatory action research as a research approach: Advantages, limitations and criticisms. *Qualitative Research Journal*, 23(3), 287–297.

DiSessa, A. A., & Cobb, P. (2004). Ontological innovation and the role of theory in design experiments. *Journal of the Learning Sciences*, 13(1), 77–103.

Fox-Turnbull, W. (2015). Contributions to technology education through funds of Knowledge. *Australasian Journal of Technology Education*, 2.

Hirsch, E. D. (2016). *Why knowledge matters: Rescuing our children from failed educational theories.* Harvard Education Press.

Lee, O., & Buxton, C. (2010). *Diversity and equity in science education: Research, policy and practice.* Teachers College Press.

Llopart, M., & Esteban-Guitart, M. (2018). Funds of knowledge in 21st century societies: Inclusive educational practices for under-represented students. A literature review. *Journal of Curriculum Studies*, 50(2), 145–161.

Louie, N., Berland, L., Roeker, L., Nichols, K., Pacheco, M., & Grant, C. (2022). Toward radical belonging: Envisioning antiracist learning communities. *Race Ethnicity and Education*, 1–21.

Nash, R. (1990). Bourdieu on education and social and cultural reproduction. *British Journal of Sociology of Education*, 11(4), 431–447

Ryan, R. M., & Deci, E. L. (2020). Intrinsic and extrinsic motivation from a self-determination theory perspective: Definitions, theory, practices, and future directions. *Contemporary Educational Psychology*, 61, 101860.

de Vries, M, Kroes, P.A., Meijers, A. (2013). *Norms in technology*. Springer.

de Vries, M., Klapwijk, R. (2023). Conclusions. In R. Klapwijk, J. Gu, Q. Yang, M. de Vries (Eds.), *Maker education meets technology education: Reflections on good practices*. Brill

Young, M., & Muller, J. (2013). On the powers of powerful knowledge. *Review of Education*, 1(3), 229–250.

CHAPTER 14

Enduring Concepts and Pluralistic Perspectives for Educating through Designerly Thinking, Doing and Acting towards Positive World Futures

Kay Stables

1 Why?

Virtually all human beings are born with the potential to think and act for themselves, to understand pasts, to envisage futures. We have the potential to believe in ourselves, to trust others, to sympathise and empathise with others. We have the potential to make things happen, for good or for bad.

2 The Potential Is Universal

Humans are born into a family, a society, a culture based on a set of beliefs about the world, the way things are and the way things could or should be. These beliefs are local and will have been established over a very long time. Inevitably the universal potential of human beings becomes controlled, or at least influenced, by the beliefs held within a family, society, culture. But there are constants across societies and a major one of these is a belief in the importance of learning and, in conjunction with this, a need for teaching. Herein lies challenges, opportunities and conflicts provoked by the extent to which people believe that their society/culture is one of *many ways, their preferred way* or *the only way*. In this chapter, I focus on the universal capabilities of humans (as designers and technologists) and explore aspects of the impact of differences in belief, challenges created by belief and possibilities and positives that emerge when *only ways, preferred ways*" and *many ways* are critiqued. My endeavour in this chapter is to do this through the lens of design per se and of design and technology education. My aim is to explore a range of possible, positive world futures that could emerge.

But first, a personal perspective on encountering difference and reflecting on hard ways I have learned to value difference that, first, challenged and then positively influenced my thinking about alternative ways of being and doing.

Differences in approaches to learning and teaching became apparent to me at a young age. I started school in 1960, aged six and a half, in Canada. Straightway I

© KAY STABLES, 2025 | DOI:10.1163/9789004710948_014

encountered the term "phonetics" as how we would learn to read. A year later my family moved to England and I found that learning to read involved the whole class chanting the letters in a word on a blackboard, then all saying the word together. I dared to ask when we would do phonetics and was embarrassed by the teacher saying that wasn't the way it was done. A painful way for a seven year old to experience autocracy! Fortunately, it didn't leave me scarred. Many moons later, once again in Canada, I was teaching a summer school with a colleague on design and technology. I found myself embarrassed by my assumptions on how the workshop would be structured and forced to confront my own ignorance. The summer school was over two weeks and attended by people from across disciplinary backgrounds. The design task was in the context of designing for ways of engaging babies and small children in being part of family meal times. After one week a participant said she would not return for week two – she saw no point in the design task we were doing, it had no place in her culture. The woman was a member of the Ojibwe Nation. We asked if there was a design task that would be more valuable. She couldn't suggest one. We then asked if there was anything she worried about for her children and she spoke instantly of the loss of the Ojibwe language, the dominance of English. Through listening to her we explored with her ways of engaging young Ojibwe in their native language and stories. Between us the idea of designing and making pop-up books of Ojibwe stories emerged. She returned for the second week and successfully took on a mission she believed in. The lesson I learned was powerful, provoking me to question the dominance of Eurocentric approaches in education.

Further learning about alternative perspectives has come with the privilege of supervising PhD students from different cultures, some of whose heritage is within the Global South. This has involved me questioning my assumptions about design and technology education, about research methods and about cultural stereotypes more generally. Each of these has led me to question the dominance of Eurocentric beliefs, to see the value of alternative perspectives and to learn from my students when, for example, the research is exploring designing from a Buddhist perspective, or 'modest dressing' and textiles from an Islamic perspective or when a student is questioning the domination of Eurocentric Design Education approaches in a Middle Eastern country rich with historic Islamic design and architecture.

3 Enduring Concepts

Stepping aside from cultural differences to contemplate the universality of potential human capabilities, there exist enduring concepts of design and technological capability of all human beings, based on innate capacities for future-facing thought and action. Nearly 100 years ago a strong proponent for this was Alfred North Whitehead who highlighted,

> The connections between intellectual activity and the body, though diffused in every bodily feeling, are focused in the eyes, the ears, the voice, and the hands. There is a co-ordination of senses and thought, and also a reciprocal influence between brain activity and material creative activity. In this reaction the hands are peculiarly important. It is a moot point whether the human hand created the human brain, or the brain created a hand. Certainly the connection is intimate and reciprocal. (Whitehead, 1929, p. 50)

Focusing on Technical Education's relation to Science and Literature, he argued for creating balance in the curriculum through the inclusion for all of technical education, which had been neglected because of "two disastrous antitheses, namely, that between mind and body, and that between thought and action" (Whitehead, 1929, p. 50).

Fast forward nearly a century to when his beliefs had the support of cognitive and neuroscience, suggesting that "The brain is designed to put 'doing' before 'seeing' or 'thinking'. We have evolved to be fundamentally active, not contemplative, creatures" (Claxton et al., 2010, p. 4) They go on to explain that

> cognitive science and neuroscience are turning those old Cartesian assumptions – that mind and body are separate, and one is clever and the other isn't – upside down. It turns out, on closer inspection, that the meat of which we are made is very clever indeed. It is no longer unthinkable that the brain, and the body of which it is an organ, could underwrite cognitive complexity, the use of language, mathematical computation and propositional reasoning. (Claxton et al., 2010, p. 8)

Even earlier than Whitehead, there is evidence that education in England was perceived to have three key learning aspects, often referred to as the 3 'Rs'. Bruce Archer highlights the incorrect assumption that these were 'reading', 'writing' and 'arithmetic'. Archer points to the inclusion of the "making and doing aspects of human activity" having a long history and reports the original conception of the 'three Rs' not being reading, writing and arithmetic but as reading and writing; reckoning and figuring; and wroughting and wrighting. Of the latter pair, he clarifies that

> Wroughting ... meant knowing how things are brought about, which we might now call technology [and] wrighting ... meant knowing how to do it, which we would now call craftsmanship. (Archer et al., 1976, p. 9)

Archer develops this insight further in focusing directly on technology and design.

> one fundamental attribute of human beings – that is, one of the attributes that define creatures as being human – is that they devise and make tools, and use these tools to adapt their environments. If Technology is 'knowing-how', then Design is 'envisaging what'. The capacity for envisaging a non-present reality, analysing it externally and modelling it externally, is the third great defining characteristic of humankind. (Archer, 1992, p. 8)

This has parallels with early craft education in Scandinavia through the concept of "sloyd" as highlighted by Kajsa Borg

> The pupils were asked to make handicraft objects that were needed or could be used in their homes. The work was expected to train the children's bodies and minds, and to support their cognitive development. The main purpose of the subject was NOT to learn the technical skills, but other aspects that were trained while working by hand, such as respect for manual work, seriousness and carefulness in any task or work. (Borg, 2006, p. 38)

The enduring concept of thought and action, mind and hand as critical components of human learning echoes through history, whether it is Elliot Eisner (in America) characterising technical education as "creative experience while you think, experience which realises your thought, experience which teaches you to coordinate act and thought, experience leading you to associate thought with foresight and foresight with achievement" (Eisner, 2002, p. 54) or Ken Baynes in England explaining that "the designer is not a special kind of person: every person is a special kind of designer" (Baynes, 2010, p. 25), or Tony Fry in Australia commenting that "We should understand that, in making a world, we largely made it by hand and, in doing so (from the perspective of ontological design), made ourselves what we are. Certainly, the development of our brain was crucial, but, without the capabilities of the hand, the brain was an agent without an actor". (Fry, 2012, p. 47)

4 Designer Perspectives on the Good and Evil Impacts of Design

There is a history of designers critiquing designing from both dystopian and utopian stances. Promoting the value that designers add to the world in relation to the increase in living standards and economic prosperity has brought with it a concern for the negative impact of consumerist desires for more and more 'stuff', more possessions. Just highlighting the views of a few indicates perspectives on this.

Papanek, one of designing's harshest critics, went so far as to comment that

> Today, industrial design has put murder on mass-production basis, By designing criminally unsafe automobiles that kill or maim nearly one million people around the world each year, by creating a whole new species of permanent garbage to clutter up the landscape, and by choosing materials and processes that pollutes the air we breath, designers have become a dangerous breed (Papanek, 1971, p. IX)

Tony Fry (2012) further suggests that unsustainable behaviour of Western civilisations leads to nihilistic de-futuring activities of humans. Focusing on social and environmental futures, Stuart Walker highlights that the idea "problems will be solved through the ingenious application of advanced, super-efficient, non-polluting technologies" is in direct conflict with "both logic and the teachings of all the major philosophical and wisdom traditions down the ages" (Walker, 2010, p. 104).

Others critique whilst also suggesting more productive approaches. Highlighting a need for an ecological world view David Orr also cites the contribution of wisdom traditions to ecological design and

> the heritage of design intelligence evident in many places, times, and cultures prior to our own ... remaking the human presence in the world in a way that honors life and protects human dignity. Ecological design is a large concept that joins science and the practical arts with ethics, politics, and economics. (Orr, 2002, p. 4)

Parodying a historic design concept that form follows function, Fleming, in Design Education for a Sustainable Future (2013), suggests that "form follows world view" highlighting differences between anthropocentric and ecocentric world views and the impact of each. Balaram, critiquing a 'hero designer' concept, foregrounds that in many instances "Much of the designing is not designing for the people but designing with the people" (Balaram, 2011, p. 195). Nan O'Sullivan (2023) commenting from an indigenous perspective, shifts ownership and capability even further even further from the 'hero designer' by highlighting the importance of designing by the people.

The above tapestry, from Whitehead through to O'Sullivan, places emphasis on alternative 'worldview' perspectives on designing. The reality of making such ideas the focus of school-level disciplinary design and technology is a challenge, given that school subjects and projects are typically packaged into bite-size pieces that are manageable within the context of the requirements of national and provincial curricula. And when the curriculum is dominated by a Eurocentric paradigm, there is a constant threat of narrowing human potential to fit into schemes that are more likely to be created around the "only way" as opposed to the many ways.

From a global perspective, it is both curious and alarming to see the extent to which Eurocentric world views dominate design and technology education. This can be seen across all levels of education and is exemplified in England by Government led control of the curriculum in mainstream schooling with an increasingly narrow focus on a knowledge-led curriculum that flies in the face of enduring concepts of humankind as thinkers and doers and, paraphrasing the warning of Tony Fry, of running the risk of the brain becoming an agent without an actor.

A common focus on repetitive 'making' projects that embody technical skills and knowledge along with the demands of formal, often high-stakes assessments, leads to a narrow curriculum that leaves little space for exploring alternative approaches and beliefs. Concerns over this are not just those expressed nearly a quarter of the way through the 21st Century. Writing in 1998, from a Taoist perspective, Jim Flowers questioned the value of and values underpinning the making paradigm that dominated technology education classrooms. He highlights, for example, that despite being set a problem-solving task, learners would highjack the task to create a problem based on something they already wished to make. His concern was that "The result of product design activities for technology students is that these students learn materialism to an extreme. They are taught that just because something can be invented or produced, it should be". (Flowers 1998. p. 21). Underpinning his concern is a lack of reality and depth in the nature of the problem-solving. He advocates a broader and deeper understanding of problem-solving and the range of ways a problem could be solved, including the possibility of a non-technological solution. Critiquing a paradigm of consumerist 'wants' over 'needs' he advocates for "students … better prepared to face the demands of global citizenry" (Flowers, 1998, p. 25).

5 Alternative Perspectives on the "the Only Way"

Different challenges of the "only way" are faced elsewhere. In a South African context, Mishack Gumbo highlights a contradiction wherein rural areas, largely populated by indigenous communities, are seen as "poorly resourced" when it comes to teaching Technology. He critiques this notion, using Ladson-Billings' concept of culturally relevant pedagogy (1995), highlighting the richness of resources available through indigenous cultures from the technologies used, to the elders who, as guardians of such knowledge, can be drawn on as educators. His point is that "Rural schools are alleged as lacking classrooms, access to services such as water and electricity, landline telephones and internet access, libraries, ... There is no regard for open classrooms where the young are taught everywhere and anytime, and the elders being the live libraries". (Gumbo, 2020, pp. 285–286) He comments on how making this visible "creates awareness in non-indigenous teachers, who may think that there are no other forms of technology besides the conventional forms. It creates a paradigm in these teachers so that they can learn along with their indigenous students who know more than they do about their local technologies. They could allow themselves to be guided by indigenous teachers about how to teach from a culturally responsive manner". (Gumbo, 2020 p. 294)

Nan O'Sullivan expresses similar ideas, but from a possibly more positive stance, when considering how design pedagogy can be guided from a Māori worldview of knowledge and values through "shifts in methods and mindsets, and ... how these can facilitate inclusion, equity, and self-determination for 'one's' individual and 'one's' collective future and towards positive change". Through critique she suggests a shift from "outmoded Euro-Anglo-American paradigms and hierarchies" still modelled as central to the discipline, to a discipline that embraces more than cooperation and equity but also accepts the intangible yet tangible tensions of commonality and contrast". She highlights Māori values that are significant for underpinning design pedagogy that "shift(s) away from the central creed of universality". This includes values such as the fluidity and longevity of reciprocity (akoranga), and authentic connections through collaboration. (whanaungatanga) and active guardianship through the recognition of responsibilities 'to' people and place (kaitiakitanga). She also shares insights into Māori concerns with the environment and circularity with ideas and understandings in the links between humans and nature in the past, present and future. Her focus is with design students in higher education but there is clearly value for mainstream schooling for, In her words "the connectivity and values imbued within Indigenous knowledge and ... the skills and courage to engage with empathy, care, respect, reciprocity, and autonomy as strategies to design by". (O'Sullivan, 2023, p. 13)

Positioning her critique explicitly on decolonising design, Danah Abdulla focuses on design education in Jordan where much of the focus in Higher education is underpinned by Eurocentric approaches. Working with research participants from design students, design academics and professional designers, she investigated the possibilities of design education curricula becoming more relevant to place and communities. Taking a future-facing stance and drawing on Escobar's focus on 'border thinking' (Escobar, 2007) her search was for design "otherwise" – exploring possibilities, not an alternative "one way" to replace a Eurocentric "one way". With an understanding of the past, her focus was to look forward and to work with her participants to develop a localised pedagogy and to imagine a decolonised "emancipatory design education that reflects on design beyond its traditional service-provider definition" (Abdulla, 2017, p. 12).

The three above examples illustrate similar and different perspectives on design and technology education. Their differences are with locality and levels of education. Their similarities reveal positive alternative perspectives on design, technology and education that have resonance with each other and point to beneficial futures.

Concerns that echo across the above perspectives are future challenges of sustainability and climate change. The Transition Design movement emerging through design academics at Carnegie Mellon University shares both concerns and approaches through their commitment to "draw on knowledge and wisdom from the past to conceive solutions in the present with future generations in mind" (Irwin et al., 2015, p. 1). With a specifically design-focused approach, they build on the citizen-based Transition Movement through which action is taken by communities at a local level towards global futures. Their strategies vary from areas such as science, economics and social theory, they seek wisdom on sustainable living from indigenous societies and postulate through design-futuring, drawing on science fiction, future-casting and speculative and critical design.

The positive potential of respect, insight and wisdom that comes from alternative beliefs and approaches is hardly yet being recognised, let alone embraced, by the majority of those concerned with design and technology education at the school level. Ideas such as those used within Transition Design are embedded in a very different culture to that of school-based design and technology.

But there are voices across the design and technology communities that support and echo the principles within the examples above and that highlight the need to move forward with global concerns and through local actions. Steve Keirl (2015) foregrounds the importance of education in creating "ethically defensible global futures". Dealing with existential challenges, for example, those provoked by climate change, is not a task to be left to overly busy design and technology teachers but to a multiplicity of agents and groups to engage in achieving this. In his words,

> It needs to be multi-located: in homes; through responsible media; in the street; through policy; and, in formal educational settings. In turn, multiple conversations are needed, both deliberative and activist, around a range of futures-oriented concepts that can inform an ethics of sustainability. (Keirl, 2015, p. 48)

6 Challenges … and Possibilities

So far in this chapter, I have commented on major challenges and major differences in world views of those aspects of life that are supported by human capabilities of designing. Attempting to not be the agent of doom, my critique has highlighted what I see as major challenges but equally has shown ways to more expansive realities and ideas that have the potential to enrich design and technology learning and capability whilst also enriching insight, empathy and understandings of alternative perspectives on living and futures.

School "disciplinary design and technology" in England (and it would seem in many parts of the globe) is driven by a Eurocentric world view. There are many reasons why things have come to be like this, but I will highlight just three drivers that are major culprits.

The first of these links simply to the prepositions used. Flowers (2010) critiques this head-on when considering the use of the word "the" in front of statements that are peppered through design and technology education where they "falsely imply uniqueness" such as "the five families of materials … the six types of material processing" (p. 2010) and, possibly most dangerous, "The Design Process" – as if there is just one. This could be seen just as a rather loose use of English, but in so much of school disciplinary design and technology, it is literally interpreted as if there is just one process. This has become a dogmatic belief rooted in English (and other) design and technology teaching and learning, despite attempts to dislodge it. We recognised this problem in the Technology Education Research Unit at Goldsmiths in the 1980s and despite establishing, through research with 10,000 15-year-olds, that designing is undertaken through iterative active and reflective processes, and despite the original National Curriculum (1989) being based on processes of designing, *The Design Process* (typically represented in a linear fashion) can still be seen on the walls of design and technology workshops across the country. This epitomises the concept of the *only way*, denying the realities of ways of designing that exist across cultures.

The second relates to a lingering, narrow group of sub-disciplines that exist in school design and technology often based around materials being used or workshop spaces they are used in. This can be contrasted with what happens beyond the classroom walls where, increasingly, long-standing areas of design such as furniture design, fashion design, and product design are either conjoined with or replaced by approaches to design such as co-design, participatory design interdisciplinary design and even postdisciplinary design. This shift doesn't fly in the face of expertise, it exploits and grows expertise as people work in multiskill-set communities. At the radical end, Bruce Mau's advice is to "Avoid fields. Jump fences. Disciplinary boundaries and regulatory regimes are attempts to control the wilding of creative life. They are often understandable efforts to order what are manifold, complex, evolutionary processes. Our job is to jump the fences and cross the fields" (Mau, 1998, p. 1).

The third relates to the content-led nature of much design and technology education. Despite a shift in National Curriculum and Assessment criteria in England to include an increased focus on user-centred, context-based designing, there is still an overbalance towards knowledge and skills such as "the use of programmable components … the impact of forces and stresses on materials … appropriate surface treatments and finishes" (DfE, 2015, pp. 6–7). This content overload has a negative downward focus on younger age groups, despite decades of research within and beyond England that has provided insight into the value of challenging and motivating context-led designing.

What chance then of supporting learners to become designers and technologists who can take on the challenges of creating and maintaining positive world futures? Are we trapped in the frameworks of school disciplinary D&T? Or are there ways to break the mould?

John Dakers (2022) makes a strong case against a dominant transmissional pedagogy to one that is based on ethnotechnological literacy that "aspires to be a transformative and liberating experience that does not stop at the school gates" (p. 3). Drawing together enduring educational concepts from progressive and radical educationalists such as Montessori, Dewey, and Freire that challenge the notion of 'curriculum-content-as-planned' in favour of 'curriculum-as-lived-experience' (Aoki, 2005), Dakers proposes a "Nomadic Pedagogy about technology" that

> requires participants to make their own decisions based upon their own experiences, variously blended with other perspectives, especially conflicting ones. These various perspectives, when considered together, have the potential to broaden out thinking about new technological possibilities by inventing solutions to scenarios that may not yet exist. The smooth space necessary for self-organised learning, means the formation of a space where the individual student's and teacher's beliefs and experiences

> meet. The varied worldviews about the technologically textured world collectively inhabited, will almost certainly conflict. However, where the teacher and the students agree that all are entitled to express what may be conflicting perspectives, and that perspectives may be challenged in a reasoned conciliatory and constructive manner, a nomadic pedagogy will have been created. The focus is no longer on providing the 'right' answer, with the concomitant fear of failure when a 'wrong' answer is given. (p. 165)

While this perspective may be seen as idealistic rather than feasible, Dakers points out that historically and currently there are schools which operate based on a Nomadic Pedagogy. He cites a current state-funded 'free' school in Northern England, the XP School, founded on this approach that despite being consummately radical and having a project-based cross-curricular structure around 'Learning Expeditions' has received the status of "Outstanding" by the School Inspectorate, an organisation known for having a rather narrow, knowledge-led view of curriculum. So, radical and brave is possible.

7 Positive World Futures?

Shifting the locus of design and technology curricula from set knowledge positioned as the pinnacle of what should be taught and learned to a locus that prioritises critique, empathy and confidence aligned with competence, autonomy and creativity requires alternative perspectives, alternative curricula and alternative pedagogies. Thinking back on my own teaching history I find myself reflecting on the Maori proverb of "Walking backwards into the future" and my own experience of teaching in an innercity co-educational comprehensive school in a deprived part of London, but within a progressive Education Authority and with a demanding but supportive head teacher. As a textiles teacher, in the 1970's it was seen as entirely acceptable to set homework for 13-year-olds to write a story about a boy landing on earth wearing a pink dress or teaching social history through the rise and fall of the hemline. A context-led curriculum through which I can view the value of walking backwards into the future.

Looking back through this chapter I am conscious of the number of designers and educators who, in one way or another, critique design, and design and technology education, highlighting the negative impacts of design and technology whilst promoting positive futures that could be achieved through more pluralistic perspectives. The wisdom, empathy, insight and knowledge that could be underpinning design and technology education at all levels and ages is staring us in the face. So, what should we do?

Between the foresight of Marc de Vries and John Dakers, I sense a 'call to arms'.

Through four decades Marc de Vries has been fundamental in creating an international Technology and Design community. Due to his leadership, we have a collective of authors contributing chapters, perspectives, expertise and experience to this and other texts. This provides an opportunity to address the challenges faced. John Dakers has provided not only the opportunity for our different perspectives to be voiced and shared but through the thinking and ideas presented in *A Nomadic Pedagogy about Technology*, insight into "a modern Agora, one in which free expression is not only encouraged but actively promoted". John Dakers' call for a liberating "free-thinking, ideal, futuristic vision" of design and technology education, feels like a door waiting to be opened.

I conclude this chapter, as I often do, with the words of Thomas Princen, who reminds us that

> humans are at their best when … they are faced with a genuine challenge … they are creative and productive … they find meaning in their own problem solving and in acts larger than themselves … they help themselves and help others … they self-organise and self-govern … they feel they are getting a fair shot at the benefits of their work. (Princen, 2010, p. 175)

The challenge is there to be addressed.

References

Abdulla, D. (2017). *Design otherwise: Towards a locally-centric design education curricula in Jordan*. University of London.

Aoki, T. T. (2005). *Curriculum in a new key: The collected works of Ted T. Aoki* (W. F. Pinar & R. L. Irwin, Eds.). Lawrence Erlbaum Associates.

Archer, L. B., Baynes, K., & Langdon, R. (1976). *Design in general Education: Part one summary of findings*. Royal College of Art, Department of Design Research.

Archer, B. (1992). The nature of research in design and design education. In B. Archer, K. Baynes & P. Roberts (Eds.), *The nature of research into design and technology education* (pp. 7–14). Loughborough University of Technology.

Balaram, S. (2011). *Thinking design*. Sage.

Baynes, K. (2010). *Models of change: the impact of 'designerly thinking' on people's lives and the environment: seminar 4 … modelling and society*. Loughborough University. Online resource. https://hdl.handle.net/2134/6092

Borg, K. (2006). What is sloyd? A question of legitimacy and identity. *Tidskrift: Journal of Research in Teacher Education* (2–3), 34–51.

Claxton, G., Lucas, B. & Webster, R., (2010). *Bodies of Knowledge: How the learning sciences could transform practical and vocational education.* Edge Foundation.

DES/WO. (1989). *Design and Technology for ages 5–16*, Department for Education and Science, HMSO.

Dakers, J. (2022). *A nomadic pedagogy about technology: Teaching the ongoing process of becoming ethnotechnologically literate.* Brill.

DfE. (2015). *Design and Technology GCSE subject content.* Department for Education, Gov.UK.

Eisner, E., W. (2002). *The arts and the creation of mind.* Yale University Press.

Escobar, A. (2007). *Worlds and Knowledges Otherwise.* Cultural Studies, 21:2–3, 179–210, https://doi.org/10.1080/09502380601162506

Flowers, J. (1998). Problem solving in Technology Education: A Taoist perspective. *Journal of Technology Education*, 10(1), 20–26.

Flowers, J. (2010). The Problem in Technology Education (A Definite Article). *Journal of Technology Education, 21*(2), 10–20.

Fry, T. (2012). *Becoming human by design.* Berg.

Gumbo, M. (2020). Teaching technology in "Poorly Resourced" contexts. In P. J. Williams & D. Barlex (Eds.), *Pedagogy for Technology Education in Secondary Schools.* (pp. 283–296). Contemporary Issues in Technology Education, Springer Nature

Irwin Irwin, T., Kossoff, G., Tonkinwise, C., & Scupelli, P. (2015). *Transition Design 2015 A new area of design research, practice and study that proposes design-led societal transition toward more sustainable futures.* School of Design Carnegie Melon University.

Keirl, S. (2015). Global ethics, sustainability, and design and technology education. In K. Stables & S. Keirl (Eds.), *Environment, Ethics and Cultures: Design and Technology education's contribution to sustainable global futures* (pp. 15–32). Sense Publishers.

Ladson-Billings, G. (1995). Toward a theory of culturally relevant pedagogy. *American Educational Research Journal, 32*(3), 465–491.

Mao, B. (1998). *Incomplete Manifesto for growth.* Bruce Man Studio. http://www.brucemaudesign.com/4817/112450/work/incomplete-manifesto-for-growth

Orr, D. W. (2002). *The nature of design: ecology, culture and human intention.* New Oxford University Press.

O'Sullivan, N. (2023). *Our Everyday: The Intangible Yet Tangible Tensions Between Commonality, Contrast, and Co-operation within Design Education.* Open Access Te Herenga Waka-Victoria University of Wellington. Conference contribution. https://doi.org/10.25455/wgtn.22650766

Papanek, V. (1971). *Design for the real world: Human ecology and social change.* Thames and Hudson.

Princen, T. (2010). *Treading Softly: Paths to Ecological Order.* MIT Press.

Walker, S. (2010). Sermons in stones: argument and artefact for sustainability. *Les Ateliers de l'ethique*, 5(2), 101–116

Whitehead, A. N., (1929). *The aims of education and other essays.* The Free Press.

CHAPTER 15

Wood Cloud: A composition of Intensity for Collaborative Learning

Maarten C.A. van der Sanden

1 Introduction

In a complex, VUCA world, systemic, holistic thinking and reasoning gain value in research and practice. A way of thinking and doing in which a variety of aspects of today and the future VUCA world are interconnected through design from various disciplines, practices and bodies of knowing such as knowledge, experience, intuition and creativity (Van der Sanden & Wehrmann, 2021). Concerning the importance of the resulting transdisciplinary learning for today and future transitions, integrating the development of technology and society, we are in need of professionals who are able and dare to see, feel and value the combinations and integration of disciplines and practices. This chapter is specifically about the *felt experience* of transdisciplinary learning. Also known as *felt knowledge* (Chappell, in press). This chapter emphasises the (felt) aesthetics of the literal composition of cognitive and affective elements of learning in complexity by using simple wooden building blocks. A so-called *wood cloud* is a tangible 3D extension of the often-used 'word cloud' (see Figure 15.1), often used in exploring processes and stages of collaboration.

This felt 3D wooden composition unveils the *intensity* of the complex problem at hand. It starts, supports and sustains a dialogue on how the complex problem at hand actually feels: dense, compact, asymmetrical, scattered, etc. The composition depicted in Figure 15.2, for example, explicates how the various domains of knowledge, skills, issues and emotions needed for a certain innovation are interconnected asymmetrically and capriciously for a team or consortium member. When team members build their wood cloud in sight of the other members around the table, the conversation will lead to what the differences are between the team members and how they experience and feel, the complex challenge ahead. By making these differences explicit, team members also may make intuitive collaborative decisions on what differences make or break for the consortium and, therefore, mitigate certain collaborative issues at the very beginning of a long and winding collaborative innovation

© MAARTEN C.A. VAN DER SANDEN, 2025 | DOI:10.1163/9789004710948_015

FIGURE 15.1 An example of a word cloud depicting the scope, importance, density and intensity of words connected with innovation (retrieved from www at Feb. 29, 2024)

FIGURE 15.2 A wood cloud (photo by the author)

road. The shared wood clouds and the dialogue thereof also unveil feelings of uncertainty, imagination, dreams and frustration when people, from their bounded rationality, think and create the wood cloud aloud. Hands, head and heart (Sennett, 2009), system 1 and system 2 thinking and doing (Kahneman, 2011), explore what one thinks and feels about the problem. Also, the amount of knowledge that one has, or a team altogether has, concerning the transitional challenge at stake measured against the knowledge and skills you or the team don't have yet becomes clear. If we bring this all together we might consider the wood cloud and the conversations as a composition of *complex problem intensity*. Moreover, since everyone at the table is uneasy with this method, it is often seen as a first and profound team experience and therefore, functions as an early start of building social capital and team safety (Wenger, 2000). Also, the aesthetics of intensity, so to speak, may encourage people just to move in complexity, simply because they would like to touch the wood cloud composition of intensity. Hence, it also helps to discover among the team, what social safety is or means for their collaboration. And honestly, the conversations have a pleasant flow of seriousness and laughter altogether. This is in contrast with the decreasing energy level of the many 'conversations' I've been observing over the years, in which professionals present their methods, and projects, or build a digital platform, as a start for long-term collaboration under high uncertainty and time pressure. Such a start is far too easy and too comfortable for true transdisciplinary collaboration and profound, radical change needed for a world in transition. Let alone, collaboration and learning together under high social and technical uncertainty and time pressure.

Moreover, the wood cloud's abstractness and simplicity support its scalability when various levels of the socio-technical system (e.g. team, organisation, society, niche, regime and landscape) are discussed by team members. The wood cloud is the 'rough diamond' version of the *essence* of the collaborative learning effort. One of the major steps in (social) design is to find the essence of the problem at hand that also resonates with the future solution, for it is a guiding principle on a human measure. The intensities or emergent, usually ambiguous, feelings concerning the essence, are hard to describe in words, though better understood when built or drawn. As, for example, the famous architect Gehry drew in 2 minutes, demonstrating the intensity of the building of the Guggenheim Museum in Bilbao, which was eventually built that way.

We might also feel the intensity of our professional challenges. For example, program managers in complex innovations report they feel unrest, are uncomfortable and at the same time are attracted by the issue as such, they feel excited and are ready to learn and to take risks. Don't we all recognise these kinds of positive, intense tensions? As the famous photographer Platon said in the Netflix series 'Abstract': "*Design that goes straight to the heart, that's what good design is*".

Design for complexity also implies that things become simple, without losing their complexity. Simplicity that enables people to move. The wood cloud is a simplification of *Form Language*[1] (Kamp,2018; Kamp & Van der Sanden, 2021), which was developed earlier for transdisciplinary learning and consists of geometrical 3D forms (e.g. pyramid, sphere, cube) that work together based on architectural principles. And of course, this all sounds abstract, as transitions are, but the wood cloud is tangible abstractness on a human measure. Its simplicity invites and challenges designers to move together into complexity in a socially safe way. Moreover, it supports team members in searching for their shared abstract essence as a guiding principle in complexity. The remainder of the chapter is about the substantiation of the wood cloud by transdisciplinary learning, art, aesthetics and STEAM's preciseness.

But before this, I would like to emphasise that Marc and I closely collaborated in the development of an integrative department of Science Education and Communication (SEC) at Delft University of Technology in the Netherlands. We share a long history of trial and error in the integration of science education and communication (Van der Sanden & de Vries, 2016), and always felt the importance and the proximity of the two domains, when it comes to something as vital as social learning for transformative innovations. We talked a lot about design thinking, triple-loop learning, and processes of critiquing as examples of critical attitudes in exploring new knowledge as a pupil, teacher, or professional in science communication and/or a team of professionals in complex innovation processes. We both strongly believe that human abilities as intuition and felt aesthetics resonate with the life-long (transdisciplinary) learning of adaptive professionals and their felt craftsmanship (Sennett, 2009). This a topic that is open for further exploration and integration in science education and communication, most certainly, when it concerns transformative innovations globally, which are the basis of transitions in energy and health.

2 Transdisciplinary Learning

To understand today's complexity or find ways to live with it, the realm of transdisciplinary learning can be helpful. And before we continue, I would like to state that there is no hierarchy or any value difference between monodisciplinary, multidisciplinary, interdisciplinary or transdisciplinary ways of thinking and doing. They all have their place when considering the problems we face, or the angles we intend to choose to tackle the challenges at stake. But for now, I would like to focus on transdisciplinary learning since it resonates with complexity and design thinking. McGregor (2017) quotes Davies (2009) on transdisciplinary learning:

> Instead of viewing disciplines as blocks of knowledge that define what to know and do, people in higher education should see disciplines as opportunities to explore different ways of thinking, through deep inquiry into complex world issues.

Next to these open spaces between disciplines, transdisciplinary learning is all about exploring and learning from the (slight) differences between the cognitive and affective elements of complex socio-technical systems through various disciplinary perspectives. The tensions between the differences and the open spaces of not knowing (yet), become explicit once we accept these disciplinary differences in form, structure and dynamics. As for silence, within conversations amongst professionals, the processes of complex innovations give sense and meaning to the conversations once they are made explicit and discussable (Verouden et al, 2016). Moreover, if we take a little side step to Hartmut Rosa's resonance,[2] little differences help us to learn, if we try to see and feel the resonance of these differences and give meaning to these (un)known and (un)felt relations through a decision-based inquiry as design thinking is (Banathy, 1996; Van der Sanden & Wehrmann, 2021). Felski (2020) connects resonance to education which it is helpful as a resource for rethinking: "[...]

as a counter concept to autonomy, it speaks to the vital role of relations in forming the self and the limits of our capacity to predict or control them". This, resonates, so to speak, with the ideas of Gibbs (2017) in a chapter on "Transdisciplinary thinking: pedagogy for complexity"[3]:

> Education creates uncertainty, so learning involves searching and struggle, and admits doubts and forms of despair. [...] These practices need to let students learn about being in the world and, indeed, change what they find. Poetic thinking[4] is transdisciplinary thinking and shifts from concepts that objectify and fall prey to reductionism to those that creatively and connectively point out difference, not to compare against but to celebrate.

Wood cloud, as a tool for collaboration, supports felt differences. Central to transdisciplinary learning is the concept of the 7 transdisciplinary habits of mind, as Mishra et al. (2011) and McGregor (2017) see it. These 7 habits of mind consist of the following consecutive steps: *perceiving, patterning, abstracting, embodied thinking, modelling, deep play and synthesising*.

1. Perceiving: is a two-layered process. People learn to observe using their five senses, and then they learn the process of *imaging* [cur. by McGregor] (calling to mind what they observed without any external stimuli).
2. Patterning: involves identifying a repeating, discernible form or plan in a seemingly arbitrary arrangement.
3. Abstracting: entails two processes. To begin, people abstract and focus on one feature of a thing to grasp its essence. Then, they use analogies (comparisons between seemingly disparate things) to explain the abstraction.
4. Embodied thinking: is also two-pronged, kinaesthetic and empathetic, wherein learning is transformed from static to tactile. Thinking is heavily influenced by body movements and sensations. Kinaesthetic thinking involves people 'thinking with their body', learning how to use the five senses to know the world around them. Thinking with the body also involves putting oneself in another person's position in order to understand them (i.e. out of one's body into another body's experience, so one can empathise).
5. Modelling: involves both abstractions and analogies [...] and *dimensional thinking* [curs. by McGregor] (space and time). When people model (or think dimensionally), they build replicas or use theories or formulas to represent and then study something to discern its purpose, nature, or composition.

6. Deep play: […] involves people intellectually playing with ideas, concepts, boundaries, or processes, so that they can open doors to new ways of thinking via unexpected breakthroughs. Beyond every superficial play, deep play is characterised as open-ended, and often leads to transformations.
7. Synthesising: involves feeling and thinking coming together into many and new ways of knowing, which could not have emerged if everything had remained separate and disconnected. Through synthesis, people develop deep, empathetic, complex connections between *each other* [curs. by McGregor] and their attendant *ideas and positions* [curs. by McGregor], the hallmark of transdisciplinary thinking.

The usefulness of wood cloud becomes most clear in the stages of abstracting, embodied thinking, modelling and deep play. By using the wood cloud, abstract issues as the intensity of the complex challenge through its composition of elements of knowledge, skills, issues and emotions become tangible and therefore mutable and moveable as Latour (2007) describes in his Actor-Network-Theory. Components of a socio-technical system are mutable, immutable, moveable or unmovable. The wood cloud supports professionals in challenging the present unmovable to be made moveable in the future, or the other way round. Through the creation of a wood cloud that fits better to one's ideas for the future, the future constellation of knowledge, skills, issues and emotions of the future network can be modelled and (deep) played by changing, and moving the current composition. It also supports professionals in developing new knowledge, since the movement of one of the pieces of wood might implicate the development of new knowledge and/or skills. Since it may leave an open space, one might like to add a new piece of wood to the existing composition.

The ability to build such an intensity is supportive of learning in uncertainty. To share deep feelings of learning and maybe understanding in a holistic way, these 7 habits of mind connect complexity, systems thinking and design thinking, by reinforcing ways to give sense and meaning to socio-technical complex reality on a human measure, since we all are capable of discovering, exploring and giving meaning to (slight) differences that help us to understand what learning tensions are in the world. Wood cloud is a means, a tool to explore the composition of differences and tensions and explicate the (collaborative) learning potential thereof.

In *SciCulture,*[5] an Erasmus+ programme,[6] we created transdisciplinary learning based on post-humanism, as a step beyond the Anthropocene. Guided by the disciplines and practices of the programme members transdisciplinary learning, art, social design and entrepreneurship. One of the outcomes of the project was *CourseKit*[7] resulting from 4 consecutive courses we designed, delivered and evaluated (Chappel et al., 2023). The CourseKit cards as part of the educational design tool, contain various card suits. Suits include the conceptual cards (see for example Figure 15.3A and B) to help you think about the bigger picture of transdisciplinary learning processes, such as Creative Pedagogies, Design Thinking Loops, Reflective Perspectives and Probe the Challenge. The other set of suits are the session-by-session cards to make your course concrete, called: Type of Teaching Block, Type of Virtual Teaching Tools, Facilitator Type, Grouping Type and Wild Cards.[8] The cards and the SciCulture canvas are designed in such a way that teaching professionals follow their design process and experience, feel the learning of the students or pupils ahead of teaching it.

Within the 7 Habits of Mind and SciCulture's CourseKit, abstracting, embodied thinking and modelling are important stages of learning under uncertainty. Steps in which art and aesthetics are important reflective sources of knowing by creating tangible items to deep play with and reflect upon. Wood cloud is such an abstract tool that supports embodied reflection and learning, modelling and deep play. Moreover, wood cloud resonates with art and the reflective power of aesthetics as is explained below.

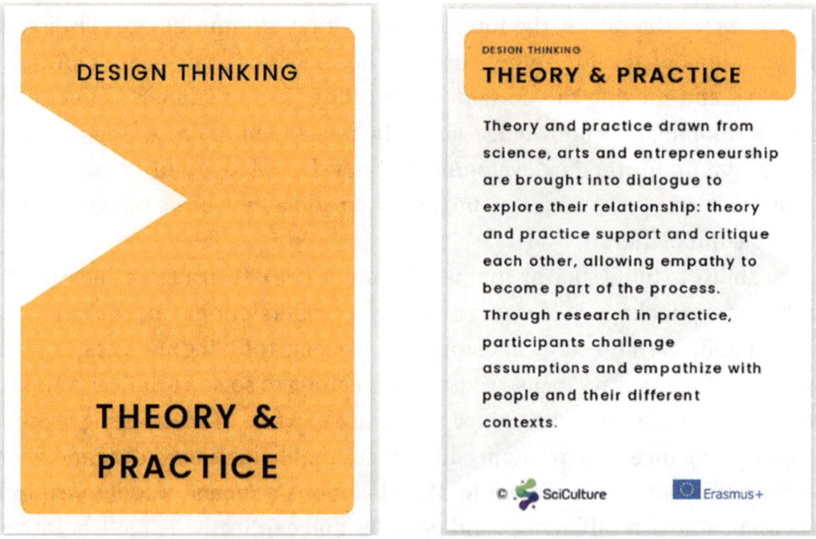

FIGURE 15.3 (A)An example of the CourseKit cards. The front side (left) indicated the suit 'design thinking' and the category 'theory & practice'. (B) The back side of the card (right) explains the idea behind the connection between theory and practice

3 Art and Aesthetics

Malevich's *architecterons* are an example of how Russian avant-garde (1890–1930) art searched for new futures for cities, citizenship and societies. Through this kind of composition, Malevich tried to explicate the city's intensity (see Figure 15.4A)., as Mondrian's Victory Boogie Woogie depicts the fuzz and movement of Broadway New York (see Figure 15.4B). Another example that I often used in lectures[9] was the 'tensions circulaire' van Jean Helion (see Figure 15.4C). To me all examples of precise asymmetry which is ambiguous in itself. Too much asymmetry will end in unrest and chaos, but just the right amount of it shows tangible, discussable and maybe exciting tensions on a human-recognisable measure. As Goodman (1976) wrote in his *Language of*

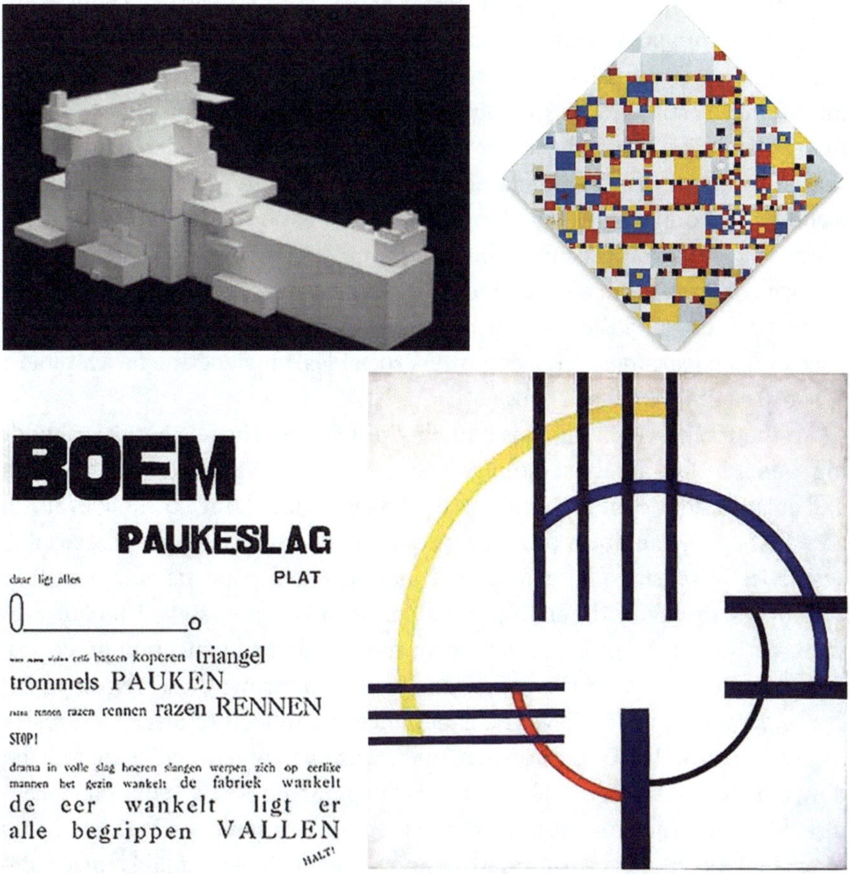

FIGURE 15.4 (A) Architecterons / (B) Victory Boogie Woogie/ (C) Jean Hélion / (D) Paul van Ostaijen

Art these kinds of precise compositions denote and describe emotions, feelings and movements that are imaginable. I do think that another great example of this kind of graphical asymmetrical precise dynamics is the poetry of Paul van Ostaijen (see figure 15.4D) in which the poem gains intensity and deeper meaning through its precise graphic presentation.

So, in the light of transformative socio-technical transitions, the composition of the intensity of the challenge is a much more tangible, and talkative starting point for learning about the value of integration of the senses, meaning and knowledge, instead of convincing each other what these components of integration content-wise at the level of cognition should be, or ought to be. Wood Cloud and its artistic precursors are much more about how the challenge feels, looks like, and is imagined. Is it a dense problem, distributed and asymmetrical? In the light of transitional uncertainty, intuitive thinking and doing might be the best way to get started and be surprised by what your intuition can do for you and within team settings (i.e. as aforementioned systems 1 and systems 2 thinking). However, much of science and technology communication is still about explaining (also in its affective approaches), rather than critically exploring deeper layers of empathic mutual understanding and shared concerns (as explained above in the SciCulture example). Building a wood cloud composition together supports professionals in getting ready to learn, as for most professionals, building compositions about the challenging intensity and the discussion thereof is something new (even awkward) to discover. By doing this, the team of professionals also develop a shared experience of deep learning, which empowers social learning and the development of social capital (Wenger, 2000).

Aesthetics, as we all know is a quality of what is. The aesthetics of buildings and interiors are most familiar next to the ones in art. But also processes and organisations contain beauty, as it has been studied since the 1990s (Strati, 2008). The way members of an organisation are stimulated and managed is aesthetically sensitive. Organisations may just feel good or energise their employees through stimulating architecture AND processes. Though, aesthetics is not often associated with communication and education processes, except sometimes from the means and tools supporting these learning processes. However, we all have the experience that a lecture feels good and a meeting went well. Good vibes, dynamics, tensions, something happened, the flow, the accelerations, the delays, the challenging things said. If one can recognise this beauty, one may also start from these recognisable aesthetic elements to explore a complex transformative innovation challenge. If all is uncertain, the talkative aesthetics of a collaboratively built wood cloud might guide you, your pupils, and your professional team in getting started together. Explicate issues that could not be extensively explained but emotionally felt by everyone, perhaps in a different way. A wood cloud compositional asymmetry that provokes thoughts, invites touch and change to the composition and therefore provokes movement of seemingly unmovable and immutable elements of knowledge, skills and emotions.

4 STEAM's Preciseness

The composition of the wood cloud and its support for transdisciplinary science education and communication resonates with the A of Art in the STEAM acronym approaches in science and technology education, and similar developments in the domain of science and technology communication. And as De Vries (2021) writes:

> As all of the STEM disciplines come together in the full complexity of reality one has to deal within design, the term integrated STEM/STEAM, or iSTEM/iSTEAM makes sense. Whereas in research one can separate out aspects of reality 'safely' as one does not change reality [...], in design one has to deal with all aspects in combination.

It is about the design aesthetics of these learning combinations and integrations of knowledge, skills, issues and emotions. Moreover, learning, and social learning, as collaboration are such fine-mazed processes based on precise interactions between knowledge, skills, issues and emotions from various disciplines, practices and system levels. Hope and comfort are precise, and if precisely written and/or well made,"it makes it possible to really cry" as Rutger Kopland (a Dutch poet) states.[10]

That makes compositions so important also in the case of learning which is a complex, sometimes chaotic but still a precise composed process. Therefore compositions of intensity work well, the making thereof may be the most important, to seek to build and enjoy e.g. precise asymmetry coming into existence. This precise aesthetics of asymmetry also refers to the cliché of perfect imperfection. A good design goes straight into the heart, whereas a too-slick design is less effective than the one that is (a bit) unpolished. Aesthetics used in this way is seen as a critical stance to what has been developed: is the educational program inviting, are there tensions and differences to explore, is there something exciting waiting for change? Strengthen learning potential? These kinds of questions guide you in needed changes which are *felt* precisely a long time before *known* precisely.

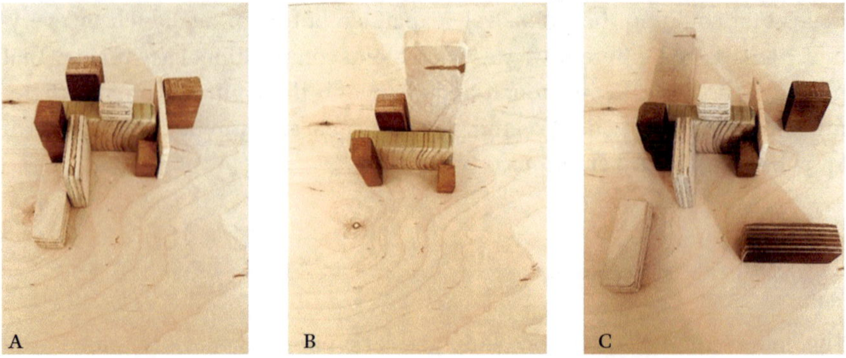

FIGURE 15.5 Small changes on wood cloud as depicted in fig. 15.2 that lead to large differences in 'look and feel' of the composition (photos by the author)

Forms and their interactions explicate this preciseness, as e.g. architecterons show the intensity of cities in quite a precise way too. It is all about the composition of flow and rhythm. If the wood cloud is too easy, or too predictable, it will not be that challenging. This exemplifies complexity, its uncertainty, its preciseness, its asymmetry and the attractiveness at the same time. Ambiguous complexity is much closer to humanity than we think and at the same time less understood than we imagine.[11]

The wood cloud as such exemplifies this one can discover how precise these kinds of compositions of intensity and collaboration are by changing them. Team members have a felt experience about how composition and interactions of knowledge, skills, issues and emotions depicted by the wood cloud radically change just by moving one piece a little bit or changing the angularity only slightly (See Figures 15.5 A, B and C: little changes make a large difference in the composition intensity).

Why all this was important for the collaboration between Marc and myself? While seeking synergy between science education and communication, and seeking the intensity of learning integration of other disciplines and practices. We both enjoy transdisciplinary education, design, design-based learning and abstract tools. Tools as an integrative, intense, precise activity supporting STEAM for lifelong social learning in complexity. As close as possible to the intense challenges we all face and recognise in real life, that is where transitions in a VUCA world start.

Notes

1 Home - VormTaalVormTaal (in Dutch)
2 Hartmut Rosa (2019) 'Resonance: a sociology of our relationship to the world'.

3 Paul Gibbs (ed.). Transdisciplinary higher education: a theoretical basis revealed in practice. Springer, 2017.
4 Gibbs (2017): Poetic thinking comes from *poiesis,* to make create.
5 Home - SciCulture
6 Home | Erasmus+ (europa.eu)
7 CourseKit - SciCulture
8 Glossary - SciCulture
9 C-Lab, a master course on designing collaboration in complex transformative innovations and decisions thereof through social design, based on the idea of living labs (community of practice).
10 Van de Schoonheid en de Troost - Rutger Kopland (2000) (youtube.com) (VPRO, 2012)
11 As a reference to Goethe's cliché: "Everything is simpler than you think and more complex than you can imagine".

References

Banathy, B. H. (1996). *Designing social systems in a changing world.* Plenum Press.

Chappell, K., Witt, S., Wren, H., Hampton, L., Woods, P., Swinford, L. & Hampton, M. (2023). Re-creating higher education pedagogy by making materiality and spatiality matter, *Journal of Creative Behaviour.* https://doi.org/10.1002/jocb.619

Chappell, K. (2024, in press). *Flowing with embodiment and materiality: Touch and time for new educational futures.* In K. Chappell, C. Turner, & H. Wren (Eds.), *Creative ruptions for emergent educational futures.* Palgrave Macmillan, in press.

Davies, D. (2009). *Curriculum is a construct.* Retrieved from In Clued-Ed website.

de Vries, M. J. (2021). *Design-based leaning in science and technology as integrated STEM.* In I. Henze & M. J. de Vries (Eds.), *Design-based concept learning in science and technology education.* Brill / Sense Publishers.

Felski, R. (2020). Resonance and education: On Education. *Journal for Research and Debate,* 3(9).

Gibbs, P. (ed.) (2017). *Transdisciplinary higher education: a theoretical basis reveales in practice.* Springer Nature.

Goodman, N. (1969). *Languages of art: an approach to a theory of symbols.* Hackett Publishing Company Inc.

Kahneman, D. (2011). *Thinking fast and slow.* Penguin Books.

Kamp, A. (2018). *Speaking architecture: a semiotic study toward understanding visual communication in architecture.* MSc Thesis, SDelft University of Technology, Delft.

Latour, B. (2007). *Reassembling the social: An introduction to actor-network-theory.* Oxford University Press.

McGregor, S. L. T. (2017). Transdisciplinary pedagogy in higher education: Transdisciplinary learning, learning cycles and habits of minds. In P. Gibbs, (Ed.), *Transdisciplinary higher education: A theoretical basis reveals in practice.* Springer Nature.

Mishra, P., Koehler, M. J., & Henriksen, D. (2011, March/April). The seven trans-disciplinary habits of mind: extending the TPACK framework towards 21st century learning. *Educational Technology*.

Rosa, H. (2019). *Resonance: a sociology of our relationship to the world.* Polity Books.

Sennett, R. (2009). *The craftsman.* Penguin Books.

Strati, A. (2008). Aesthetics of organization. In S. Clegg & J. R. Bailey (Eds.), International Encyclopedia of Organization Studies. Sage.

Van der Sanden, M. C. A. & de Vries, M. J. (Eds.). (2016). *Science and technology education and communication: Seeking synergy*. Sense Publishers.

Van der Sanden, M. C. A. & Kamp, A. (2021). *Discussable complexity in science communication through form language.* PCST-virtual conference.

Van der Sanden, M. C. A. & Wehrmann, C. (2021). *A mind set called design thinking.* In I. Henze, & M. J. de Vries (Eds.) *Design-based concept learning in science and technology education.* Brill Sense Publishers.

Verouden, N. W., Van der Sanden, M. C. A. & Aarts, N. (2016). Silence in interdisciplinary research and collaboration: Not everything said is relevant, not everything relevant is said. *Science as Culture*, *25*(2), pp. 264–288.

Wenger, E. (2000). Communities of practice and social learning systems. *Organization*.

CHAPTER 16

Food as an Aspect of Technology Education in the 21st Century

Marion Rutland

1 Introduction

The chapter begins by drawing on the paper 'Food in the school curriculum in England: Its development from cookery to cookery' (Owen-Jackson, 2016, p. 70). It explored:

- the teaching of food in the English school curriculum and considered its position and possible future role in the school curriculum.
- It drew on 10 years of research by university staff, teacher educators, teachers and pupils into rhetoric and lived experiences of those involved in teaching about food, from a range of literature and personal research.
- The paper explored the changing nature of food in the school curriculum.

The research concluded 'that schooling is political, and from its conception, food teaching has been used for political purposes, for preparing girls for their future roles addressing the health concerns of the nation'. However, it was argued that the study of food can be much more than this. It relates to pupils' lives and can contribute to their scientific learning as well as developing their understanding of ethical, social and political issues. Issues that still in 2024 relate to the status and position of food education in the English school curriculum.

Food was first introduced into the English elementary school curriculum in the mid to late 19th Century for philanthropic or utilitarian reasons. This coincided with a population moved from a rural to an industrial, urban factory-based setting. Cookery had a low status and was intended to teach working class girls basic cooking skills to improve their standard of living, improve their family health and prepare them for low-paid employment. It was not until early 20th Century that it appeared as 'domestic science' in girls' grammar schools for more academically able girls. Even then it focused on nutrition within an essentially practical subject, with little attempt to teach underlying scientific principles (Rutland, 1997, 2006; Rutland & Owen-Jackson, 2015a). Similarly, manual instruction, later known as craft design technology (CDT) for boys was met with hostility partly because of its association with the working

© MARION RUTLAND, 2025 | DOI:10.1163/9789004710948_016

class (Penfold, 1988). Both subjects focused on developing useful 'practical skills and developing of routine mechanistic skills' or 'training', rather than 'education' as a body of knowledge and concepts. This influenced the status of both subjects for more academically able pupils in the school curriculum up to the early 1970s (Rutland, 2006).

The Sex Discrimination Act (1975) made sexual discrimination unlawful, and schools were required to enable boys and girls to have equal curriculum access for both CDT and cookery, then known as home economics which included cooking, needlework, budget management, child development, health and hygiene. The Nuffield Home Economics project (1982) introduced a more scientific, investigative approach to practical food activities. Though, many home economics teachers were less qualified or interested in these aspects of the subject (Davies, 1981). A situation that has continued up the present day, as indicated by a mixed range of views expressed by current food teachers (Davies & Ballam, 2023, pp. 4–5). These include 'many food teachers expressing dissatisfaction with 'limited options available', 'current course being criticised for being too focused on food science and nutrition, lacking practical skills, and not providing a suitable academic qualification'. Yet, wanting an 'A Level option to provide academic rigour and enhance future prospects' and on the other hand 'more focus on 'heathy eating and nutrition, practical cooking skills and food preparation'.

Documents written by the Her Majesty's Inspectors (of Schools) (HMI) (DES, 1985, p. 1) saw the primary aims of home economics as helping prepare 'boys and girls for some aspects of everyday living and the adults' responsibilities of family life' and concerned with hygiene, safety, health, and diet'. This was not the case with HMI guidance for CDT (DES, 1987), where an emphasis on designing practical solutions and creative problem solving was highlighted, a philosophy much closer to the central philosophy of the design process when the National Curriculum Technology document (DES, 1990) arrived (Rutland, 2017).

2 The English National Curriculum 1990

Prior to the introduction of the National Curriculum in 1990 the teaching of food in home economics included a wide range of content ranging from learning to cook, feeding the family in a domestic context, consumer awareness and the sources and uses of foods in our diet. A pattern that can be found in the teaching of food in countries such as Australia (Turner & Wilks, 2016), New Zealand and the USA. It is still taught today within home economics in countries such as Southern Ireland, Scotland, the Caribbean, Malta and Cyprus.

Following the introduction of the National Curriculum in 1990 and the amalgamation of the National Association of Home Economics and Technology (NATHE) and the Design and Technology Association (DATA) in 1993, the inclusion of food within D&T as 'food technology', was developed by the members of DATA, the new combined organisation. This required a change of emphasis for food not only from the domestic arena but to a wider commercial food product development approach. This was not easy for food teachers, many were confused and alienated by the terminology used in official documents (DES, 1990). However, with a determined effort, the production of resources and the introduction of external examinations for pupils aged 16 years, called the General Certificate of Education (GCSE) Food Technology and an A Level Food Technology examination for 18-year-olds, progress was made. On the positive side, the status and associated gender related issues did improve, and the subject was taught to the full ability and age range. The intended knowledge content was rigorous and required pupils to combine "thinking and doing" with an ability to make informed decisions in food product development with a learning style based on problem-solving rather than on rote learning and practice (Rutland, 2017).

Factors affecting the development of Food Technology in the curriculum.

HMI recommended that the nature of food technology should be clarified (Ofsted, 2006) and that learning in food technology should be more intellectually challenging to include "designing, product development, empirical testing and applying maths and science" (Ofsted, 2008: 35), with more in-depth nutritional knowledge and greater scientific understanding and technical rigor (Ofsted, 2011). Unfortunately, this clarification did not occur. Other factors included a shortage of well-qualified food technology teachers and some examination boards placing too high an emphasis on "industrial practices" and the use of larger-scale equipment to the detriment of knowledge, understanding and basic skills needed in food product development (Rutland, 2006).

The introduction of food technology coincided with a changing lifestyles for many people in England, with a move away from the domestic production of food in the home to an increased consumption of industrially designed and manufacture of foods. This reflected the increasing consumption of ready meals and highly processed foods alongside the availability of large, well-stocked food supermarkets (Rutland, 2018). A further major concern was the interpretation of "designing" in food, misinterpreted as the requirement to "draw" when designing food products. There was a lack of understanding that designing with food is essentially a hands-on-activity where pupils foster and use their knowledge and understanding of the physical, chemical, and nutritional properties of foods by exploring and experimenting when developing their food products (Rutland & Owen-Jackson, 2015b).

Concerns were raised regarding the relationship between pupils 'learning to cook' and food technology. An HMI Report (Ofsted, 2006: 5–6) noted confusion about the basic aims of food technology 'in essence, a tension exists between teaching about food to develop skills for living and using food as a means to teach the objectives of design and technology'. Ideally, food technology should embrace an understanding of the properties of the food materials and apply this to developing food products. This illustrates a fundamental difference between those who believe that children should be taught the 'life skills' of cooking and those who believe that a wider perspective is required (Rutland, 2018). In England 'Licence to Cook' for pupils aged 11–16, a non-statutory government initiative funded by a Department for Education concerned about rising incidents of obesity in the community was introduced (DES, 2008). Its aim was to provide all pupils with the opportunity to learn how to cook, as well as make informed decisions about diet and nutrition, health and safety and wise food shopping. However, it was confusing for food teachers as it required all pupils to learn basic cooking skills through dedicated lessons in food preparation techniques, diet and nutrition, hygiene and safety and wise food shopping (Rutland, 2008) and appeared to question what was taught in food technology.

3 Developments in England 2013–2016

In the 2013 English review of the National Curriculum for D&T, these dual and sometimes conflicting views of the teaching of food became key issues. Food was retained within D&T for pupils aged 5–14 years (DfE a, 2014) and included terms such as 'ingredients and food' with pupils expected to design and make with food ingredients, working in the home and wider industrial contexts. However, despite views expressed such as 'food technology education must be more than just cooking' (Goodburn, 2013, p. 2), a new separate 'cooking and nutrition' section was included, and 'learning to cook' was described as a 'crucial life skill'.

These themes continued in the review of the National Curriculum for D&T when all GCSE and A Level subjects (DfE a, 2014) were reformed, and it was decided to develop a combined GCSE Cooking and Nutrition, with a name change later to GCSE Food Preparation and Nutrition (DfE, 2015). The new qualifications focused on ensuring students acquired a good understanding of food and nutrition together with excellent cooking skills (ibid: pp. 6–7). The draft GCSE Subject Content for D&T (DfE b, 2014) did not include food as material. All the range of former food courses, such as home economics, hospitality, catering and food technology were to be addressed through just one new examination course, the GCSE Food and Nutrition. A new A Food Technology level course, providing a pathway for pupils aged 18 years and progression to higher education degree courses, was not developed as it was argued, that there were several high quality vocational qualifications such as 'confectionary' and 'butchery' available (DfE a, 2014).

4 Current Situation in 2024

In England the GCSE Food Preparation and Nutrition examination has been the only food related course available for school pupils aged 14–18 since 2014 though 'there are talks of it being reviewed shortly' (Davies & Ballam, 2023, p. 3). The Design and Technology Association in their latest Report 'Reimagining D&T' show 'some reluctance to let 'Food and Nutrition go from the D&T suite' (Design and Technology Association, 2023, p. 6). 'The Food Teachers Centre' founded by Lousie Davies in 2013 a former Deputy Chief Executive of the D&T Association, is a UK based self-help group supported by experienced Associates and volunteers. She was involved in the development of the new D&T National Curriculum in 2014 and the new GCSE Cooking and Nutrition and later GCSE Food and Nutrition, though no food related members of DATA were involved. Food remained within D&T for pupils aged 5–14 years but not for 14–18-year-olds. A new GCSE Food and Nutrition was available for pupils aged 14–16 years but there was no food course available for pupils wanting to progress to higher education and food related courses. The Food Teachers Centre Report (Davies & Ballam, 2023) noted that only 23% of those respondents (teachers) offering any post 16 vocational courses and out of 306 responses 93% felt there was not sufficient post 16 level choices available. Some of the options available include WJEC level 3 Food Science and Nutrition (being the most available) but others include Hospitality, Leith's Professional Cookery Level 3, International Baccalaureate (IB) Food Science and Technology and Business & Technology Education Council (BTEC) Home Cooking Skills (Rutland & Seabrook, 2023).

International, food education is alive and well with varying approaches and styles to the teaching of food. For example, in Wales, food is taught as a material area within D&T; in Northern Ireland, it is taught as home economics, while in Scotland, food is in both the Technologies and Health and Well-being curriculum areas, the former focusing on design and make work with food and the latter on food and health. In Australia, there are similar approaches. Turner (2013, p. 483) noted that "many educators consider food preparation and safe handling essential life skills". However, she saw an urgent need for the repositioning of food technology as a rigorous study in food science and innovation as practiced in industry with clarification of the content of hospitality courses. In New Zealand the recent focus in on 'food literacy', in the 'Technology' curriculum clearly teaches food technology and 'food literacy'. In many countries, with increasing concerns over health, food education sometimes known as food literacy (Food Literacy Center, 2023), includes an understanding the impact of your food choices on your health, environment and economy – and that these impacts are not experienced equitably are key issues. Today, in many countries, there are concerns about health issues relating to diet, obesity (particularly in children), cancer, heart disease, hypertension, diabetes, and undernutrition, though food is not taught in any form in some countries.

5 The Future?

I return to the paper Food in the school curriculum in England: Its development from cookery to cookery. 'Teaching about food in the school curriculum is more that the transmission of practical skill or preparing young people for work in the food industry, it should ensure that our children become informed and responsible consumers of food' (Owen-Jackson & Rutland, 2016, p. 70). Food teaching has the potential to be the vehicle for informing and enriching the lives of children in the 21st century in practical, intellectual and social domain. 'It can make an important and major contribution to the preparation of our pupils for their life in the 21st Century and how this should be achieved should be a matter of continuing debate in England and internationally' (Rutland, 2017, p. 12).

Essentially, just teaching children 'to cook', though important is not sufficient in the mid-21st century, a wider perspectives is needed. Lifestyles in many countries have changed since food education was first introduced in the mid-19th century. Women's roles are not longer just in the home looking after the family and preparing food that the family has grown or bought locally. The food industry is a powerful, worldwide organisation in many countries and produces an increasing amount of the foods we eat. It needs employees that have the knowledge, understanding and skills related to designing and making a wide range of food products suited to the needs and health of the current and future generations. The increased availability and intake of ultra-processed foods (UPF), that are highly refined and contain industrial substances such as preservatives, emulsifiers, sweeteners, and artificial colourings and flavours are a cause for concern (BNF, 2023). These foods are cheap, available, easy to store and quick and easy to prepare for busy people and families.

There is also the issue that more children in the UK are experiencing food poverty compared with a decade ago. Food poverty or 'household food insecurity' can be caused by many economic factors and affects children the most. The enormous strain it puts on families, their health and wellbeing, impact on children's development, educational attainment and their social mobility are clear. Food poverty can have a significant impact on health, both in terms of nutritional deficiency affecting growth, mental health and wellbeing, but also long-term health problems (Seabrook & Rutland, 2024).

This chapter argues that food education requires a robust, theoretical framework that addresses socio-cultural issues together with scientific theory (food science); technological understanding, environmental issues, product design, nutritional knowledge and the development of cooking skills, taught through an experimental, sequential and integrated methodology. This would enable children to develop the important food related knowledge, understanding and the skills required for their future health and wellbeing. This would also provide a pathway to a wide range of food related career opportunities in the food and hospitality industries, teaching and food related professional occupations such as the hospitality industry, healthcare and nursing.

There is a very urgent need in England to review the situation regarding the GCSE and A Level Food examination courses to enable pupils to follow such a pathway. This should involve teachers, the Office for Standards in Education, Children's Services and Skills (Ofsted), food related lecturers and admissions officers in higher education, the D&T Association and other professional organisations related to food education. As for the future name of such an examination course? If food technology or new GCSE courses with a similar name and content for pupils aged 14–18 years is revised and updated, there also must be other school food-based examination courses available for pupils interested in more practically oriented food activities related to developing cooking skills for the hospitality industry and other similar food related employment.

References

BHF. (2023). Are ultra-processed foods linked to early death? Accessed May 11, 2023, from https://www.bhf.org.uk/informationsupport/heart-matters-magazine/news/behind-the-headlines/ultra-processed-foods

Davies, M. (1981). An investigation of home economics teachers' interest in the contributory areas of the subject. *Journal of Consumer Studies & Home Economics, 5*, 148–155.

Davies, L., & Ballam, R. (2023). *Food Education: fit for the future?*. Food Teachers Centre. Accessed February 19, 2024, from https://foodteacherscentre.co.uk/wp-content/uploads/2023/05/Food-Education-fit-for-the-future.pdf

Design and Technology Association. (2023). *Reimagining D&T: Because design and innovation matter.* Banbury, Oxfordshire, UK. Accessed February 27, 2024, from https://www.designtechnology.org.uk/media/4843/reimagining-dt-our-vision-v9.pdf

DES. (1985). *Home economics from 5–16 curriculum matters 5.* HMSO.

DES. (1987). *Craft, design and technology from 5–16.* HMSO.

DES. (1990). *Technology in the National Curriculum.* HMSO.

DES. (2008). *Licence to Cook.* Accessed February 26, 2024, from https://webarchive.nationalarchives.gov.uk/ukgwa/20150220014223/http:/www.nationalstemcentre.org.uk/elibrary/technology/collection/1577/licence-to-cook?view=grid

DfE. (2014a). *Reforming GCSE and A Level subject content consultation.* Department for Education.

DfE. (2014b). *Design and Technology: Draft GCSE subject content.* Department for Education.

DfE. (2015). Food preparation and nutrition GCSE subject content. London: Department of Education.

Food Literacy Center. (2023). What is food literacy?. Accessed August 30, 2023, from https://www.foodliteracycenter.org/what-food-literacy

Goodburn, K. (2013). The growing opposition to the draft D&T programme of study. *D&T news Issue*, 2, 2.

Nuffield Home Economics Project. (1982). Hutchinson & Co (Publishers) Ltd. https://doi.org/10.1079/PNS19840051

Ofsted. (2006). *Food technology in secondary schools (HMI 2633).* HMSO.

Ofsted. (2008). *Education for a technologically advanced nation design and technology in schools 2004–07 (Reference no. 070224).* HMSO.

Ofsted. (2011). *Meeting technological challenges? Design and technology in schools 2007–10 (Reference no. 100121).* HMSO.

Owen-Jackson, G., & Rutland, M. (2016). Food in the school curriculum in England: Its development from cookery to cookery. In *Design and Technology Education: An International Journal, 21*(3), 63–73. Design and Technology Association.

Penfold, J. (1988). *Craft, design and technology: Past, present and future.* Trentham Books

Rutland, M. (1997). *Teaching food technology in secondary schools.* David Fulton Press

Rutland, M. (2006). The inclusion of food technology as an aspect of technology education in the English school curriculum. In M. J. de Vries & I. Mottier (Eds.), *International handbook of technology education: The state of the art* (pp. 273–284). Sense.

Rutland, M. (2008, July 2–4). Licence to cook: The death knell for food technology? In E. W. L. Norman & D. Spendlove (Eds.), *The Design and Technology Association International Research Conference*). Design and Technology Association, Loughborough University.

Rutland, M. (2017). Food in the School Curriculum: A Discussion of Alternative Approaches. In M. J. de Vries (Ed.), *Handbook of Technology Education, Springer International Handbooks of Education.* https://doi.org/10.1007/978-3-319-38889-2_25-2

Rutland, M. (2018). Food Education in the School Curriculum: A discussion of the Issues, Influences and Pressures on the Teaching of Food. In *PATT36 Research and practice in technology education: Perspectives on human capacity and development* (pp.461–467). Athlone Institute of Technology, Co. Westmeath, Ireland 18–21st June 2018.

Rutland, M., & Owen-Jackson, G. (2015a). Food technology education: preparation for life and work? In *Plurality and Complementarity of Approaches in Design and Technology Education. PATT 29* (pp. 349–356). Marseille: France 7–10th April.

Rutland, M., & Owen-Jackson, G. (2015b). Preparing to teach food technology. In G. Owen-Jackson (Ed.), *Learning to teach design and technology in the secondary school* (3rd ed.). Routledge.

Rutland, M., & Seabrook, R. (2023). Tackling food poverty: The role and importance of food education in the United Kingdom schools. In *PATT40 Pupils' Attitude towards Technology Educational Research Conference* (pp 709–717), Liverpool John Moores University, Liverpool, UK. 31st October to 3rd November 2023.

Seabrook, R., & Rutland, M., (2024). Food Poverty and how it affects UK children in the long term. In G. Singh Lalli, A. Turner, M. Rutland (Eds.), *Food Futures in Education and Society*. Routledge, Abingdon, Oxon, UK.

Turner, A. (2013). The utility of genre theory in technology education: A case study into food technology teaching. In J. Williams & D. Gedera (Eds.), *Proceedings of Technology Education for the Future – A Play on Sustainability: 27th Annual Pupils Attitude Toward Technology Conference* (pp. 482–490), December 2–6. Christchurch: Southern Cross University.

Turner, A., & Wilks, J. (2016). A place for food in Australian schools: a socio-historical review of food education. *International Journal of Technology and Design Education.* https://doi.org/10.1007/s10798-016-9377-9. Published online 31st August 2016.

CHAPTER 17

Why It Is Essential That (Ethno)Technological Literacy Is Embodied into STEM and STEAM Education

John R. Dakers

> If all we do is bring to light what we already know, then what is the point of what we are doing?
>
> IAN BUCHANAN

1 Introduction

It has been said that a camel is a horse designed by a committee. Essentially, in terms of technology education, I will argue, similarly, that the reason the concept of STEM or STEAM education is proving ineffective in its delivery, is because of the ill-conceived incorporation of too many historically incompatible subject domains into a single project. In other words, STEM and STEAM have been, and continue to be, designed by a variety of independent committees, usually governmental, who, unlike the committee of designers of the camel, literally never communicate with each other. I offer my slightly facetious interpretation of how one of these committees might have thought through the creation of STEAM from STEM thus:

Well into the meeting, the chairman came to discussion point seventeen on the agenda.

"Now", he said, "we come to the business of changing the concept of STEM. It is clearly a great idea, certainly from my point of view, but, alas, teachers do not seem to accept it. Also, as I am reliably informed, it appears to lack any creative element. So, let us rethink it into something everyone will love. Any suggestions?

One man who had a thick beard put his hand up and said enthusiastically, "Why don't we add the letter 'C' for creativity?"

© JOHN R. DAKERS, 2025 | DOI:10.1163/9789004710948_017

"Yes, but how does that fit into the acronym STEM?" questioned the man opposite.

Looking towards the chairman, the secretary suggested, "Maybe we should think of other letters to add that would enhance the acronym".

"Yes, let's begin with the letter 'A'." said the chairman authoritatively. "It is, after all, the first letter of the alphabet. 'A' not only transforms STEM into STEAM, which is pretty cool, but also adds the potential for another subject to be included in the STEM family".

"Algebra perhaps?" said one rather keen, but somewhat shy committee member.

"Absolutely not". said the chairman. "That already forms part of STEM located, as it is, in the mathematics box".

"Well, the only other school subject that begins with 'A' is Art". said the lady with the splendid hat.

"Or The Arts". Interrupted a man who had been lost in a book for most of the meeting.

"The Arts", exploded the chairman, "but The Arts subjects taught at school can include History, Geography, English (In English speaking countries), Philosophy (in France), Music and Drama. That would be", he paused for a moment, 'that would be STEHGEPMDM? That is almost every subject taught at school, it is almost the entire curriculum. That is just silly.

"It is also a bit difficult to pronounce", said the shy committee member.

"Yes", said the chairman, it is impossible to pronounce, and so it will not do". After another pause, he stated with some authority, "Let us agree that the 'A' should be Art. Art is, after all, a subject that oozes creativity, and that is what we want STEM to have: more creativity. STEAM it is then".

Everyone agreed and retired to have some tea.

2 Multiple Perspectives on What Defines STEM or STEAM Education

There is a plethora of available information regarding the constitution of STEM or STEAM education today. Marc Sanders, for example, outlines the difficult process that resulted in the genesis of STEM. "In the 1990s, the National Science Foundation (NSF) [in the USA] began using 'SMET' as shorthand for 'science, mathematics, engineering, and technology'. When an NSF program officer complained that 'SMET' sounded too much like 'smut', the STEM acronym was born" (Sanders, 2009, p. 20). Sanders, and many others, I suspect, thought however, that this had something to do with the science of STEM cell research. As STEM evolved over time, moreover, it continued to confuse and perplex. As Sanders suggested:

> Most, even those in education, say 'STEM' when they should be saying 'STEM education,' overlooking that STEM without education is a reference to the fields in which scientists, engineers, and mathematicians toil. Science, mathematics, and technology *teachers* are STEM *educators* working in STEM *education*. It's an important distinction. In addition, there is the common misconception that the 'T' (for technology) means computing, thereby distorting the intended meaning of the STEM acronym. Suffice it to say, STEM is often an ambiguous acronym, even to those who employ it. (*ibid,* p. 20) (italics in original)

Other examples of the lack of agreement as to what constitutes STEM education can be found in the introduction to the book *STEAM Education: Theory and Practice*. The editors borrow from a 2019 report by Tsupros et al., which defines STEM education thus:

> STEM education is an interdisciplinary approach to learning where rigorous academic concepts are coupled with real-world lessons as students apply science, technology, engineering, and mathematics in contexts that make connections between school, community, work, and the global enterprise enabling the development of STEM literacy and with it the ability to compete in the new economy. (Khine & Areepattamannil, 2019, p. 3)

According to this definition, STEM education is a form of interdisciplinary education involving science, mathematics and technology/engineering subjects. It advocates also that a connection must be made between the academic world of the school and the 'real' world outside the school. This definition, it is claimed, manifests through the development of STEM literacy, a literacy that forms a portal between the school and the 'new' economy. A preferred outcome in this definition favours learning to compete in the 'new' economy. A preference for competition is endorsed over cooperation. In addition, the linking of STEM school subjects to the economy suggests a link with industry, thus implying vocational education. This forms the basis of a recurring theme throughout the 2019 edited collection referred to above.

Another edited collection, published in 2015 by Ge et al., entitled *Emerging Technologies for STEAM Education*, has, likewise, emphasised this link to vocationalism. The preface opens with the following:

> There has been a great deal written about the need for addressing the 21st-century digital literacy in curricula at all levels and the particular concerns about failing to prepare college graduates for careers in science, technology, engineering, the arts, and humanities (STEAM), as well as other career areas. (Ge et al., 2015, p. v)

STEAM education is championed as preparing college (university in the USA) graduates for the world of digital literacy, but is this the same as technological literacy, and does it apply to school-based STEAM education? The emphasis in the quotation suggests that STEAM education will prepare college graduates for careers in science, technology, engineering, the Arts and Humanities, essentially, pretty much every career imaginable. However, once again, does this apply to school-based STEM or STEAM education? Indeed, can it? Universities have different Faculties for Science, Engineering, and The Arts and Humanities, separating each into specific disciplines, as is the case in secondary schools. Faculties of technology, however, are extremely rare. Technology universities such as MIT do exist, especially in the United States of America. However, the subject matter on offer does not articulate much with those offered in technology education.[1]

Whether the concept of creativity can be grafted onto STEM education also raises various complications. For example, adding Art (A) to STEM education implies that those who support STEAM over STEM, evidently consider the concept of STEM education as lacking any creative element: hence the need to add Art. This further implies that the individual subjects that constitute STEM education must also lack any creative element. A keen supporter of STEAM education, on the other hand, advocates for the addition of The Arts thus:

> According to the Rhode Island school of design, (RISD), the goal is to foster the true innovation that comes with combining the mind of a scientist or technologist with that of an artist or designer. Former RISD President John Maeda, an early influencer for STEAM, has emphasised the idea that design thinking and creativity are essential ingredients for innovation ... The addition of the 'A' (The Arts) to the original STEM discipline to create STEAM is important, in part, because practices such as modelling, developing explanations and engaging in critique and evaluation (argumentation) have too often been underemphasised in the context of math and science education. (Lathan, 2022)

Note above that Lathan prefers the catch-all concept of 'The Arts', leading us back to STEHGEPMDM, which, as our expert committee above concluded, is unlikely to work. This addition, Lathan claims, will develop students' ability to critique and evaluate. He also mentions that this appears to have been under-emphasised "in the context of math and science education". Yet surely the emphasis on Science and Maths, together with 'The Arts' (A) would be something other than STEAM; it would be SAM?

de Vries (Hendrika) offers an intriguing conception of STEAM by arguing that:

> Art was added and thought to engage students to foster inclusive and gender-equal classrooms, and therefore helping to achieve success and promote inclusive and critical and creative thinking of all students. (de Vries, 2021, p. 1)

Notice that de Vries, unlike Lathan above, considers the 'A' to represent Art, a subject she thinks fosters gender equality, inclusivity, and criticality. Creativity in Art is unarguable. Whilst no one would object to pursuing gender-equal classrooms, I am not sure, however, that the addition of Art alone would foster gender equality, inclusivity and criticality in STEAM.

In the second edition of their book *Teaching STEM in the Secondary School: Helping Teachers Meet the Challenge,* Frank Banks and David Barlex offer a comprehensive set of arguments as to why STEM education, now including STEAM education, is so vital in today's technologically textured world. In the forward, Jim Al-Khalili, a distinguished professor of physics at the University of Surrey, makes the argument that there is more to STEM than just:

> learning scientific facts or solving algebraic equations – or indeed to prepare the next generation of technologists and engineers ... When we talk today about STEM-based industries, we mean much more than just those in the engineering and technology sector, as it encompasses more broadly almost all aspects of modern life. Thus, if school science exams do little more than test pupils' abilities to recall facts then they will be ill-prepared to enter the modern workforce. (Al-Khalili & Barlex, 2021, p. VII)

Al-Khalili's argument that testing students' ability to recall facts will ill-prepare students to enter the modern workforce, echoes Semetsky's argument that where education follows a prescribed curriculum, it necessitates a "model of teaching reduced to the unidirectional transmission of pre-given content" (Semetsky, 2017, p. 11).

Banks and Barlex offer a research-based account, offering several examples from around the world and a variety of clearly articulated pedagogical models considered necessary for delivering a STEM education-based integrative set of subjects. They not only promote a more critically orientated approach as being vital for teaching these subjects, they also highlight the need to help teachers understand how to teach these integrative projects. In addition, they indicate that policymakers in England regard STEM education to be good for the needs of industry and the economy. However, despite policymakers' excitement in this respect, the curriculum in England remains much the same as before STEM, with no significant changes in school delivery being evident. Importantly, Banks and Barlex add to the various definitions offered thus far, the concept of values as an essential element in STEM or STEAM education, primarily related to sustainability issues such as climate change.

What is concerning, however, is that the concept of values already represents a significant component in the teaching and learning of technological literacy. Some mention of technological literacy is given by Banks and Barlex, but not in great depth. This raises some critical questions. Since technological literacy continues to be considered a vital component of education about technology, how does technological literacy morph into STEM literacy?

It appears that our colleagues across the pond have the answer to this. In 2003, the ITEA produced a new set of assessment protocols for promoting technological literacy. They defined technological literacy thus:

> We live in a technological world. Living in the twenty-first century requires much more from every individual than a basic ability to read, write, and perform simple mathematics. Technology affects virtually every aspect of our lives, from enabling citizens to perform routine tasks to requiring that they be able to make responsible, informed decisions that affect individuals, our society, and the environment. Citizens of today must have a basic understanding of how technology affects their world and how they exist both within and around technology. Technological literacy is as fundamentally important to students as traditional core subject area knowledge and abilities. It has often been left for individuals to gain through their daily activities. However, technological processes and systems have become so complex that the happenstance approach is no longer effective. A massive, coordinated effort is needed that should involve schools, mass media and entertainment outlets, book publishers, and museums. Schools, in collaboration with the community, must bear the bulk of this effort, because the educational system can provide the most comprehensive study of technology. (ITEA, 2003, p. 4)

Arguably, for the first time, the term 'affect' enters the conversation, a conversation that discusses a rationale for technological literacy. "Technology", they say, "affects virtually every aspect of our lives'" This highlights an important additional concept for inclusion in scientific, technological, engineering or mathematical literacy. It is important to note, however, that affect is not effect, although it is generally held that it is affect that causes effect. Affective knowledge "serves as the necessary condition for effective learning, defined as the creation of concepts and assigning meaning to experience (Semetsky, 2009, p. 444; ITEEA, 2020, 2021). Buchanan defines affect thus:

> It should be noted that affect is something that we experience passively rather than actively. One may be affected both by internal stimulus, from the imagination, the instincts, or more generally the unconscious … or external stimulus, which may take a huge variety of forms. (Buchanan, 2018)

Affect, considered in terms of technological or STEM literacy, must include the development of understanding the very palpable impact that technology has upon individuals, families, cultures, societies and the environment.

Since 2003, ITEEA subsequently updated their Standards to include STEM education. In the preface to the new Standards, the lead investigators explain some potential consequences of the human/technology interface and how it affects people, societies, cultures, and the natural environment, whether knowingly or unwittingly. "Many of our current global problems were created by our technological choices. This increases the need for technologically literate citizens who participate in decision making" (ITEEA, 2020, 2021, p. VIII). In the opening pages of chapter one of the publication, a justification is given as to why ITEEA consider technological and engineering literacy to be such a crucial part of technology/STEM education.

> Technology and engineering are pervasive in all aspects of our lives. Every human activity is dependent upon the products, systems, and processes created to help grow food, provide shelter, communicate, work, and recreate. As the world grows more complex, it is increasingly important for everyone to understand more about technology and engineering. People need to understand technology's impacts on their lives, society, and the environment, as well as how to use and develop technological products, systems, and processes to extend human capabilities. These understandings are all important elements of technological and engineering literacy. (ITEEA, 2020, 2021, p. 1)

Once again, we see, albeit implicitly, that the way technology affects humans, societies and the environment, highlights the need for young people to learn the ongoing process of becoming technologically literate and where better to engage with that than in technology education?

I have reconceptualised the concept of becoming technologically literate into becoming ethnotechnologically literate in my book entitled: *A Nomadic Pedagogy About Technology: Teaching the Ongoing Process of Becoming Ethnotechnologically Literate* (Dakers, 2022). The prefix 'ethno' denotes this form of literacy's social, cultural and affective aspects, aspects that I argue, are distinctly lacking in technology and STEM curriculum delivery. Furthermore, in the ITEEA rationale written in 2019 for the much-needed revision of the Standards for Technological Literacy, the authors acknowledged that the concept of technological literacy is difficult to define and is subject to change, depending upon cultural conditions.

Citing Zollman (2012), the ITEEA rationale acknowledges that "there is a difference between literacy and being literate. STEM literacy should, therefore, not be viewed as a content area but as a shifting, deictic[2] means (composed of skills, abilities, factual knowledge, procedures, concepts, and metacognitive capacities) to gain further learning". They go on, now citing Krupczak et al. (2016), to articulate the concept of technological literacy as a fluid construct that is "constantly evolving or changing" and also, citing Gagel (1997), that "to maintain relevance, its content must evolve as a function of changing cultural traditions. The utility of such a literacy would depend on its ability to adapt and keep pace with constant change" (Buelin et al., 2019, p. 5) In other words, (ethno)technological literacy is an ongoing process that "is irreducible to a static recognition but demands an experiential and experimental encounter that would force us to think and learn, that is, to create a singular meaning for a particular experience" (Semetsky, 2009, p. 444).

The many perspectives, some of which have been discussed above, appear unable to co-exist. But why is this? The many variations on the theme of STE(A)M are processes that, whilst looking good on paper, are extremely difficult to achieve, particularly in a secondary school setting. Secondary school curricula comprise competing subject domains, each with their own specialist 'expert' teachers, whether from the sciences, technologies, engineering or the arts, who, with few exceptions, compete with one another for student enrolment. Moreover, Science and Mathematics are prioritised over Technology education by policymakers. Often, Science and Mathematics subjects are compulsory in upper secondary school settings, whereas technology subjects are voluntary in the later stages. This emphasis is given to science and maths at the expense of technology education within STE(A)M education. In a published academic paper written in 2018, issues related to common themes regarding STEM education across different countries were discussed. The authors offered local perspectives from across five nations: the USA, Australia, Scotland, France and Germany. They concluded, amongst other things, that common recurring problems in the delivery of STEM education across nations:

> STEM [has] become a catch-all acronym for science and mathematics education in schools and in much educational discourse. However, the T (technology) and E (engineering) aspects often become lost or seen as subsidiary to science and mathematics. This matters because STEM education is seen as an area of concern in several nations: STEM education policy is shaped by, and responds to, the perception that STEM knowledge and innovation drives economic growth and recovery in a post-industrial globalised world, despite the evidence for this being mixed. (Wright et al., 2018)

This research further demonstrates the low status of the technology component in STE(A)M education. Furthermore, it casts serious doubt on the perspective that policymakers continue to champion, that STE(A)M education drives economic growth in a globalised economy.

The negative status highlighted in Wright et al., further suggests that technology education, especially craft-based technology education, centres around the commonly held perceptions discussed above: It has a trades-based vocational orientation. Whilst science and mathematics are perceived as academic subjects, students and the public continue to perceive technology education as a subject domain for less academically able students, particularly boys.

Other areas of concern have become evident, particularly in relation to STE(A)M education delivery. A recent BERA-commissioned report,[3] for example, highlighted several issues related to pedagogy and curriculum. These issues reveal the task of uniting of these subjects as problematic:

> A difficulty in critically reviewing the literature, and understanding the perceptions and practices, as gleaned through conversations with practitioners, was the lack of conceptual clarity with respect to the key terms used. Whilst STEM has currency as an essentially economic term identifying those educational areas that have the greatest impact on a developed country's GDP, the pedagogical and curricula implications are less obvious. STEAM retains this lack of educational clarity, indeed adds to it, firstly: by being itself a portmanteau term; secondly, by having varied modalities and associated purposes; and thirdly, because the terms' art' and 'arts' are also used interchangeably, and often uncritically.

> Thus, it is unclear if 'art' or 'arts' refer to, for example, the arts, or specific art forms/practices, or pedagogies used in teaching art subjects or arts and humanities. It is also unclear whether STEAM is intended to imply a reconfiguration of disciplinary relationships, interdisciplinary and/or transdisciplinary conceptions of the subject. Further, it is unclear whether an arts-infused or arts-integrated approach is implied. Whilst this poses no difficulties for individual case studies, either in educational or professional practices, or reported in the academic literature, the lack of conceptual clarity limits the possibilities for a comprehensive account to be developed. The conceptual issues are further complicated by an apparent conflation of STEAM with creative approaches to teaching in the STEM subject areas, which the commission recognises as encompassing significantly different purposes. (Colluci-Gray et al., 2017, p. 8)

A two-year longitudinal European research study looking into ways to increase participation in STEM subjects across Europe, particularly by women, was carried out with the European Commission between 2001 and 2003. One consensus reached in the concluding report was that the delivery of science, and particularly technology subjects at the school level, was in much need of significant reform. This was reported in a paper written by Dow entitled: *The Need to Change Pedagogies in Science and Technology Subjects: a European Perspective* (Dow, 2006). Dow reported, as the title of the paper suggests, that it was essential that new forms of pedagogy be implemented across Europe in science and technology subjects at the school level. Dow goes on to outline the problem of the extant pedagogical models involved in the delivery of craft-based technology education across Europe, all as concluded in the study:

> The mechanistic processes which underlie the dominant existing model have the effect of reducing technological knowledge to small discrete components which are learned, mostly through drill and practice, and subsequently tested in situations completely devoid of any meaningful context. This was supported from the research which indicated that technology education across Europe had a vocational emphasis which tended, for the most part, to be directed towards skill acquisition in the service of trades occupations, and was taught in whole class settings with children carrying out essentially the same tasks. This type of instruction in turn results in a depersonalisation and fragmentation of the child's experience, and in its "whole class", assembly line production of skills, estranges children from not only the subject and each other, but from themselves as well. (Dow, 2006, p. 311)

It becomes apparent that the social construction of craft-based technology education as a component part of STEM education is recognised by students, parents and particularly females, as being an initial training ground designed to meet the perceived needs of industry, an industry gender-biased towards males. This has implications for the recruitment of females into technology education. These recurring problems continue to situate the delivery of technology education stubbornly as being a subject that maintains content and pedagogy in desperate need of change.

A distinction begins to emerge between the practical aspect of craft-based technology education and an understanding of the socio-cultural aspects of technology. Aristotle differentiates between skill and action, stating that they are generically different. He tells us that while "[craft] production [skill] has an end distinct from itself, this could not be so with action, since the end here is acting well itself (Aristotle, 2002, p. 107). This infers that acting well is seeing beyond what is good for oneself by considering what is good for all. This calls into question the practicability of forming a workable union between craft-based technology education and technological literacy. It also calls into question the practicability of forming a relationship between art and STEM.

In general terms, perhaps C. P. Snow got it right in his seminal and enduring Rede lecture entitled; *The Two Cultures and the Scientific Revolution*. He delivered this lecture at Cambridge University in 1959. Now in book form, he argued that the sciences and humanities (The Arts) were incompatible (Snow, 1961). This 'two cultures' argument has never been resolved, especially in schools. The field of architecture, for example, has made links between the creative endeavours of architects in association with the more quantitively orientated endeavours of structural engineers. However, the design of the building is, for the most part, in the hands of the architect, whilst the design of the structural elements is the engineer's responsibility. The architect's designs and the engineer's structural calculations make it possible. Occasionally, the two elements come together when one individual can merge both the architectural design with the structural: Santiago Calatrava is a supreme example of this rare genre.

Feenberg highlights the distinction between STEM and The Arts, between what he implies thinking quantitatively and qualitatively means. He postulates, "Once function is selected, the engineer has the last word on its implementation and the humanist interpreter is out of luck. This is the view of most engineers and managers; they are at home with 'function' but have no place for meaning" (Feenberg, Questioning Technology, 1999, p. 84). Regarding the dangers of supporting and teaching technology in purely quantitative terms, Feenberg reveals the potentially dystopic outcomes that may be made manifest because of this reductive form of pedagogy. A world, he suggests, "in which human individuality has been completely suppressed. Here people become, as Marshall McLuhan once said, the 'sex organs of the machine world" (Feenberg, 2006, 2014, p. 17).

One important issue that emerges from these capricious descriptions of what does or does not constitute STEM or STEAM education, is that nothing ever seems to change. In 2006, Marc de Vries concluded that in order to build up a sound technology educational theory and practice:

> We need to work on a technology education practice that convinces governments, school administrators and school boards, teachers, parents, and pupils that teaching and learning about technology is not only fun but also a necessary component of every (future) citizen's general education. (de vries, 2006, p. 9)

Fast forward twelve years when Marc was editor of the Springer *International Handbook of Technology Education*, and nothing had significantly changed. In his chapter on the history of technology education, he lamented that "hopefully in due time, there will be a second international handbook of technology education ... showing that technology education has been able to overcome the hurdles of survival and flourishes in many countries worldwide" (de Vries, 2018, p. 83).

3 What Is the Purpose of STEM or STEAM?

Given the above descriptions and perspectives on the concepts of STEM and STEAM education, it is not easy to differentiate between them and the concept of technological literacy. STEM and STEAM education are likely to be condemned to similar futures if there continues to exist a vocational element in school-based technology education, one that follows the logic of Adam Smith's concept of the division of labour.

Education, in all its guises, must be a democratic and transformative process for all individuals, whatever their place in society. This must start with governments recognising that school-based technology education, like all education, is not an apprenticeship leading to what they consider useful employment in the interests of the economy. It involves instead, as Rousseau postulated in 1762, one of two choices "When instead of raising [individuals] for [themselves], one wants to raise [them] for others? [ ] Forced to combat [between] nature or the social institutions, one must choose between making a [person] or a citizen, for one cannot make both at the same time" (Rousseau, 2002, Book 1, p. 39).

"Few would disagree that the world as we know it is sinking into an economic, political, social and ethical abyss of previously unimaginable depths" (Jun, 2011, p. 1). STEM education, in all its forms, is considered by policymakers from around the world as a subject area to be given a high priority to help address this. STEM education is continually being lauded as a crucial set of subjects, the graduates of which will go on to support future economic growth and technological progress. Why, then, does STEM education appear to continue to resist being born into classroom practice? This, despite a plethora of research in support of STEM as part of school-based education. Jun considers one of the reasons might be wider than STEM education alone, it might also be socio-cultural.

Perhaps the most tragic and frightening aspect of contemporary life is the systemic lack of imagination – the hopeless acquiescence of the powerless to those in power, coupled with the latter`s insistence that everything is the way it is because, in some sense, it could not be otherwise. For Deleuze and Guattari, the ethical question isn`t `What ought we do?` but `What might we do?` or `What could we do?` The reason that we are living in decidedly [bad] times isn`t just that people aren`t asking the ethical question, but that they are routinely denied the ability to ask it or, worse, are placed in situations where the desire to ask it never emerges on its own (Jun, 2011, p. 4).

STEM education and technology education before it, lack, in any real sense, a way to empower students by giving them a forum in which they can consider the ethical dimensions, good or bad, that pervade all technologies. A forum such as this could facilitate the formation of what (Aoki, 2004) terms, a-curriculum-of-lived-experience,' one in which the affect that technologies have upon each student's own lived experience, can be discussed and debated openly and democratically; one in which no individual is in charge of the knowledge that is co-produced and one that is subject to constant variation in world views. A pedagogical framework such as this is guided, for example, not by 'What ought or should we do about mobile phones in classrooms?', but 'What might, or what could we do about mobile phones in classrooms?'. The former seeks answers and guidance from those in authority, the latter requires students "to bring to mind the things that are not immediately present [to their] senses. With [their] imaginations [they] can step out of the here and now and … speculate, visualise and suppose" (Robinson & Robinson, 2022, p. 4). It is, after all, our collective imaginations that have created the world as we know it.

Perhaps we should add Imagination and Design to the STEM acronym. We should also lose the 'A' because it does seem to confuse the situation. That would give us STIMED, but that doesn't really work either. Once again, our committee is STYMIED for the lack of a school subject that starts with Y.

Notes

1 See, for example, http://catalog.mit.edu/subjects/
2 A deictic expression is a word or phrase (such as this, that, these, those, now, then, here) that points to the time, place, or situation in which a speaker is speaking.
3 BERA: British Education Research Association

References

Aoki, T. T. (2004). *Curriculum in a new key: The collected works of Ted T. Aoki.* (W. F. Pinar, & R. L. Irwin, Eds.). Routledge.

Aristotle. (2002). *Nicomachean Ethics* (R. Crisp, Trans.). Cambridge University Press.

Buchanan, I. (2018). *A Dictionary of Critical Theory SECOND EDITION Buchanan, Ian. A Dictionary of Critical Theory (Oxford Quick Reference). OUP Oxford. Kindle Edition.* Oxford University Press.

Buelin, J., Daughterty, M. K., Hoepfl, M., Holter, C., Kelley, T., Loveland, T., & Sumner, A. (2019). *ITEEA standards for technological literacy revision project: Background, rationales and Structure.* ITEEA.

Dakers, J. R. (2022). *A Nomadic Pedagogy About Technology: Teaching the ongoing process of becoming ethnotechnologically literate.* Brill Publishers.

de Vries, H. (2021, March). Space for STEAM: New Creativity Challenge in Education. *Frontiers in Psychology, 12*, 1–6.

de vries, M. J. (2006). Two decades of technology education in retrospect. In M. J. de Vries, & I. Mottier (Eds.), *International handbook of technology education: Reviewing the past twenty years* (pp. 3–11). Sense Publishers.

de Vries, M. J. (2018). Technology education: An international history. In M. J. de Vries (Ed.), *International handbook of technology education* (pp. 73–84). Springer.

Feenberg, A. (1999). *Questioning technology.* Routledge.

Feenberg, A. (2006/2014). What is philosophy of technology? In J. R. Dakers (Eds.), *Defining technological literacy: Towards an epistemological framework* (Revised and Updated Second Edition ed.). Palgrave MacMillan.

ITEA. (2003). *Standards for technological literacy.* International Technology Education Association.

ITEEA. (2020–2021). *Standards for Technological and Engineering Literacy: The role of technology and engineering literacy.* International Technology ane Engineering Educators Association.

Jun, N. (2011). Introduction. In N. Jun & D. W. Smith (Eds.), *Deleuze and Ethics* (pp. 1–4). Edinburgh University Press.

Khine, M. S., & Areepattamannil, S. (2019). Introduction. In M. S. Khine, & S. Areepattamannil, *STEAM Education: Theory and Practice* (pp. V–VII). Switzerland: Springer.

Robinson, K., & Robinson, K. (2022). *Imagine if ... Creating a Future for us All.* Penguin Random House UK.

Sanders, M. (2009, December/January). STEM, STEM Education, STEMmania. *The Technology Teacher*, 20–26.

Semetsky, I. (2009). Deleuze as a philosopher of education: Affective knowledge/effective learning. *The European Legacy,, 14*(4), 443–456.

Semetsky, I. (2017). Introduction: A Primer on Edusemiotics. In I. Semetsky, *Edusemiotics: A Hanbook* (pp. 1–15). Springer.

Snow, C. P. (1961). *The two cultures and the scientific revolution.* Cambridge University Press.

CHAPTER 18

Technology Education in Context

Beyond Consumer Cultures

Mike Martin

1 Introduction

In developing a technology curriculum for the future it is essential that we consider our current context. This is particularly important at the moment as the global context in which we find ourselves becomes increasingly significant at a personal social and political level. For this reason, it is important that all our thinking and action about the future of the subject is foregrounded by a consideration of climate change and other global issues. As Vincent Block (2019) writes - Nothing else matters!

Given that technology education is by its very nature involved in the use of resources, we should recognise that the impact of technology education needs to be 'net zero' or better. So what form of technology education will be fit for purpose in our current era? What are the criteria that it needs to meet in order for it to be a key component in the general education of young people?

In order to answer these questions this chapter will consider the relationship between technology and context, as well as the relationship between technology education and context. The nature of consumer cultures will be considered along with the characteristics of our currently product-centred technology education. Re-imagining technology education as a context-driven curriculum is proposed. The chapter concludes with a conceptual model of what such a curriculum would look like and highlights key action that needs to be undertaken for schools, teachers and learners to adopt a new way of thinking for the future.

2 Technology and Contexts

All technology is influenced by both culture and the environment. One only needs to look at housing solutions in different countries to see how they are shaped by environmental influences and how cultural practices can affect the structures that we live in. It is also the case that technology, and the products

© MIKE MARTIN, 2025 | DOI:10.1163/9789004710948_018

of technology, influence our cultural practices and shape the environment around us (Pacey, 1986). The work of many contemporary philosophers of technology such as Coeckelbergh (2020) relate to postphenomenology and an exploration of what things do (Verbeek, 2005) to us, our relationship with technology and indeed our relationships with each other.

Technology is always conceived produced and eventually used in a particular context and the consideration of this is a key factor in the success of products.

3 Technology Education and Contexts

Formal technology education of the kind that we find in schools is, by its very nature, simplified to help understanding. In doing so, technology becomes rarified (Feenberg, 2017) and is usually taken out of context so that its evaluation can be simplified and consequently more accessible to learners. As Dewey (1902) says, this is different from the normal everyday experiences of young people at home or indeed anywhere outside of school. Technology education is also influenced by the context in which it is done and the cultural practices of teachers. Whilst there are always national curriculum documents that suggest content and processes there is inevitably variation in both teacher's interpretation and learners experience. Locally available materials and machine-tool restrictions inevitably constrain what pupils can make. In the long term, it is also the case that technology education affects the environment and cultural practices as learners use their knowledge to make decisions about technology and the products of technological practice.

In considering the current Design and Technology curriculum in England (DfE, 2013) and its association with contexts, it is worth going back to the origins of technology education and the thinking of those who initially brought it about as a subject in the National Curriculum (DfE, 1990). The Working Group Interim Report (DES, 1988) placed all technological activity in context and there was considerable emphasis on the development of generic skills and design processes. Enabling young people to develop, for example, skills of cutting in one material and making the connection to cutting another material was seen as acceptable practice and something to be encouraged. Soon after becoming statutory, the curriculum was fought over by a number of different interest groups. This resultant social-political shaping of the curriculum was documented by Layton (1992) and the use of the term products became synonymous with practice in technology education. At the time Martin and Riggs (1999) provocatively wrote about lost contexts and the tyranny of products suggesting that such a focus on products as the outcome of lessons, and units of work, was done at the expense of contexts.

4 Consumer Cultures

So far the chapter has considered the global situation in which we find ourselves, the link between technology and context and the similar link between technology education and context. Before a critique of the current curriculum, and day-to-day practice is undertaken, it is important to consider the nature of consumer cultures (Lury, 2011).

Our relationship with technology and the products of technology needs to change. It could be considered a generalisation to say that much of what we buy helps to define who we are. The marketing of new products is based on the assumption that we would like better products that make life easier and enable us to do more things. In this sense, Papanek (1995) asks the right question in his book *The Green Imperative*, is convenience the enemy?

A current example to consider is the uptake of air fryers in England. As a new way to prepare food we are told that an air fryer is more energy efficient and a convenient way to cook. If, however, we already have a means of cooking that is capable of meeting our needs then why should we buy another one? Saving money is a long-term consideration but what of the resources and energy used to make the fryer in the first place? Are those who have purchased air fryers planning to recycle or give away their conventional ovens and microwaves – or do they just have a more crowded kitchen?

It is also the case that the large majority of things that we buy are made elsewhere and we have no connection or even information about the means of production. Whilst we have this disconnection there is no way of knowing about the impact that our consumer choices have had, and will have, on the environment and the people involved in the production process.

5 Product-Centred Technology Education

I would suggest that the current curriculum for Design and Technology in England (DfE, 2013) can be considered product-centred with units of work defined as projects to make three-dimensional objects from resistant materials, textiles etc. With the focus on the product, any consideration of context is inevitably marginalised. In school, existing products are often seen as useful starting points with the aim of producing something better, or perhaps, more specifically suited to the need that has been identified or indeed created. Processes of designing may well be limited or seen in narrow terms as what pupils will be making can be pre-determined. One advantage of conceiving the curriculum in this way is that there is relevance to a modern technological society and learners are introduced to the world of goods and how they are made. There is often an end result to take home and this in itself is seen as a positive aspect to the subject. The factual knowledge base that learners have access to and can be tested on is constrained and this can create a degree of equality of experience across different schools.

One of the disadvantages of a product-centred technology education curriculum is that it promotes the notion that addressing technological issues (human needs and wants) necessarily involves making. Creativity can be limited as a result of the limitations of what can be physically made with defined materials and processes. Often design process skills do not need to be exercised in order to get a good result. This can limit opportunities for decision-making and the exercise of value judgements. Another disadvantage of the inward-looking nature of the subject is that there is little concern with some of the most significant forms of technology that affect our lives such as renewable energy sources and electric vehicles. If the focus is on learners being engaged in things that they can personally make then this limits their exposure to technology and technological practices outside of the classroom (Martin, 2023). In addition, it may promote increased consumption and the deterministic view that technology needs to continuously improve.

6 Technology Education Reimagined

The technological world in which we live has changed rapidly since the National Curriculum (DfE, 1990) came about some 35 years ago. To some extent, the practice of making products in school has kept up with the changes in manufacturing processes and materials. The broader context of depleting resources and global warming is, however, yet to be completely factored into the day-to-day practice of technology education. It is not just about increasing the awareness of sustainability and sustainable design but a shift in the nature of the subject that is needed at this time. Technology can no longer be seen outside of the context in which it is conceived developed used and disposed of. Instead, the consideration of wider contextual issues needs to pervade every aspect of the day-to-day practice of teachers in our schools. To support this basic human needs and wants should be the driver for activities in the subject. Systems thinking that takes account of the origins of materials and considers their disposal needs to be developed. In addition, the outcomes of learning activities need to be focused more on the appropriation of attitudes towards technology as much as on developing skills and knowledge.

There are of course a number of challenges to be faced when moving away from a product-centred technology education to one focused on contexts. One of these is the potential reduction in making activities for which, I believe, there is a strong case for their continued inclusion (Martin, 2022). A second concern is the issue of motivation is a key concern as one of the appeals of the subject is the time spent working hands-on with materials to produce something that can be taken home. One of the other concerns is in subject of maintaining its unique contribution by providing a window into the technological world and keeping up with technological change.

Avoiding the promotion of consumption and the values of consumption will be a further challenge and learners need to be led by teachers who have adopted a different mindset about what is important in the subject.

7 Context-Driven Technology Education

In developing an understanding of what technology education could look like in the future, it will be helpful to develop a model that is able to capture, in diagrammatic form, how such a curriculum might be enacted.

7.1 *Global Context*

The ring represented in Figure 18.1 provides the framing for all activities and thinking in the future technology education context. It is based on human needs which need to provide the backdrop for the subject and activities undertaken as part of it. No hierarchy is presented here but there are obvious human needs such as the need for food, water, shelter and energy that can usefully support genuine technological activity that will make a difference to the human and natural environment in which we live. Considering the types of things that have been done in technology education over the last three decades it was felt useful to include communication and recreation.

7.2 *Context User Purpose*

Figure 18.2 provides further elements that help to define genuine design and technological activity with the consideration of context user and purpose. Making without any defined context or user is essentially craft activity. These are valuable and can be a significant part of a future curriculum but opportunities

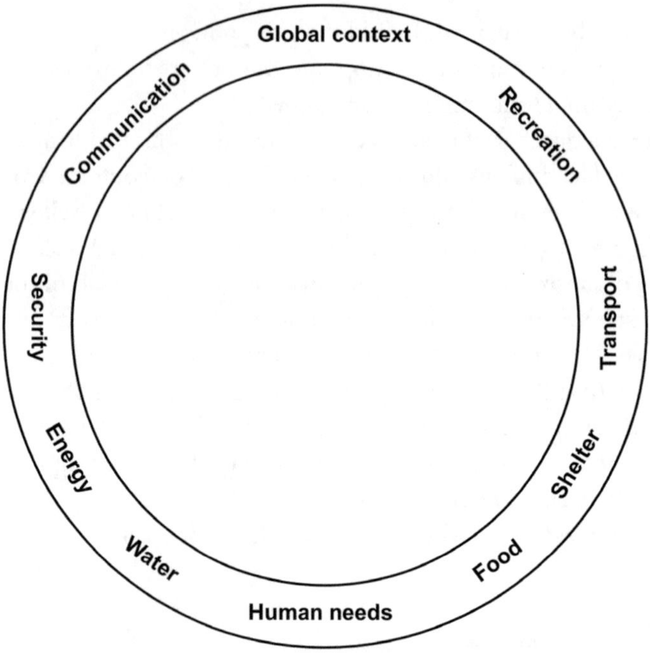

FIGURE 18.1 Framing for all activities and thinking in the future technology education context

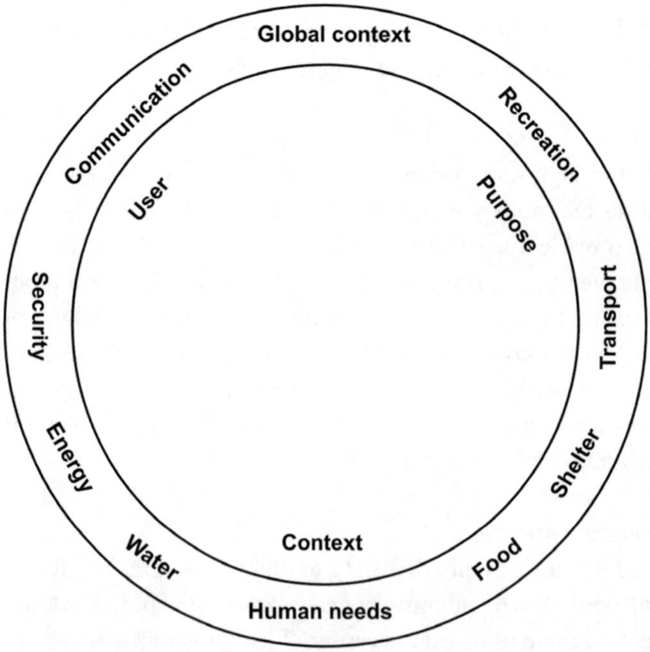

FIGURE 18.2 In order to explain the use of these terms, let's consider a collection of cups

for learners to engage in activities that require consideration of these three elements are essential.

All of the cups shown have a similar purpose in that they are designed to contain liquids for drinking. Depending on the user and, more significantly the context in which it is to be used, the design will vary. The materials used and the overall shape are determined by these factors and clearly illustrate the significance of context, user and purpose when considering design and technological activity that explores human needs.

8 Valuing Making and Designing

In terms of the day-to-day activities that learners might be engaged in, the model presented in Figure 18.3 has overlapping circles of valuing making and designing. These will be explored individually.

8.1 *Valuing*

From an early age, children are constantly experiencing the physical world, having daily contact with human-made and natural materials. Over time they

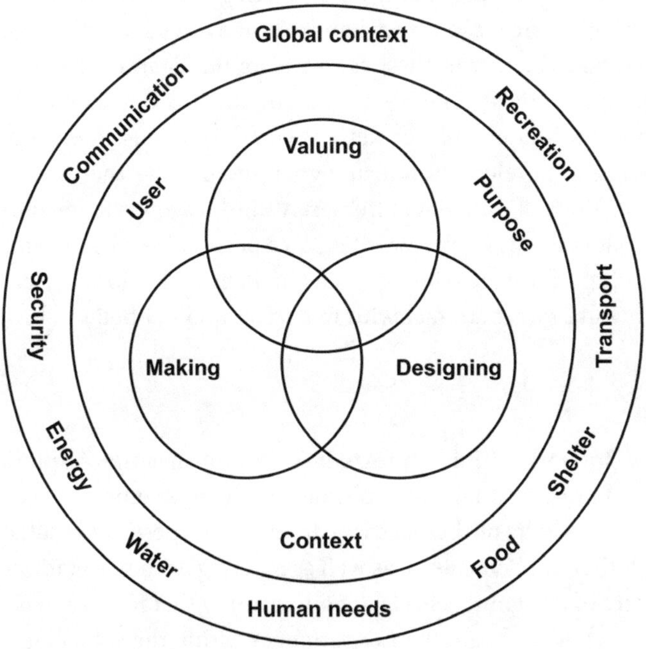

FIGURE 18.3 Valuing making and designing

become discerning about what they like and dislike and begin to make judgments about the world around them. Consequently, valuing experiences are some of the first that children have linked to technology education (Benson & Lawson, 2017). As a part of an exploration of contexts for design and technological activity, such valuing can be very useful in considering existing solutions that have met human needs. As learners progress through school the nature of valuing activities needs to change so that learners begin to provide some justification for their views and the choices that they make. In this sense valuing can be seen as a skill to be acquired, and improved, as learners work their way through the curriculum.

8.2 *Making*

Making activities have always been at the heart of technology education but the nature of them has changed over time. When the subject became part of general education there was a great deal of emphasis on the use of hand tools and the expectation that some learners would use the skills developed in related professions. As I have written in earlier work (Martin, 2013) the requirements for hand skills, using machine tools and cadcam resources have changed over time and a number of eras can be defined. Whilst it is currently possible to make many products without personally manipulating materials and using hand tools, there are legitimate reasons why hands-on activities should be part of any future curriculum. The first of these is to ensure that learners have an appreciation of where materials come from - making the link between technological activity and the shaping of our natural world as a consequence of it. The second reason is to develop our fine and gross motor skills that we use throughout life for all kinds of different activities. A third reason is to encourage learners to see that making is very rewarding on a personal level and helps our well-being. Creating a future version of the technology education curriculum that does not involve direct contact with materials would, I believe, be a mistake.

8.3 *Designing*

In enabling learners to work within contexts it is natural that they explore possible ways of developing technological solutions to the problems that they encounter. Ensuring the results of their activity are fit for purpose, and match the context in which they are being used, as well as meeting user expectations is an important aspect of designing. This is the opportunity that learners have to make decisions. As children begin their experiences within the subject it is legitimate to first focus on valuing and making activities. Indeed, an understanding of material properties and ways of working with them is essential for designing activities to have any significant meaning. As learners progress through the curriculum the demands on their designing need to increase so that they are able to make decisions that they can justify.

9 Theory into Practice

So far the presentation of a future technology education curriculum that is context-driven has been theoretical. In order to better explain the model outlined above it will be useful to provide an example from experience. Below is the description of a valuing, designing and making unit of learning that was undertaken by first-year undergraduate primary students on a teacher education course.

The global context selected for valuing, making and designing was security and particularly focused on keeping things safe. The specific context was a day trip such as a school class going to a museum or the zoo. As this was aimed at learners who had not had much experience with design and technological activities, the learner themselves was selected as the user. The purpose was to carry essential items including a snack, water and money and be made in such a way that the user would have their hands 'free'.

In introducing the unit of learning to those involved it was important to consider the language used to describe the different elements involved. For example, if the word 'bag' was used then it has often been the case that learners focus on something that can be held by hand and the solutions that they develop to carry the items tend to be predictable and similar to each other. If, on the other hand, the word carrier is used, and the learners first engage in a valuing activity about carrying things in different ways, then the outcomes are generally more diverse and reflect the creative ideas that they have. Here we see the significance of how we present the context for learning.

One specific valuing activity that can be linked to this unit of learning relates to the joining of materials, specifically the handles and straps the top part of existing products designed for carrying. The example from Nuffield Primary Solutions (2024) explains this in more depth.

In terms of designing, this example of a unit of work lends itself to full-size model making (or prototyping) with paper, card and recycled materials. Such authentic modelling is very important to allow genuine evaluation to take place where users can be involved.

In order to make this a genuinely useful making activity, the carrier could be made by reusing or upcycling existing materials and tested on an actual trip to the zoo.

10 Advantages

One advantage of situating technology education activities in a global context, such as security, is that it links directly to human needs And avoids the generation of products that are not needed. As such it reflects best practices in contemporary manufacturing where an account is taken of the impact on people in the environment. Another additional match is the development of generic valuing-making and designing skills that are, to some extent, independent of materials. The subject would also be more outward-facing and engage learners and issues of social responsibility and technology justice.

11 Considerations

One of the disadvantages of a context-driven technology education would be less emphasis on products that can be taken home. Another consideration would be the increased challenge for learners as they become more responsible for decision-making. There would also need to be a change in thinking by all involved, particularly teachers who will be attached to two activities that have worked well for them over time. With such a curriculum it would need to be recognised that the coverage of materials and processes would vary from school to school, making the emphasis on generic skills and processes even more important.

12 Conclusion

As was said in the introduction, the current curriculum is out of touch with the global context in which we find ourselves and is consequently not fit for practice. A new curriculum that focuses on technology in context is, I believe, essential if we are to help young people manage living in the rapidly evolving world of the future.

References

Benson, C., & Lawson, S. (2017). *Teaching design and technology creatively*. Routledge.

Blok, V. (2019). Nothing Else Matters: Towards an Ontological Concept of the Materiality of the Earth in the Age of Global Warming. *Research in Phenomenology*, *49*, 65–87.

Coeckelbergh, M. (2020). *Introduction to philosophy of technology*. Oxford University Press.

DES. (1988). *National Curriculum Design and Technology Working Group: Interim Report*. HMSO.

Dewey, J. (1902). *The child and the curriculum*. University of Chicago Press.

DfE. (1990). *Technology in the National Curriculum*. London: HMSO.

DfE. (2013). *National curriculum in England: Design and technology programmes of study*. https://www.gov.uk/government/publications/national-curriculum-in-england-design-and-technology-programmes-of-study

Feenberg, A. (2017). *Technosystem: The social life of reason*. Harvard University Press.

Idhe, D. (1990). *Technology and the lifeworld: From garden to earth*. Indiana University Press.

Layton, D. (1992). *Values and design and technology*. Design Curriculum Matters: Occasional Paper No. 2. Loghborough: Loughborough University.

Lury, C. (2011). Consumer Culture (2nd ed.). Polity Press.

Martin, M. (2023). Extracurricular Perspectives: Valuing Technology beyond the Classroom. In D. Gill, D. Irving-Bell, M. McLain, & D. Wooff (Eds.), *The Bloomsbury Handbook of Technology Education* (pp. 401–409). London: Bloomsbury Academic.

Martin M. (2013). Five eras of making and designing. In PATT27 Technology Education for the Future: A play on sustainability. Hamilton New Zealand: University of Waikato.

Martin M. (2022). The role of making in D&T. In A. Hardy (ed.), Debates in Design and Technology Education (2nd ed.). Routledge.

Martin, M. C., & A. Riggs. (1999). Lost contexts and the tyranny of products. IDATER 1999 Conference, Loughborough: Loughborough University.

Nuffield Primary Solutions. (2024). https://dandtfordandt.wordpress.com/resources/nuffield-primary-dt-resources/

Pacey A. (1986). The Culture of Technology, Cambridge: MIT Press.

Papanek, V. (1995). *The green imperative: ecology and ethics in design and architecture*. Thames and Hudson.

Verbeek, P. (2005). *What things do: Philosophical reflections on technology, agency and design*. Penn State Press.

CHAPTER 19

Who Are We Fooling with Schooling?

Why We Must Break from a Nineteenth Century System and Educate for the Twenty-First Century

Philip A. Reed

If someone who is removed from the educational enterprise were to look at compulsory schooling, they may conclude that mastery of language arts and mathematics was the goal. Without questioning the importance of these foundational disciplines, however, it is clear the world is much more transdisciplinary than disciplinary or even multi-disciplinary. Why then is teaching and learning still organised in historically limiting ways? Additionally, certain subjects are still labelled "academic" while others are labelled as "practical" even though they all meet defining elements as academic disciplines. Efforts to elevate transdisciplinary/convergence education will fail if these issues are not addressed. We see this in educational movements such as STEM/STEAM education where endeavours are often stymied due to the historically rigid perceptions and structures of the educational enterprise. Perhaps the solution to our current state of compulsory education requires a multi-pronged approach: First, a look at the literacies needed in our contemporary world and, second, a structural reformation of compulsory education. Recommendations are made on the knowledge, skills, and dispositions of citizens in the twenty-first century by considering core and under-represented disciplines in compulsory education. An argument is also presented to change the structure of schooling from content-focused silos to contextual learning and metacognition.

Reflecting on the career of Marc J. de Vries (2024) immediately makes me consider the evolution of science and of science education. Whereas Marc started his career teaching physics, mathematics, and technology, he used didactics and ultimately became a renowned philosopher of technology, focusing on design and technology education. This is telling since it is in opposition to the development of science as a discipline. The field of science is recognised as emerging from philosophy, more specifically Aristotle's doctrine of "impetus" (Butterfield, 1957). Marc's career path, however, seems more a parallel to modern science education, which emerged from the academies in the nineteenth century to foster thinking (DeBoer, 1991).

© PHILIP A. REED, 2025 | DOI:10.1163/9789004710948_019

While a goal of compulsory education may be to foster thinking, the current state of siloed curriculum, a focus on standardised assessments, and structural implications in compulsory education seem to be in opposition to this goal (Reed, 2007, 2018). This is why de Vries' 'come-full-circle' career is so poignant. In *Teaching About Technology: An Introduction to the Philosophy of Technology for Non-philosophers* he provides a compelling rationale why educators should embrace philosophy (de Vries, 2016). Structured as a textbook (in my opinion, *the* textbook), *Teaching About Technology* provides engaging lessons and exercises on how to teach the philosophy of technology. This approach is sound if we truly want to foster thinking. As an example, Dakers (2023) has taken a detailed look at the history and philosophy of technology education to recommend a nomadic pedagogy as a way to break from the craft tradition in technology education.

The educational enterprise at the compulsory level, however, maintains historical barriers that illustrate schooling and educating are not the same. There are traditions on what constitutes a discipline that results in which disciplines continue to receive the most emphasis in schools. This chapter makes recommendations on the knowledge, skills, and dispositions citizens need in the twenty-first century by considering core and under-represented disciplines in compulsory education. Technology education is used as an expanded disciplinary example. Additionally, an argument is presented to change the structure of schooling from a focus on content silos to one that supports contextual learning and metacognition. In the Festschrift spirit, I provide references but also take liberty in making some unsupported claims, generalisations, and pose questions to foster dialogue and, hopefully, action towards educating as opposed to schooling.

1 Current State of Schooling

The current model of compulsory schooling is similar in many countries with a focus on language (reading and writing), mathematics, physical science, and history. In the United States, a relatively young country, these four disciplines were established over a century ago and became the core subjects for different reasons. The study of language was introduced by church schools so citizens could learn the Bible. Mathematics and history were added through a legislative bill from Thomas Jefferson in 1778 so citizens would learn to "manage their affairs" (Urban & Wagoner, 1996, p. 72). The most recent core subject added was science, which was recommended in 1893 by the National Education Association's (NEA) Committee of Ten (DeBoer, 1991).

It is clear the four core disciplines remain firmly entrenched when reviewing assessments such as The Nation's Report Card in the U.S. (National Center for Education Statistics, 2024) or the Programme for International Assessment (PISA; OECD, 2024). There is no question these four disciplines are important, but should they continue to dominate primary and secondary education? Humanity continues to progress and so have many disciplines needed to live a fulfilling life in a global, contemporary society. Yet, we continue offering a nineteenth-century education in the twenty-first century.

There are perceived and real impediments to change in compulsory schooling. My use of language in the preceding paragraphs is an example of a perceptual barrier. I deliberately refer to the four ensconced disciplines as core, not academic, yet the media, the public, and even educators refer to them as the academic disciplines. My choice of language is not to insinuate they are not academic disciplines, rather, it is to say they are the fundamental disciplines selected for compulsory schooling and, yes, other disciplines are academic as well. However, the use of the phrase "academic disciplines" when referring to the four core disciplines (for over a century!) makes every other discipline less than.

Technology education is a part of vocational education/career and technical education (CTE) in many countries and even some CTE scholars compare programs with an "occupational" orientation to those with an "academic" orientation (Gordon & Schultz, 2020). Globally, CTE is an elective for most students, or a different secondary education pathway altogether, so studies comparing CTE and required core disciplines is common and necessary. The use of language that positions one discipline over another, however, is divisive. Such language has ingrained an either/or mentality when, in reality, there are common characteristics that define every discipline: content, epistemology, and a history of practice, inclusive of curriculum, teaching, and research (Reed, 2018). Technology as an academic discipline has been clarified through the work of, among others, Warner et al. (1947), Olson (1963), DeVore (1966), Savage and Sterry (1990), Volti (2017), and most recently in *Standards for Technological and Engineering Literacy* (ITEEA, 2020).

Real impediments to change in compulsory schooling are mostly related to the structure of the school day. Postman (1993) describes how lower schools emerged during the widespread dissemination of print brought on by movable type in the fifteenth century. Text has remained the dominant communication medium in education for four hundred years and heavily influenced the shape of compulsory education. Real disruption, however, has been occurring for the past forty years through instructional technologies and our penchant for being entertained across all aspects of society (Postman, 1995, 2005). The scheduling in compulsory schooling remains an impediment defined by periods of study in a subject (i.e., bells) organised around factory work structures from the nineteenth century (Urban & Wagoner, 1996).

Other real impediments to change in compulsory schooling centre around content and motivation. Olson's (2013) review on the nature of technology content in *A framework for K-12 science education* (National Research Council, 2012) sums up the current state of schooling in several ways. First, she details at least fifteen points where the *Framework* either downplays or omits aspects of the nature of technology. This supports ideas that there is too much emphasis on practices/skills and, that as outlined, the *Framework* does not aid students in obtaining a truly liberal education. In fact, she concluded that "denying students access to the nature of technology, including its assumptions, limitations, and consequences, is nothing short of educational malpractice" (Olson, 2013, p. 242).

Secondly and equally powerful, Olson positions her *Framework* review around the story of a former sixth-grade student who was perpetually disengaged. One day when she asks, "Why are you here?" he responds, "I don't know, you tell me" (p. 217). As educators, we have all likely experienced similar students, and we need to ask ourselves whether we are educating them or merely treating them as a cog in the machinery of schooling. Olson (2013) admitted that she did not have a compelling answer for this student. Overall, I am not convinced that we, the schooling enterprise writ large, provide a compelling answer for students' "why" in compulsory schooling.

2 Current State of Education

The "why" question highlights the stark differences between schooling and educating. This leads to the question of "what" education should include in compulsory schooling. Globally, there have been monumental advances in the breadth and depth of post-secondary education, yet primary and secondary education persist as large, immovable objects. There are government and private sector grants to foster advancements in education but often these programs end when the funding is over and long-term impacts are difficult to measure (Sun et al., 2021).

An additional issue with compulsory education is that it does not reflect the advances in society over the past century. Areas such as instructional/educational technology are being integrated into education and teacher training (ISTE, 2024) but even advocates for the core disciplines claim there is not enough time for instruction (see National Science Teaching Association, 2024). To address the "what" in compulsory education, we must recognise there are more relevant ways to teach the core subjects and newer literacies that are vital for all students in the twenty-first century. Note that I make no effort to define these literacies individually, rather, I use the term literacy in reference to the basic knowledge, skills, and dispositions each discipline uniquely adds to the educational enterprise. Let's look at a few examples.

Most schools organise mathematics around arithmetic, algebra, geometry, trigonometry, and other pure mathematics disciplines. Unless you aspire to be a mathematician, doesn't it make more sense to study mathematics from an applied standpoint? In the United States, for example, many states implemented a policy requiring all students to take one financial literacy class prior to graduation. (Yes, many, not all states and, yes, one class). This was a reaction to the 2007–2008 global financial crises. Shouldn't such contextual classes have been happening since 1778 when Thomas Jefferson sponsored that bill supporting mathematics to help citizens with their affairs (Urban & Wagoner, 1996)? Would students be more engaged and mathematics more relevant if courses focused on a context such as consumer mathematics or employment areas such as construction, manufacturing, or information technology (IT)?

The COVID pandemic provided numerous lessons in what did and did not work in compulsory education. There were many lessons on the use of instructional technology such as equitable access to hardware and online access. A deeper understanding of pedagogical implications is still emerging but given the drop in basic skills on large-scale assessments (National Center for Education Statistics, 2024; OECD, 2024), more work is needed on the effective use of instructional technologies in primary and secondary education. Organisations such as the International Society for Technology in Education (ISTE) provide standards for instructional technology (ISTE, 2024) but the pandemic and the reliance on online communications have demonstrated a need for the broader area of media and information literacy as defined by organisations such as the American Association of School Librarians (AASL). The AASL standards (2024) go beyond the use of instructional technology by providing crosswalks to ISTE standards, ASCD's Whole Child Tenets, coding, and other initiatives that help students locate, evaluate, and use valid information in a broad range of contexts.

A second pandemic lesson surrounds health and wellness literacy. A pandemic naturally causes uncertainty regarding physical and psychological well-being, but the amount of misinformation and confusion during the COVID-19 pandemic demonstrated a severe lack of health literacy. Science addresses some of these issues in compulsory education yet, like mathematics, compulsory education science content is primarily organised around the work of scientists (i.e., classes in earth science, biology, chemistry, and physics). Wouldn't students be better served with more contextual courses such as personal wellness or global health? Contextual courses could include content from the physical sciences as well as other disciplines such as psychology, kinesiology, nutrition, and human services, among others.

Numerous other literacies vital for global citizens today are apparent by simply following international news. An understanding of information technology, including data science and artificial intelligence, is becoming important for work, communication, financial well-being, and maintaining our privacy. Politics, business, and religion provide numerous examples of the need for a basic understanding of ethics. Wars, immigration, and online behaviours are a few of the justifications for cultural literacy, including strategies for diversity, equity, and inclusion. Global warming, pollution, and the depletion of natural resources offer compelling reasons for environmental literacy.

Each of the preceding examples highlights three important issues that address the "what" in contemporary education. First, compulsory schooling is not providing the education needed for global citizens in the twenty-first century. Second, the literacies mentioned are all intimately tied to technology and engineering education since that is the discipline that focuses on the designed world and touches every aspect of the human experience (de Vries, 2017; ITEEA, 2020; Petrina, 2007). Finally, these examples demonstrate the integrative nature of education today. Educators such as Dewey (2011) have long advocated for education to improve society and as society has become more complex, we need to focus on the "how" in order to make meaningful changes to compulsory education.

3 Opportunities

Any proposal on how to educate runs the risk of over generalising since each country and state/province has very different education models (Tucker, 2019; Williams, 2006). There are curriculum and structural changes, however, that can be made no matter what the existing system entails. Most of the recommendations offered here are evolutionary, not revolutionary, and scale up existing practices or offer them in a slightly different way. The foremost change is to dethrone the core disciplines from occupying so much dedicated time in schooling to make way for the many new literacies needed in compulsory education. We know the importance of the core disciplines for providing a knowledge base so students can move from knowledge lean to knowledge rich forms of learning (National Academies of Sciences, Engineering, and Medicine, 2018). Therefore, core subject knowledge still needs to be offered in the early years and other disciplines should be integrated. Technology and engineering educators have learned that integrating content with core subjects in primary and elementary education is a successful strategy (Kirkwood & Foster, 1997; Virginia Children's Engineering Council, 2024). Thus, research and practice support a model to build basic skills at lower levels and move to a model that increases depth and breadth as student progress into secondary education.

In the United States, the National Center on Education and the Economy (NCEE, 2008) proposed a reinvention of secondary education, primarily the last two years of high school. The proposed model would have students finish common school in grade ten (age 16) and provide students with two options: two years of upper secondary study (e.g., advanced placement or International Baccalaureate) or two to three years of vocational school or community college. Students under each path in this model would earn a high school diploma and have options to earn certificates and associate degrees depending on their interests and goals. Adoption of the NCEE model by school systems is unknown, although there is evidence that more high school students in the United States are using their last two years to simultaneously earn their diploma and an associate degree (Barshay, 2020).

The NCEE model provides more student autonomy in the final years of secondary education and there are several other strategies to make secondary education more student-centered. Career academies are schools that focus all coursework around a career area such as those found in the National Career Clusters Framework (AdvanceCTE, 2024). Core subjects in career academies focus on reading, writing, and numeracy lessons around the academy theme and there are topical courses, integrated projects, and often work-based learning options for students. Career academies are expanding and showing positive outcomes for students (National Career Academy Coalition, 2024; Stern et al, 2010). In school systems that have career academies, students have choices to apply to academies they are interested in or to continue their secondary education in a traditional comprehensive high school.

One reason career academies are becoming popular is their use of contextual learning. Teachers that actively engage the student in contextual learning utilise five strategies:

- Relating – learning in the context of one's life experiences or preexisting knowledge.
- Experiencing – learning by doing, or through exploration, discovery, and invention.
- Applying – learning by putting the concepts to use.
- Cooperating – learning in the context of sharing, responding, and communicating with other learners.
- Transferring – using knowledge in a new context or novel situation—one that has not been covered in class (Crawford, 2001, p. 111).

The Center for Occupational Research and Development (CORD) has demonstrated the effectiveness of these REACT strategies in mathematics and science (CORD, 1999a, 1999b) and continues to broaden their work in integrated contexts such as science, technology, engineering, and mathematics (STEM).[1]

Technology and engineering education has a long history of using contexts to organise teaching and learning (Reed, 2017). ITEEA's *Standards for Technological and Engineering Literacy* (2020) uses standards, practices, and contexts as organisers and is designed for flexibility so curriculum developers and teachers can use contexts for full courses or merely for units or individual lessons. Regardless of how contexts are implemented, it is important to use authentic contexts whenever possible such as the United Nations Sustainable Development Goals (2024), the Grand Challenges for Engineering (National Academy of Engineering, 2024), or a local context. The use of authentic contexts is supported by the CORD model for contextual learning and the broader research on how people learn (National Academies of Sciences, Engineering, and Medicine, 2018) because they are likely to connect to students' experiences and prior knowledge more readily than contrived contexts.

The COVID-19 pandemic, as noted above, provides many authentic contexts for the needed literacies in compulsory education as well as some structural opportunities. New ways of delivering technology and engineering education emerged and demonstrated that we do not need to be tied to physical laboratories (Mentzer, 2022; Sain, 2022). Even before the pandemic, however, Custer (2000) stressed the importance of finding new ways of schooling using authentic contexts and had some pointed recommendations for technology and engineering education:

- Increase breadth and depth of content knowledge.
- Address issues of image and recognition.
- Share the load of teaching technology with other disciplines.
- Actively participate with other disciplines.
- Become more adept with educational and political policy.
- Work with leaders in colleges to integrate disciplines and break down barriers. (pp. 129–130)

All these recommendations have been addressed to varying degrees since Custer (2000) introduced them almost a quarter century ago. STL (ITEEA, 2007) and STEL (ITEEA, 2020) have helped with content, image, and recognition. The proliferation of elementary technology education (Kirkwood & Foster, 1997; Virginia Children's Engineering Council, 2024), working with prominent policy groups (i.e., National Assessment Governing Board, The National Academies), and the rise of makerspaces, technological competitions, and engineering design in science education (Reed, 2018) have aided disciplinary, multi-disciplinary, and transdisciplinary efforts.

Politics is the one area recommended by Custer (2000), Sanders (2000), and Starkweather (2015) that has largely gone unaddressed. This may be due to the shrinking number of technology educators (Moye, 2009; Volk, 1997) or from the continuance of the core disciplines to dominate the narrative on compulsory education. Despite a lack of overt political action by technology educators, documents such as *Charting A Course For Success: America's Strategy for STEM Education* (National Science and Technology Council, 2018) highlight many ways technology and engineering is integrated into compulsory education. Additionally, the publication *Convergence Education: A Guide to Transdisciplinary STEM Learning and Teaching* (National Science and Technology Council, 2022) details the importance of engineering design and CTE in convergence education.

Technology and engineering educators need to continue working in policy and political circles despite the entrenchment of the core disciplines. Integrating content across the curriculum ensures more students will receive the literacies they need in the twenty-first century. There is no room for siloed content or posturing to claim turf. Efforts in education often take a long time before their impact is noticed so sustained efforts are needed. Research can impact policy and practice and should be shared broadly through international activities such as the PATT conference (de Vries, 2015). Incredibly, Marc de Vries has organised the PATT conferences for over three decades and continues to serve a long tenure as editor of the *International Journal of Technology and Design Education*. His consistent leadership in global scholarly efforts is rare, not to mention his own prolific contributions, and clearly puts him atop a very distinguished list of scholars who have solidified technology and engineering education as a discipline for all students. Such scholarly efforts within journals, conference proceedings, and periodic futuring (e.g., de Vries & Tamir, 1997; Martin, 2000; Williams et al, 2015) must be continued to have meaningful impacts on policy and politics.

We are clearly at a historical deflection point in education with COVID-19, climate change, globalisation, and the proliferation of communications technologies. It is imperative to continually reflect on how to educate when we face pervasive change, are so connected, and have so much knowledge at our fingertips. For example, with the emergence of artificial intelligence (AI), how do we teach writing and even artistic endeavours (e.g., graphics, audio) when it can be done for us? One solution, perhaps the most humanistic solution, is to teach disciplinary foundations in contextual ways at the elementary level and to move progressively into ways of convergence thinking and metacognition (National Science and Technology Council, 2022). In technology and engineering education, de Vries (2016) demonstrated how this can be done through philosophy. Dakers (2023) articulates how the craft-based model in technology education has run its course, is not sustainable, and the viable path forward is through a model based on ethnotechnological pedagogy. Additionally, Well's (2016) provides a process for teaching design using inductive and deductive questioning. These models, among others, transcend historical calls for technical and socio/cultural curriculum (e.g., DeVore, 1966) by providing the "how" for technology and engineering education to foster the knowledge, skills, and dispositions students need in the twenty-first century.

Note

1 https://www.cord.org/

References

AdvanceCTE. (2024). *The National Career Cluster's Framework.* https://careertech.org/what-we-do/career-clusters/

American Association of School Librarians. (AASL). *Standards crosswalks.* https://standards.aasl.org/project/crosswalks/

American Montessori Society. (2024). Who was Maria Montessori? https://amshq.org/About-Montessori/History-of-Montessori/Who-Was-Maria-Montessori

Barshay, J. (2020). Research on early college high schools indicates they may pay for themselves in long run. *The Hechinger Report.* https://hechingerreport.org/research-on-early-college-high-schools-indicates-they-may-pay-for-themselves-in-the-long-run/

Butterfield, H. (1957). *The origins of modern science* (revised ed.). The Free Press.

Center for Occupational Research and Development (CORD). (1999a). *Teaching mathematics contextually.* CORD Communications, Inc.

Center for Occupational Research and Development (CORD). (1999b). *Teaching science contextually.* CORD Communications, Inc.

Crawford, M. L. (2001). *Teaching contextually: Research, rationale, and techniques for improving student motivation and achievement in mathematics and science.* CCI Publishing, Inc.

Custer, R. L. (2000). Blurring the boundaries. In G. E. Martin (Ed.), *Technology education for the 21st century.* Council on Technology Teacher Education 49th Yearbook. https://hdl.handle.net/10919/5531

Dakers, J. R. (2023). *A nomadic pedagogy about technology: Teaching the ongoing process of becoming ethnotechnologically literate.* Brill.

DeBoer, G. E. (1991). *A history of ideas in science education: Implications for practice.* Teachers College Press.

DeVore, P. W. (1966). *Structure and content foundations for curriculum development.* (ED020429). ERIC. https://eric.ed.gov/?q=%22Structure+and+content+foundations+for+curriculum+development.%22&id=ED020429

de Vriees, M. J., & Tamir, A. (Eds.). (1997). *Shaping concepts of technology: From philosophical perspective to mental images.* Kluwer Academic Publishers.

de Vries, M. J. (2015). Research challenges for the future. In P. J. Williams, A. Jones, & C. Buntting (Eds.), *The future of technology education.* Springer.

de Vries, M. J. (2016). *Teaching about technology: An introduction to the philosophy of technology for non-philosophers* (2nd ed.) Springer.

de Vries, M. J. (ed.). (2017). *Handbook of technology education.* Springer International. DOI 10.1007/978-3-319-38889-2_9-1

de Vries, M. J. (2024). Marc de Vries. https://www.tudelft.nl/en/tpm/our-faculty/departments/values-technology-and-innovation/people/full-professors/profdr-m-marc-de-vries

Dewey, J. (2011). Democracy and education. Simon & Brown.

Gordon, H. R. D., & Schultz, D. (2020). *The history and growth of career and technical education in America* (5th ed.). Waveland Press.

International Technology and Engineering Educators Association (ITEA/ITEEA). (2007). *Standards for technological literacy: Content for the study of technology* (3rd ed.). https://www.iteea.org/technology-for-all-americans-project

International Technology and Engineering Educators Association (ITEEA). (2020). *Standards for technological and engineering literacy: The role of technology and engineering in STEM education.* www.iteea.org/stel.aspx

International Society for Technology in Education (ISTE). (2024). *Standards.* https://iste.org/standards

Kirkwood, J. J. and Foster, P. N. (Eds.). (1997). *Elementary school technology education.* Forty-sixth yearbook of the Council on Technology and Engineering Teacher Education. McGraw-Hill. http://hdl.handle.net/10919/19146

Martin, G. E. (Ed.). (2000). *Technology education for the 21st century.* Council on Technology Teacher Education 49th Yearbook. Glencoe McGraw-Hill.

Mentzer, N., & Mohandas, L., (2022). Student experiences in an interactive synchronous HyFlex design thinking course during COVID-19. *Interactive Learning Environments.* DOI: 10.1080/10494820.2022.2124423

Moye, J. J. (2009). The supply and demand of technology education teachers in the United States. *The Technology Teacher, 69*(2), 30–36.

National Academies of Sciences, Engineering, and Medicine. (2018). *How people learn II: Learners, contexts, and cultures.* The National Academies Press. https://doi.org/10.17226/24783

National Academy of Engineering. (2024). *NAE Grand Challenges for Engineering.* https://www.engineeringchallenges.org/challenges.aspx

National Career Academy Coalition. (2024). *Data and research.* https://www.ncacinc.com/category/data-research/

National Center for Education Statistics. (2024). *National Assessment of Educational Progress.* https://nces.ed.gov/nationsreportcard/

National Center on Education and the Economy. (2008). *Tough choices or tough times: The report of the New Commission on the Skills of the American Workforce* (Revised and Expanded Edition). Jossey-Bass.

National Research Council. (2012). *A framework for K-12 science education.* National Academies Press. https://nap.nationalacademies.org/catalog/13165/a-framework-for-k-12-science-education-practices-crosscutting-concepts

National Science and Technology Council. (2018). *Charting A Course for Success: America's Strategy for STEM Education.* https://www.energy.gov/sites/default/files/2019/05/f62/STEM-Education-Strategic-Plan-2018.pdf

National Science and Technology Council. (2022). *Convergence Education: A Guide to Transdisciplinary STEM Learning and Teaching. https://www.whitehouse.gov/wp-content/uploads/2022/11/Convergence_Public-Report_Final.pdf*

National Science Teaching Association (NSTA). (2024). Position statement: Elementary school science. https://www.nsta.org/nstas-official-positions/elementary-school-science

Olson, D. (1963). *Industrial arts and technology.* Prentice Hall.

Olson, J. K. (2013). The purposes of schooling and the nature of technology. In Clough, M. P., Olson, J. K., & Niederhauser, D. S. (Eds.). *The nature of technology: Implications for learning and teaching*, 217–248. Rotterdam, Netherlands. Sense Publishers.

Organisation for Economic Cooperation and Development (OECD). (2024). *Programme for International Student Assessment (PISA).* https://www.oecd.org/pisa/

Petrina, S. (2007). *Advanced teaching methods for the technology classroom.* Hershey, PA: Information Science Publishing.

Postman, N. (1993). *Technopoly: The surrender of culture to technology.* Vintage.

Postman, N. (1995). *The end of education: Redefining the value of school.* Vintage.

Postman, N. (2005). *Amusing ourselves to death: Public discourse in the age of show business* (20th anniversary edition). Pinguin Books.

Reed, P. A. (2007). The journey towards technological literacy for all in the United States: Are we there yet? *Technology and Engineering Teacher, 67*(3), 15–22. http://www.iteea.org

Reed, P. A. (2017). Technology education standards in the United States – history and rationale. In M. de Vries (ed.), *Handbook of Technology Education*, Springer International Handbooks of Education, DOI 10.1007/978-3-319-38889-2_9-1. Springer International Publishing, Inc.

Reed, P. A. (2018). Reflections on STEM, standards, and disciplinary focus. *Technology and Engineering Teacher, 71*(7), 16–20. http://www.iteea.org

Sain, J. (2022). Using targeted online learner analytics to enhance design-based learning. Paper presented at the 108th Mississippi Valley Technology Teacher Education Conference. https://the1909conference.org/archives-2

Sanders, M. E. (2000). Power play or no play? In Martin, G. E. (Ed.). (2000). *Technology education for the 21st century*. Council on Technology Teacher Education 49th Yearbook. https://hdl.handle.net/10919/5531

Savage, E., & Sterry, L. (1990). *A conceptual framework for technology education.* International Technology Education Association.

Starkweather, K. N. (2015). Politics and policy. In P. J. Williams, A. Jones, & C. Buntting (Eds.), *The future of technology education.* Springer.

Stern, D., Dayton, C., and Raby, M. (2010). *Career academies: A proven strategy to prepare high school students for college and careers.* https://www.acteonline.org/wp-content/uploads/2018/02/Career_Academies_a_Proven_Strategy.doc

Sun, M., Kennedy, A. I., & Loeb, S. (2021). The Longitudinal Effects of School Improvement Grants. *Educational Evaluation and Policy Analysis, 43*(4), 647–667. https://doi.org/10.3102/01623737211012440

Tucker, M. S. (Ed.). (2019). *Vocational education and training for a global economy: Lessons from four countries.* Harvard Education Press.

United Nations. (2024). *Department of Economic and Social Affairs, Sustainable Development: The 17 Goals.* https://sdgs.un.org/goals

Urban, W. & Wagoner, J. (1996). *American education: A history.* McGraw-Hill.

Virginia Children's Engineering Council. (2024). *Resources.* https://vachildrensengineering.org/resources

Volk, K. (1997). Going, going, gone? Recent trends in technology teacher education programs. *Journal of Technology Education, 8*(2), 66–70.

Volti, R. (2017). *Society and technological change* (8th edition). Worth Publishers.

Warner, W. E. [with Gary, J. E., Gerbracht, C. J., Gilbert, H. G., Lisack, J. P., Kleintjes, P. L., and Phillips, K.]. (1947, April). *The new industrial arts curriculum.* Paper presented at the first meeting of the American Industrial Arts Association, Columbus, OH.

Wells, J. G. (2016). PIRPOSAL model of integrative STEM education: Conceptual and pedagogical framework for classroom implementation. *Technology and Engineering Teacher, 75*(6), 12–19.

Williams, P. J. (Ed.). (2006). *International technology teacher education.* Fifty-fifth Council on Technology and Engineering teacher Education Yearbook. McGraw-Hill. http://hdl.handle.net/10919/19155

Williams, P. J., Jones, A., & Buntting, C. (Eds.). (2015). *The future of technology education.* Springer.

CHAPTER 20

The Genesis of Contemporary Technology Education

From Attitude Studies to a Mature Field

Piet Ankiewicz

1 Introduction

1.1 *Purpose and Research Methodology*

Marc de Vries has been associated with pupils' attitudes toward technology (PATT) and the related PATT conferences to such an extent that he has become synonymous with it. However, his far-reaching impact on the broader field of technology education should not be restricted to PATT only. The purpose of this chapter is to illustrate how de Vries' initial involvement in attitude studies in the Netherlands has promoted attitude studies and given rise to the maturation of the broader field of technology education internationally. Furthermore, it indicates how his philosophical disposition has informed and steered attitude studies in technology education.

The research methodology pursued a systematic literature review followed by a meta-synthesis of a selection of literature on students' attitudes toward technology (cf. Buckley et al., 2022; Strimel et al., 2020). Firstly, a systematised literature review was conducted of the PATT conference proceedings since their inception in 1986, the field's most productive source of research papers (Williams, 2016), using the search terms 'attitudes', 'beliefs', 'interest', 'perceptions', 'conceptions', 'behaviour', 'attitude measurement' and 'instruments'. Secondly, the result of more than 40 PATT papers directed me to more than 60 articles in the International Journal of Technology and Design Education (ITDE), Journal of Technology Studies, Journal of Technology Education, Design and Technology Education: An International Journal, Research in Science and Technology Education, and International Journal of Science Education, and a book chapter which focuses on attitudes studies (cf. Buckley et al., 2022). Parts of this chapter have been published elsewhere in a different format and genre (cf. Ankiewicz, 2018a, b, 2019a, b).

© PIET ANKIEWICZ, 2025 | DOI:10.1163/9789004710948_020

1.2 *The Contribution of the Pupils' Attitudes toward Technology (PATT) Foundation and Its Studies to Instruments for Ascertaining Students' Attitudes*

Before the 1980s, research related to students' attitudes towards technology was unusual (Yu et al., 2012). In the 1980s, several countries introduced technology education as a successor to some form of craft or technical education and it began to develop its distinct research area. Studies into students' attitudes towards and concepts of technology mostly contained information on students' ideas when entering technology education (Kőycű and de Vries, 2016).

The most noted study of students' attitudes towards technology has probably been the work pioneered by Marc de Vries and his PhD supervisor, Prof. Jan Raat, as part of 'Project Physics and Technology' in the Department of Physics Education at Eindhoven University of Technology in the Netherlands in 1984 (de Vries, 1988; Volk & Yip, 1999). The first part of the research was done among students of ages 13 to 14 in secondary general education regarding their attitudes as well as how they conceptualised technology.

de Vries played a pioneering and pivotal role in conceptualising and developing the PATT instrument used in the Netherlands, referred to as PATT-NL, as the first instrument specifically designed for this purpose. Results in the Netherlands were so significant that an international extension of the research was the logical next step (Ardies et al., 2013). In 1986, ten countries participated in pilot studies with the aim of increasing the reliability and validity of the PATT-NL instrument. It was Newton's Principia (1687) that impacted on the development of celestial mechanics, and the theory of gravitation only a century later (18th Century). In a similar vein, I believe that the PATT 1 workshop report (Raat and de Vries 1986) will be regarded as technology education's Principia in the centuries to come (see Figures 20.1 and 20.2). In 1987, 12 countries from across the world (e.g. Australia, India, Kenya, Mexico, Nigeria and European countries like Belgium, France, Italy, Poland and the UK) started using PATT-NL in survey studies with the aim to ascertain and describe the attitudes of students towards technology (De Klerk Wolters 1989a, 1989b; Dugger 1988).

The subsequent development of related international surveys led to workshops and PATT conferences, which have brought scholars involved in technology education together for almost four decades to provide a discussion platform for PATT-related issues, as well as an exchange of ideas and information to contribute towards the development of technology education (Jones et al., 2013; Kőycű & de Vries, 2016; Mottier et al., 1991; Volk & Yip, 1999). Initially, PATT studies aimed to investigate secondary school students' attitudes and the concepts they had of technology (De Klerk Wolters, 1988). PATT studies were a means of generating theoretical knowledge with practical implications

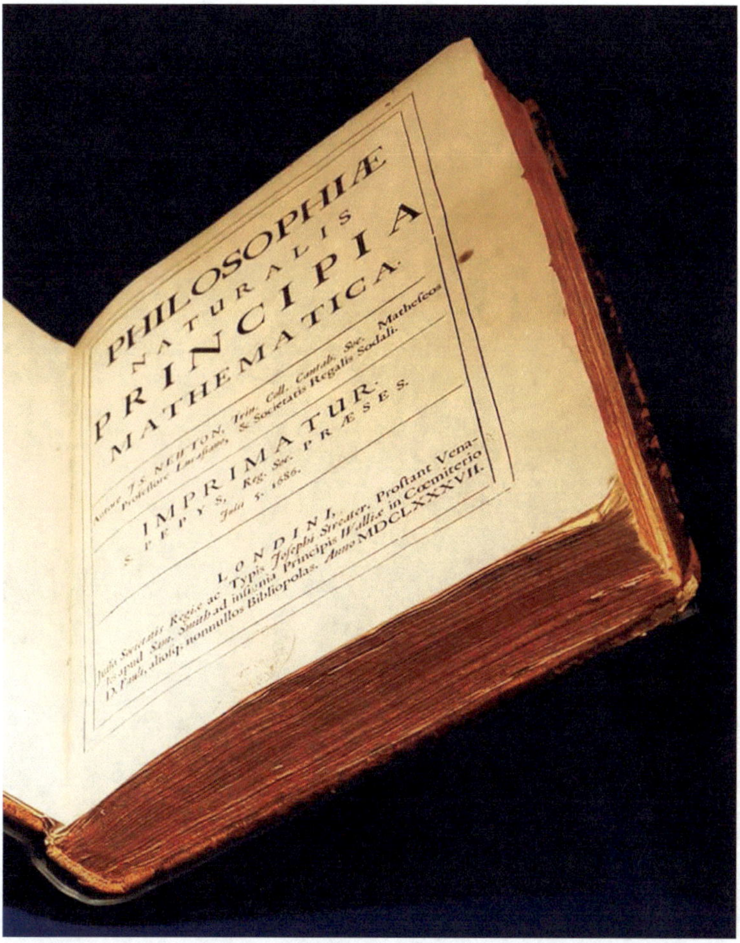

FIGURE 20.1 Newton's Principia (1687)
SOURCE: PHOTOGRAPH TAKEN BY WERNER ENGELBRECHT.

for the development and assessment of technology education and were not aimed primarily at curriculum content (De Klerk Wolters, 1989a), although they valued the link between research and curriculum development (Raat, 1988). PATT studies and the PATT Foundation, of which de Vries is the chair, have played an international leadership role in the maturation of the field of technology education for almost four decades. PATT studies and the 40 related PATT conferences, which de Vries has co-organised, have been instrumental in determining the research agenda and establishing an international research fraternity in technology education. PATT conferences have also become an international discussion forum for all aspects of technology education: curriculum development, research, teacher education, assessment and pedagogical

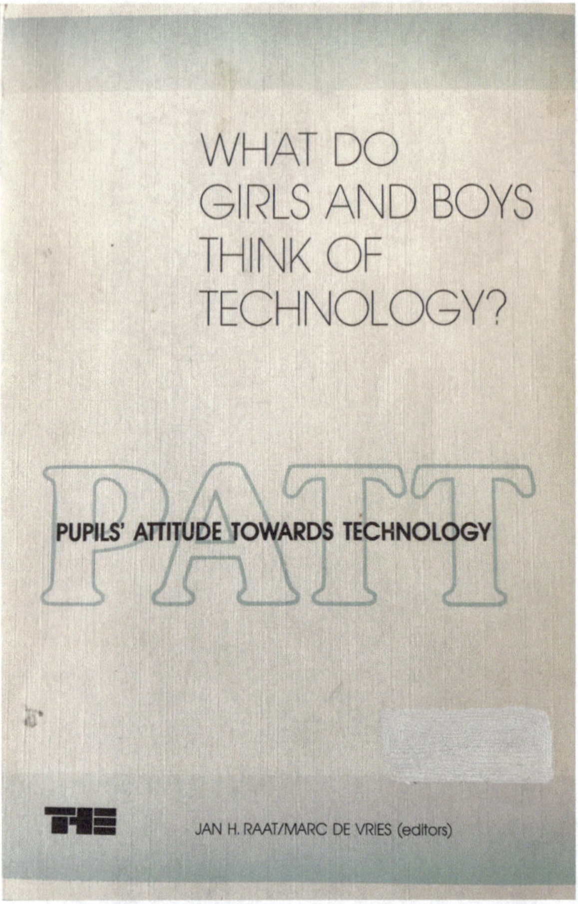

FIGURE 20.2 The PATT 1 workshop report (1986)
SOURCE: PHOTOGRAPH TAKEN BY THE AUTHOR.

issues in schools and higher education institutions (HEIs) (Jones et al., 2013; Mottier et al., 1991).

de Vries' involvement in PATT-NL has not only given rise to PATT conferences but also to a top-tier journal in the field, ITDE, which was established in 1990, and of which de Vries has been editor-in-chief since 1999. ITDE has been the most productive journal in the field over the period 2006–2013 (Williams, 2016). Under de Vries' editorship, ITDE's status has advanced to Category A with an impact factor of 2,177 in 2022, covering a vast range of research topics. ITDE is an important platform for the exchange of research results in the field of technology education and has a significant influence. de Vries was the most influential and prolific author in ITDE for the period 2000–2018. Delft University of Technology to which de Vries is affiliated, was the most prolific institution in the ITDE. Furthermore, the main research themes of ITDE refer to problem-solving, technological literacy, project-based learning, primary education, cognition, community of practice, engineering design and education, STEM education, computer innovation performance, collaborative learning and curriculum development (Xu et al., 2020).

PATT conferences, because of their frequency, are the most productive source of research papers in the field of technology education (Williams, 2016). The number and ranking of classical PATT studies focusing on students' views of technology at these conferences have however declined over time in favour of technological literacy, which is still the most common category of papers presented (Volk & Yip, 1999; Williams, 2013, 2016). Hence, it may seem as if research questions concerning the PATT survey in its current form have been exhausted and that there is a need to fundamentally rethink and revitalise PATT studies. In my view, the decline in the use of the PATT survey in contemporary research may be ascribed to the maturation of technology education over the past three decades (de Vries, 2018). Initially, there was a substantial number of pilot and survey studies using the mainstream PATT instruments in countries around the world to ascertain students' ideas when entering technology education (Kőycű & de Vries, 2016), which provided researchers with a good understanding of students' ideas when entering technology education (Ankiewicz 2018a). To better understand what has been measured by the various instruments it is necessary to analyse the concept of 'attitudes'.

1.3 *The Traditional/Psychological Approach to Attitudes*

Attitude is a broad concept with different definitions and interpretations, as well as controversies regarding the dimensionality of attitudes, with various models comprising one to three dimensions (Ankiewicz et al., 2001; Ardies et al., 2013). The traditional approach (also known as the psychological approach) is that attitudes have an integrated three-dimensional nature, consisting of cognitive, affective and behavioural components (Breckler, 1984; Fishbein & Ajzen, 1973; Ostrom, 1969).

The cognitive component of attitudes includes a person's ideas or opinions that express the relationship between situational and attitudinal objects (Gagné, 1977). Statements that reflect a person's perception and knowledge of the attitudinal object are part of the cognitive component (Corsini & Ozaki, 1984). The affective component refers to a person's 'feeling' or emotion concerning an attitudinal object (Heaven, 1982). The behavioural component includes a person's predisposition or readiness for action, as well as his or her actions concerning the 'behavioural object' (Ankiewicz et al., 2001; Gagné, 1977). For Bagozzi and Burnkrant (1979) attitude is primarily the interplay of affect and cognition, with the behavioural tendency as a secondary consequence (Ankiewicz et al., 2001; Van Rensburg et al., 1999).

According to the traditional approach, an attitude towards a concept such as technology thus is the person's collection of beliefs about it (cognitive component) and associated episodes linked with emotional reactions (affective component). The stimulation of these reactions results in decisions to engage in behaviour (behavioural component), such as choosing to take a technology course, to read about technological matters or to adopt a technology-related hobby (Ankiewicz et al., 2001; White, 1988). Scholars in technology education often acknowledge, either implicitly or explicitly, the traditional approach to attitudes (De Klerk Wolters, 1988, 1989a; Metsärinne & Kallio, 2015; Rohaan et al., 2010; Tseng et al., 2013; Volk & Yip, 1999). A possible means to fundamentally rethink and revitalise PATT studies is to identify which of the components of the traditional approach to attitudes have been measured by the various instruments and then to focus on those components which may have received less or no attention. The components of attitudes that are measured by the mainstream PATT instruments, namely PATT-NL, PATT-USA and PATT-SQ will be identified in the next section.

2 Instruments for Measuring Students' Attitudes toward Technology

2.1 *The Mainstream PATT Instruments*

De Klerk Wolters (1988, p. 41), who was involved in the design of PATT-NL, described attitudes as "a certain negative or positive feeling towards technology based on certain knowledge of and ideas about technology that may lead to a certain behaviour with reference to technology". In this context, attitude was used as a collective term for someone's affinity, behaviour and conceptualisation in relation to technology (Rohaan et al., 2010), in accordance with the traditional approach of attitudes (Ankiewicz, 2019a; De Klerk Wolters, 1989a).

PATT-NL consisted of two questionnaires, measuring the affective component and the cognitive component. The attitude questionnaire comprised Likert-type items containing a five-point response format, with six subscales measuring affective components of attitude such as interest, gender, consequences, difficulty, curriculum and careers in technology (Ankiewicz, 2019a; De Klerk Wolters, 1988; Rennie & Jarvis, 1995; Rohaan et al., 2010; Van Rensburg et al., 1999). Based on his own research and parallel international research, de Vries (1988) defined five different dimensions, namely interest in technology; aspirations for a technological career; perceived consequences of technology; perceived difficulty of technology; and students' perception of technology as a subject suitable for both genders. These dimensions of attitude served as the basis for the six subscales of the affective component (Ardies et al., 2015; De Klerk Wolters, 1988).

One of the aims of technology education is the development of a positive concept of technology; therefore, students' concepts have always been an important element in PATT studies (de Vries, 2005). The concept questionnaire with four subscales measuring the cognitive or knowledge component of attitude towards technology, based on the five general characteristics of technology which were identified by de Vries in the PATT 1 Workshop report (Ankiewicz, 2019a; Bame & Dugger, 1989; Becker & Maunsaiyat, 2002; De Klerk Wolters, 1989a, 1989c; de Vries, 1992; Jeffrey, 1993; Raat & de Vries, 1986; Rennie & Jarvis, 1995; Rohaan et al., 2010; Van Rensburg et al., 1999). Thus, de Vries' dimensions of attitude as well as his general characteristics of technology laid the conceptual foundation for the affective and the cognitive components respectively of attitudinal studies.

In its early form, PATT-NL included an essay (qualitative) section (Luckay & Collier-Reed, 2014). This read: *Technology can mean different things to different people. When you read the word 'technology' what comes into your mind?* to ascertain students' cognitive views of technology (Ankiewicz, 2018a, 2019a).

The original PATT-NL was translated and modified by Bame et al. (1993) for use in the USA (Ankiewicz, 2018a, 2019a; Volk & Yip, 1999). PATT-USA was a one-page instrument consisting of four parts. The first was a short, written description of technology; then 11 questions to gather demographic data and information about the technological climate of students' homes; 58 statements to assess students' attitudes towards technology (affective component); and 31 statements to assess students' concept of technology (cognitive component). The PATT-NL essay question was replaced with a brief statement of what the students thought technology was (Ankiewicz, 2018a; Bame & Dugger, 1989; Boser et al., 1998; De Klerk Wolters, 1988, 1989a).

The attitude questionnaire (affective component) of the PATT-USA instrument was recently reconstructed and revalidated by Ardies et al. (2013) (Ankiewicz, 2018a, 2019a). This resulted in the shorter PATT-SQ instrument with six sub-factors and 24 items of attitude towards technology to address the issue of length with PATT-NL and PATT-USA (Ardies et al., 2013; Volk & Yip, 1999).

PATT-NL and PATT-USA, in accordance with the traditional approach to attitudes, only ascertained students' technological concepts (cognitive component) and attitudes (affective component) as crucial prerequisites for technological activities (behavioural component) (Ankiewicz et al., 2001; De Klerk Wolters, 1988; Van Rensburg et al., 1999). PATT-SQ, as well as its translated, Swedish version, labelled PATT-SQ-SE (Ankiewicz, 2019a; Svenningsson, 2024; Svenningsson et al., 2018, 2022) and Slovenian version, PATT-SQ-SI (Rupnik &

Avsec, 2019), only ascertained students' attitudes (affective component). The three mainstream PATT instruments did not assess the behavioural component of attitude (cf. Tzeng & Yu, 2023).

Several related instruments, which do not seem fundamentally different from the initial mainstream PATT studies, were derived from the mainstream PATT-USA. Subsequently, PATT-HK, PATT-Taiwan, PATT-Turkey, PATT-ELEM for elementary school students, PATT-G4 for fourth graders (Papadopoulos, 2021) and the PATT-STEM instrument for measuring upper secondary school students' attitudes towards and concepts of engineering as part of Science, Technology, Engineering and Mathematics (STEM) education followed suit. The Technology Attitude and Concept Scale (TACS-Thai) and TEC-Lab questionnaire were developed similarly (Ankiewicz, 2019a).

2.2 *New and Non-Related Instruments to PATT Studies*

New instruments, which were non-related to the PATT studies, such as the Attitudinal Technology Profile (ATP), the Test of Technical Attitudes (TTA), The Finnish National Board of Education (FNBE) attitude test, the online AFT-Switzerland and Human Being and Technology (HBT) questionnaires were also developed (Ankiewicz, 2019a).

To resolve some of the contextual and formulation problems experienced with PATT-USA in South Africa, Van Rensburg et al. (1999) designed the ATP questionnaire to be used in the lower secondary school (ages 13 to 14). In Part A of this instrument, students were familiarised with the construct of technological products to avoid misconceptions. In Part B, 24 items were included on a five-point Likert-type scale to assess students' attitudinal technology profile (Ankiewicz, 2018a, 2019a; Ankiewicz et al., 2001). The items were designed and formulated as descriptive propositions linked to the affective components of the content of technology and attitude. By using descriptive propositions, it was also possible to integrate the affective component of attitude to some extent with the behavioural component (only students' readiness for action) (Ankiewicz, 2018a; Ankiewicz et al., 2001; Van Rensburg et al., 1999).

The HBT questionnaire was devised to examine students' learning related to the HBT cross-curricular theme among grade 9 Finnish students. It was divided into three sections, namely questions on students' knowledge about technology (cognitive component), their attitudes towards technology (affective component) and their activity know-how (behavioural component) (Järvinen & Rasinen, 2015). The HBT questionnaire appears to be the first instrument to include a section that measures students' actions or technological activities (thus the behavioural component) directly as part of the behavioural component of attitudes, and not only students' readiness for action as with the various mainstream PATT instruments. This was an extension of the ATP questionnaire, which measured the readiness for action as part of the behavioural component to some extent (Ankiewicz, 2019a; Ankiewicz et al., 2001). The discussion turns now from a psychological approach to attitudes to a philosophical approach.

3 A Philosophical Approach to Attitudes

Two decades ago, de Vries introduced me to Mitcham's (1994) fourfold philosophical framework according to which technology is manifested as object, knowledge, activity and volition. Technological knowledge and volition, with their origin within human beings, give rise to technological activities expressed as concrete technological objects. Based on his philosophical disposition and insight we superimposed the four components of general philosophy, namely ontology, epistemology, methodology and volition onto the four modes of manifestation of technology respectively (Ankiewicz et al., 2006; de Vries, 2017), as illustrated in Figure 2. By superimposing general philosophy onto Mitcham's framework was a 'philosophy modulation' of Mitcham's framework – a construct which I explicitly coined in Harvey and Ankiewicz (2023) for the first time.

Mitcham's (1994) philosophical framework of technology yielded fresh insights into the perceptions and attitudes of students towards technology and instruments for measuring these. The traditional approach to attitudes resembles Mitcham's philosophical framework (refer to Figure 20.3) in which technological knowledge (epistemology) and volition are prerequisites for technological activities (methodology), which result in technological objects (ontology). Thus, de Vries and his co-workers' departure from a psychological

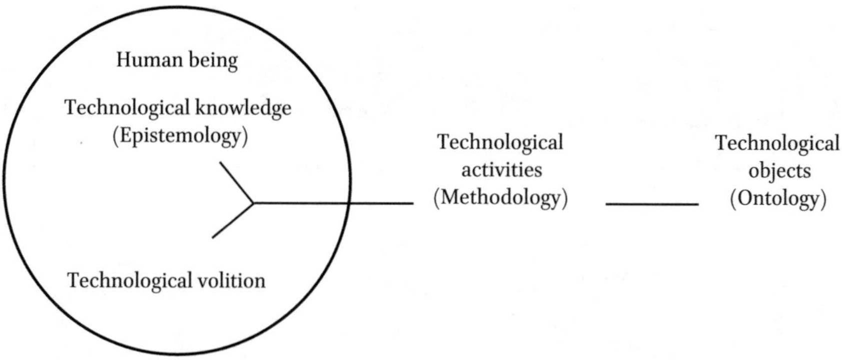

FIGURE 20.3 A 'philosophy modulation' of Mitcham's framework (Mitcham 1994:160, as modulated by Ankiewicz, De Swardt and de Vries 2006)

approach before Mitcham's conception of his philosophical framework was visionary, and most probably aligned with de Vries' aptitude for philosophy (Ankiewicz, 2019b).

Viewed from a philosophical perspective, PATT-NL and its derivatives (i.e., PATT-USA and PATT-SQ) ascertained students' technological concepts (epistemology; cognitive component) and attitudes (volition; affective component) as crucial prerequisites for technological activities (methodology; behavioural component) that would result in technological objects (ontology) (Ankiewicz et al., 2001; Van Rensburg et al., 1999). However, these instruments did not ascertain the behavioural component of students' attitudes. Except for the HBT questionnaire the closest that other instruments came to ascertaining the behavioural component (methodology or activities) was to measure readiness for action (e.g. the ATP questionnaire) (Ankiewicz, 2019b).

Figure 20.4 – as a 'philosophy modulation' of attitudes – depicts the superposition of technology as knowledge (epistemology), volition, and activity (methodology), according to Mitcham's (1994) fourfold philosophical framework (Figure 20.3) onto the cognitive, affective and behavioural components of attitudes, respectively according to the traditional approach (indicated with dotted lines). A general empirical finding from the PATT studies (Ankiewicz, 2018a) was that students' concepts of technology appears to influence affect and not the other way around (refer to the vertical arrow pointing from the cognitive to the affective component in Figure 3) (De Klerk Wolters, 1989c; Rohaan et al., 2010).

As a result of the behavioural component being neglected, its relationship with the other two components of attitudes remains unknown. Considering the 'philosophy modulation' of attitudes, PATT studies have also provided

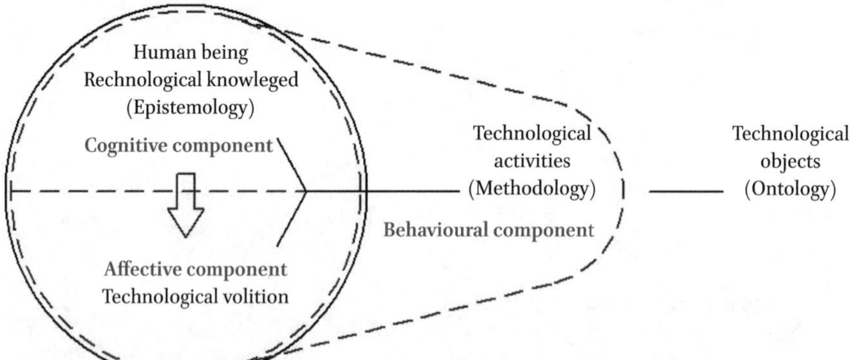

FIGURE 20.4 A 'philosophy modulation' of the traditional approach to attitudes (Ankiewicz 2019b)

some insights into the relationship between technological knowledge (epistemology) and volition, both having their origin within human beings. However, the relationship between them and the other two components that manifest outside human beings, i.e., activities (methodology) and objects (ontology), is also still unknown (Ankiewicz, 2019b).

While being cognisant of the neglected behavioural component of attitudes, Svenningsson et al. (2022) explored the relationships between the cognitive, affective and potential behavioural components of students' attitudes in a Swedish context by using PATT-SQ-SE. In addition to the empirical finding from the PATT studies that students' concepts of technology appear to influence their affect (Ankiewicz, 2018a), Svenningsson et al. (2022) identified relationships among the remaining attitude components, however, without being able to discern the directions of the relationships. Unlike the HBT questionnaire, Svenningsson et al. (2022) have only studied behavioural intention – and not the behaviour itself – which is the immediate motivational factor for behaviour itself.

4 Conclusion

Historically, de Vries laid the conceptual foundation for attitude studies from a psychological view, specifically for the mainstream PATT instruments and their derivatives. Stemming from his initial interest in attitude studies, PATT conferences are still being held almost four decades later, covering a vast range of research topics now. His prolific contributions as an author to ITDE, as well as his editorship, have been instrumental in establishing and maintaining a top-tier journal, which one should expect for a mature field. Furthermore, his aptitude for the philosophy of technology has empowered scholars to pursue philosophical studies, specifically in the field of attitude studies.

From a psychological view, the mainstream PATT-NL instrument and its derivatives (i.e., PATT-USA and PATT-SQ) were aligned with the traditional approach to attitudes. The mainstream PATT instruments ascertained students' technological concepts (cognitive component) and attitudes (affective component) as crucial prerequisites for technological activities (behavioural component) (Ankiewicz et al., 2001; Van Rensburg et al., 1999). However, these instruments did not ascertain the behavioural component of students' attitudes. Except for the HBT questionnaire, the closest that other instruments came to ascertaining the behavioural component was to measure readiness for action (e.g. the ATP questionnaire). From a philosophical view – which was inspired by de Vries' philosophical aptitude – the 'philosophy modulation' of attitudes also reinforced the missing behavioural component of students' attitudes. A shift from a psychological to a philosophical approach to attitudes might also strengthen the cognitive component of attitudes which has disappeared gradually after PATT-NL and PATT-USA, as the traditional approach to attitudes has vanished in later iterations of PATT studies in exchange for a more general conception of attitudes (Svenningsson, 2024).

To fundamentally rethink and revitalise PATT studies, students' attitudes towards technology should perhaps in future be ascertained more innovatively, imaginatively and holistically by also focusing on the behavioural component as one of the three components of attitudes. If the behavioural component of students' attitudes were to be ascertained in future, the results might lead to insights into the relationship between the components or manifestations inside human beings and those outside them. This might also better support more holistic teaching and learning practices in technology education in which the integrated three-dimensional nature of students' attitudes towards technology is emphasised in future technology education curricula.

References

Ankiewicz, P. (2018a). Perceptions and attitudes of pupils toward technology. In M. J. de Vries (Ed.), *Handbook of technology education, Springer International Handbooks of Education.* Switzerland: Springer.

Ankiewicz, P. (2018b). Rethinking Pupils' Attitudes Towards Technology (PATT) studies. In *PATT 36 conference proceedings*, (pp. 1–8). Athlone, Ireland.

Ankiewicz, P. (2019a). Perceptions and attitudes of pupils towards technology: In search of a rigorous theoretical framework. *International Journal of Technology and Design Education.* *29*(1), 37–56. https://doi.org/10.1007/s10798-017-9434-z.

Ankiewicz, P. (2019b). Alignment of the traditional approach to perceptions and attitudes with Mitcham's philosophical framework of technology. *International Journal of Technology and Design Education.* *29*(2), 329–340. https://doi.org/10.1007/s10798-018-9443-6.

Ankiewicz, P., De Swardt, E., & de Vries, M. (2006). Some implications of the philosophy of technology for science, technology and society (STS) studies. *International Journal of Technology and Design Education, 16*(2), 117–141.

Ankiewicz, P., Van Rensburg, S., & Myburgh, C. (2001). Assessing the attitudinal profile of South African learners: A pilot study. *International Journal of Technology and Design Education, 11*(2), 93–109. https://doi.org/10.1023/A:1011210013642.

Ardies, J., De Maeyer, S., & Gijbels, D. (2013). Reconstructing the pupils' attitude towards technology-survey. *Design and Technology Education: An International Journal, 18*(1), 8–19.

Ardies, J., De Maeyer, S., Gijbels, D., & Van Keulen, H. (2015). Students' attitudes towards technology. *International Journal of Technology and Design Education, 25*(1), 43–65.

Bagozzi, R. P., & Burnkrant, R. E. (1979). Attitude organization and the attitude-behaviour relationship. *Journal of Personality and Social Psychology, 37*, 913–929.

Bame, E.A., & Dugger, W.E. (1989). Pupils' attitude towards technology: PATT-USA. In *PATT 4 conference proceedings*, (pp. 309–319). Eindhoven, The Netherlands.

Bame, E., Dugger, W., de Vries, M., & McBee, J. (1993). Pupils' attitudes toward technology - PATT-USA. *Journal of Technology Studies, 19*(1), 40–48.

Becker, K. H., & Maunsaiyat, S. (2002). Thai students' attitudes and concepts of technology. *Journal of Technology Education*, 6–19.

Boser, R.A., Palmer, J.D., & Daugherty, M.K. (1998). Students' attitudes toward technology in selected technology education programmes. *Journal of Technology Education, 10*(1), 4–19.

Breckler, S. J. (1984). Empirical validation of affect, behaviour, and cognition as distinct components of attitude. *Journal of Personality and Social Psychology, 47*(6), 1191–1205.

Buckley, J., Adams, L., Aribilola, I., Arshad, I., Azeem, M., Bracken, L., ... , & Zhang, L. (2022). An assessment of the transparency of contemporary technology education research employing interview-based methodologies. *International Journal of Technology and Design Education, 32*(4), 1963–1982.

Corsini, R. J., & Ozaki, B. D. (Eds.). (1984). *Encyclopaedia of psychology*, Vol. 1. New York: John Wiley.

De Klerk Wolters, F. (1988). PATT research in 1987/88. In *PATT 3 conference proceedings*, (pp. 39–46). Eindhoven, The Netherlands.

De Klerk Wolters, F. (1989a). The PATT-project, An overview of an international project in technology education. In *PATT 4 conference proceedings* (pp. 290–308). Eindhoven, The Netherlands.

De Klerk Wolters, F. (1989b). Summary of theme 2. In *PATT 4 conference proceedings*, (pp. 331–334). Eindhoven, The Netherlands.

De Klerk Wolters, F. (1989c). A PATT study among 10 to 12-year-olds. In *PATT 4 conference proceedings*, (pp. 324–330). Eindhoven, The Netherlands.

de Vries, M. (1988). Technology in physics education. In *PATT 3 conference proceedings* (pp. 182–188). Eindhoven, The Netherlands.

de Vries, M. J. (1992). Dutch technology education developments: A comment. *International Journal of Technology and Design Education, 2*(3), 58–60.

De Vries, M. J. (2005). The nature of technological knowledge: Phidosophical reflections and educational consequences. *International Journal of Technology and Design Education, 15*(2), 149–154.

de Vries, M. J. (2017). Philosophy as critique. In Williams, P. J. & K. Stables (Eds.), *Critique in design and technology education.* Dordrecht, NL: Springer. https://doi.org/10.1007/978-981-10-3106-9.

de Vries, M.J. (2018). Preface. In M. J. de Vries (Ed.), *Handbook of technology education, Springer International Handbooks of Education*. Switzerland: Springer.

Dugger, WE. (1988). Technology – The discipline. In *PATT 3 conference proceedings*, (pp. 255–271). Eindhoven, The Netherlands.

Fishbein, M., & Ajzen, I. (1973). Attitudes and opinions. *Annual Review of Psychology, 24*, 487–544.

Gagné, R. M. (1977). *The conditions of learning*. New York: Holt, Rinehart & Wilson.

Harvey, N., & Ankiewicz, P. (2023). A principle-based philosophical framework for fashion design praxis and education in new design landscapes. *International Journal of Technology and Design Education, 33*(5), 1883–1900. https://doi.org/10.1007/s10798-022-09794-7.

Heaven, P. C. C. (1982). *Houdings*. Pretoria: Academia (Afrikaans).

Järvinen, E. M., & Rasinen, A. (2015). Implementing technology education in Finnish general education schools: Studying the cross-curricular theme 'Human being and technology'. *International Journal of Technology and Design Education, 25*(1), 67–84.

Jeffrey, T. J. (1993). Validation of a technology attitude scale for use by American teachers at the middle school level. In *PATT 6 conference proceedings*, (pp. 357–364). Breukelen, The Netherlands.

Jones, A., Bunting, C., & De Vries, M. J. (2013). The developing field of technology education: A review to look forward. *International Journal of Technology and Design Education, 23*(2), 191–212.

Kőycű, Ü., & De Vries, M. J. (2016). What preconceptions and attitudes about engineering are prevalent amongst upper secondary school pupils? An international study. *International Journal of Technology and Design Education, 26*(2), 243–258.

Luckay, M. B., & Collier-Reed, B. I. (2014). An instrument to determine the technological literacy levels of upper secondary school students. *International Journal of Technology and Design Education, 24*(3), 261–273.

Metsärinne, M., & Kallio, M. (2015). How are students' attitudes related to learning outcomes? *International Journal of Technology and Design Education, 26*(3), 353–371. https://doi.org/10.1007/s10798-015-9317-0.

Mitcham, C. (1994). *Thinking through technology.* Chicago: The University of Chicago Press.

Mottier, I., Raat, J. H., & De Vries, M. J. (Eds.). (1991). Report on the PATT 5 conference. In *PATT 5 conference proceedings* (Vol. 1, pp. 9–13). Eindhoven, The Netherlands.

Ostrom, T. M. (1969). The relationship between the affective, behavioral, and cognitive components of attitude. *Journal of Experimental Social Psychology, 5*(1), 12–13.

Papadopoulos, J. (2021). *Pupils' Attitudes Towards Technology–Grade 4 (PATT-G4): Establishing a valid and reliable instrument.* (Doctoral thesis, Virginia Tech).

Raat, J.H. (1988). Introduction to the theme of the conference. In *PATT 3 conference proceedings*, (pp. 30–38). Eindhoven, The Netherlands.

Raat, J. H., & De Vries, M. J. (1986). What do girls and boys think of technology? In *PATT workshop report*, (pp. 33–36). Eindhoven, The Netherlands.

Rennie, L., & Jarvis, T. (1995). Three approaches to measuring children's perceptions of technology. *International Journal of Science Education, 17*(6), 755–774.

Rohaan, E. J., Taconis, R., & Jochems, W. M. (2010). Reviewing the relations between teachers' knowledge and pupils' attitude in the field of primary technology education. *International Journal of Technology and Design Education, 20*(1), 15–26.

Rupnik, D., & Avsec, S. (2019). The relationship between student attitudes towards technology and technological literacy. *World Transactions on Engineering and Technology Education, 17*(1), 48–53.

Strimel, G. J., Kim, E., Grubbs, M. E., & Huffman, T. J. (2020). A meta-synthesis of primary and secondary student design cognition research. *International Journal of Technology and Design Education, 30*(2), 243–274.

Svenningsson, J. (2024). Having an attitude toward technology: Rethinking PATT studies from a theoretical perspective to study students' attitudes toward technology (Doctoral thesis, Linköping University).

Svenningsson, J., Höst, G., Hultén, M., & Hallström, J. (2022). Students' attitudes toward technology: exploring the relationship among affective, cognitive and behavioral components of the attitude construct. *International Journal of Technology and Design Education, 32*(3), 1531–1551. https://doi.org/10.1007/s10798-021-09657-7.

Svenningsson, J., Hultén, M., & Hallström, J. (2018). Understanding attitude measurement: exploring meaning and use of the PATT short questionnaire. *International Journal of Technology and Design Education, 28*(1), 67–83. https://doi.org/10.1007/s10798-016-9392-x.

Tseng, K. H., Chang, C. C., Lou, S. J., & Chen, W. P. (2013). Attitudes towards science, technology, engineering and mathematics (STEM) in a project-based learning (PjBL) environment. *International Journal of Technology and Design Education, 23*(1), 87–102.

Tzeng, S. Y., & Yu, K. C. (2023). Rethinking measures of attitude toward technology in technology education. In S. Davies, M. McLain, A. Hardy & D. Morrison-Love (Eds.), *The 40th International Pupils' Attitudes Towards Technology Conference Proceedings 2023, 31 October–3 November, Liverpool John Moores University, Liverpool, UK*, 809–820. https://doi.org/10.24377/PATT40.2023.

Van Rensburg, S., Ankiewicz, P., & Myburgh, C. (1999). Assessing South African learners' attitudes towards technology by using the PATT (Pupils' Attitudes Towards Technology) questionnaire. *International Journal of Technology and Design Education, 9*(2), 137–151.

Volk, K. S., & Yip, W. M. (1999). Gender and technology in Hong Kong: A study of pupils' attitudes toward technology. *International Journal of Technology and Design Education*, 9(1), 57–71.

White, R. T. (1988). *Learning science*. Basil Blackwell.

Williams, P. J. (2013). Research in technology education: Looking back to move forward. *International Journal of Technology and Design Education, 23*(1), 1–9.

Williams, P. J. (2016). Research in technology education: Looking back to move forward... again. *International Journal of Technology and Design Education, 26*(2), 149–157.

Xu, M., Williams, P. J., Gu, J., & Zhang, H. (2020). Hotspots and trends of technology education in the International Journal of Technology and Design Education: 2000–2018. *International Journal of Technology and Design Education, 30*, 207–224.

Yu, K.-C., Lin, K.-Y., Han, F.-N., & Hsu, I. Y. (2012). A model of junior high school students' attitudes toward technology. *International Journal of Technology and Design Education, 22*(4), 423–436.

CHAPTER 21

Design Learning at the Limits of Pedagogy

Richard Kimbell

1 A Personal Context

As a pupil in school in the 1960s I had been provided with a rich diet of craft courses and was proficient in woodwork, metalwork and technical drawing. I was not only proficient, but was also confident in my ability. But when I entered Goldsmiths College I was taken aback to be faced with the challenge of doing more than just making stuff. I had to imagine it in advance of its existence and manage that imagination through designing and modeling processes up to the point where my 'real' (making) skills could kick in.

The 1st year summer examination (1967) was a Road to Damascus experience. We were to design and make a museum/gallery portable/collapsible stool. Something to collect at the entrance and carry around to view the displays. There were 24 of us in the group and after a week of designing and 2 days of making all the work went on display. We all then migrated to the Union bar. But after an hour or so I felt compelled to go back to the display of our work and wander around it. The products were of course all different. Some were clever, some beautiful, some complicated, some interesting, or degrees of all of these. And I just stood there in amazement thinking 'where did all this stuff come from?' Obviously it came out of our heads … and for the first time that I can consciously recall, I began to glimpse the power of ideas.

When I left Goldsmiths College in 1970 and began my teacher career in Devon, I was a committed design teacher and found myself working with a terrific group of colleagues. But by then I understood that we were doing utterly different things. The tradition of the department was of high quality craftwork. We all had high expectations of learners, and standards of work were high. But we expected (and got) utterly different things from learners. Not just different in the sense of using different materials or making different products, but different in the sense that whilst my colleagues' priorities were primarily with the tool skills to be imparted through set class projects (in wood/ textiles/ metal /food), mine were with trying to find ways to liberate learners' imagination, so they too could create ideas and manage them into artefacts; to become capable and confident in thinking for themselves. In those ROSLA[1] years (1971–74) my colleagues and I spent a lot of time thinking – and arguing - about the

© RICHARD KIMBELL, 2025 | DOI:10.1163/9789004710948_021

contradictory pedagogies that we were espousing and equally about the contradictory epistemologies in which those pedagogies were rooted. From these debates I wrote my first book 'Design Education – the foundation years' that John Eggleston helped me to get published with Routledge in 1982, and chapter one was all about this conundrum. The chapter title was 'craft or design – what's the difference'. We all used the same tools and the same materials. We used the same workshops and had the same timetable framework, but the youngsters were learning quite different things.

I was aware, however, that there was one central commonality to it all. We – the teachers – were in charge. What the learners did was what we allowed /encouraged instructed them to do. The studio classroom workshop was our domain.

2 Coercive Pedagogy

The concept of pedagogy has ancient roots and is often described as pertaining to the theory of teaching and the study of teaching methods (Wang & Huang, 2018). This involves the teacher or "pedagogue" being responsible for adapting and transforming curriculum content and holding the central role in transforming learners' understandings (Dobson, 2012). In those debates with colleagues in the 1970s it bothered me that underlying the idea of pedagogy was that we (the teachers) were in effect saying to learners … 'You have to know this stuff … I know it and you don't, so just do as I tell you'.

As Delahunty and Kimbell (2021) outline

> inherent still in the concept of pedagogy is the teacher's central role and arguably hold on the learning process (McAuliffe, Hargreaves, Winter, & Chadwick, 2009).The teacher-centric nature of pedagogy brings to the fore the concept of Foucauldian power where the teacher is the holder and translator of knowledge for their students. Consequently, knowledge represents the values of those with the power and the ability to circulate them (Foucault, 1990; Levitt, 2008) which aligns with the shaping power that knowledge has on pedagogy.

One of the uncomforable features of STEM curricula emerges in this knowledge issue, and in particularly with science. Science, with its high status conceptual structure [science as knowledge] (Young & Muller, 2013), tends often to overpower any sense of science as a useful human pursuit [science as investigation]

(Appleton, 1995]. And this knowledge-centred view of science education emerged clearly in Osborne and Collins' report published by King's College London; *Pupils' and Parents' Views of the School Science Curriculum* (2000).

- Pupils found it hard to make the connections between school science and their everyday lives.
- For many topics, the value of the knowledge lay simply in its instrumental value for passing an exam.
- Teachers acknowledged that much of the pupils' dissatisfaction was legitimate but felt that they were the product of a content-dominated and overloaded curriculum (Osborne & Collins, 2000, p. 6).

Beyond the knowledge issue Bonawitz et al (2011) explored the shaping (limiting) power of pedagogy. They conducted an experiment in which preschool children viewed an adult demonstrating the functionality of a novel toy in three different conditions. Only one of those conditions would be classed a pedagogical demonstration. And the children in that 'demonstrated' group restricted their exploration of the toy to only those functions demonstrated by the adult. The other two groups tended to explore and uncover more un-demonstrated functions of the toy.

When Pink Floyd (Roger Waters) wrote *Another brick in the wall*, it was a nihilistic protest against what he saw as the rigid and abusive power of teachers. But you don't need to be a Pink Floyd fan to find coruscating critiques of the duality of pedagogy, and perhaps this is most forcibly expressed by Ranciere when he describes …

> the myth of pedagogy, the parable of a world divided into knowing minds and ignorant ones, ripe minds and immature ones, the capable and the incapable, the intelligent and the stupid. (Rancière, 1991, p. 5)

Ranciere's pedagogic myth sees the world divided in two, with 'explication' (from the knowing to the not-knowing) as the stultifying weapon of the pedagogue. And Ranciere's antagonism to this view of pedagogy leads him to conclude that the teacher's role should rather lie in ensuring the equality of teachers and learners (p. 35)

> The ignorant person will learn by himself what the master doesn't know if the master believes he can and obliges him to realise his capacity: a circle of power homologous to the circle of powerlessness that ties the student to the explicator of the Old Method. (p. 15)

The image that Ranciere was raging against was that of the teacher as the master and student as the apprentice; a remnant of values that evolved from the medieval craft guilds and later the industrial revolution (Blaschke & Hase, 2016). In the context of this chapter it is hard to avoid the obvious association with craft teaching and its instructional pedagogy.

Moreover we have all witnessed the inexorable process of neo-liberal reform in the UK and other industrialised nations. In education in England this has seen Local Education Authorities emasculated and their teacher advisor/support teams abolished – with the remnants housed within new privatised organisations. And ironically this trend happened alongside the centralised imposition of a new, nationalised, hegemonic curriculum along with its ludicrous assessment regime (DES, 1988) and its all-powerful police force OFSTED.[2] This unholy alliance (nationalising the curriculum whilst privatising [decimating] teachers support services) perhaps justifies the cynicism and despair that was so brutally expressed by Pink Floyd (Evans, 2004; Shore & Wright, 1999; Slaughter & Leslie, 1997). And teachers' despairing response in 1992/3 was a total national boycott particularly of the assessment processes.

Pedagogy cannot avoid being seen as part of this inexorable and problematic process when the language of teachers and teaching has adapted to accommodate it. 'Delivery' is the official watchword, with teachers as the postal service delivering a curriculum as if it is a known and finite thing, and in fact nothing more than another Amazon product on a delivery van. Direct from the Department of Education to your head – via a few classrooms and a whole load of explication (Department of Education, 2021).

3 Alternative Conception of Teaching and Learning

The emergence of a new view of learning depends upon what Ranciere saw as the 'equality' of teacher and learner, but what more recent writing describes as learner agency. Students are now expected to be critical thinkers with the abilities to access, collate, critique and synthesise knowledge and understanding, and inevitably this foregrounds the necessity for increased agency and autonomy in the teaching and learning process (Delahunty & Kimbell, 2021).

This has arisen naturally in the sphere of adult learning, where the label andragogy has gained traction. Sharifi et al. (2017) present a contextual differentiation between pedagogy, as children focused, and andragogy, as adult focused education that emphasises ownership of the learning process under the guidance of the teacher.

> Knowles discusses the idea that as learners mature their self-concept evolves from a mode of dependency to independency informed by societal roles and life experiences (Knowles, 1972; Knowles et al., 2005). In the model of andragogy the focus becomes more metacognitive in nature actualised in the negotiation of learning processes (Lee & McLoughlin, 2010; Luckin, 2010). (Delahunty & Kimbell 2021 p. 753)

But it is difficult to argue that age is the determining issue in any move towards a view of learning as self-determined. So perhaps it was inevitable that heutagogy has emerged, particularly in web and technology-mediated contexts, in the last 20 years. It is a theory concerned with self-determined learning, with control over the learning process placed on the student to make his/her own decisions on what and how to learn (McAuliffe et al., 2009). The articulation of heutagogy as distinct from pedagogy – or perhaps as a refinement of pedagogy – has deep resonance in the world of design learning.

3.1 *Design Learning*

I did not realise at the time that my struggles to articulate (in the 1970s) a view of design education as distinct from craft education was just one small part of a much bigger picture. That bigger picture was the emerging international realisation of the importance of learner autonomy, and my small part of that lay in trying to find teaching and learning strategies that would result in imaginative, skillful and thoughtful young designers without drowning them in endless explication. Such explication abounded at the time in the form of skills teaching and technological knowledge-banks. 'You will need these skills ... you need to know this ...'

In the world of design learning everything is changed, and not least the view of knowledge. It might well be that schools are premised on the idea that learners start from a state of NOT knowing, but gradually they grow through various degrees of half-knowing into a state of 'rather-more' knowing. In short – in a learning context - we are wrong (I think) to use the term 'knowledge' as though it represents a *state* of being, and we should rather think of 'knowing' as a process – as representing a state of growth – a state of emergent transition, as I argued in the DATA journal 'reflection' in 2010

> The importance of this for design is that our students are constantly getting into design tasks that require them to find out about things that are (at the outset of the task) unknown – or maybe half-known. Rather than saying that they can't do that sort of designing until they have learned the text-book – what we do is formulate a view of knowing that empowers

> learners to take action with provisional knowledge - and that encourages them to refine and deepen that knowledge in response to the demands of the task. So we have deliberately transposed the issue of 'knowing' stuff into the business of 'finding-out-about' stuff. (Kimbell, 2010)

As the DES pointed out thirty years ago ...

> The designer does not need to know all about everything so much as to know what to find out, what form the knowledge should take, and what depth of knowledge is required for a particular purpose. (DES, 1981, p. 5)

It follows of course that both learners and teachers must be able and willing to operate in this indeterminate zone of activity where hunch, half-knowledge and intuition are essential ingredients. And this is exactly the heutagogic challenge for learners and teachers alike. Hicks (1983) saw the challenge as one for teachers.

> Teaching facts is one thing; teaching pupils in such a way that they can apply facts is another, but providing learning opportunities which encourage pupils to use information naturally when handling uncertainty, in a manner which results in capability, is a challenge of a different kind. (Hicks, 1983, p. 1)

But it equally applies to learners. Anyone who has taught design will know how unsettling it is for learners when they find that their teacher does not simply supply all the direction that they think they need in order to achieve a design outcome. The heutagogic challenge for learners is to grow confidence and capability in their own decision-making. Not ignoring the teacher – but being prepared to pick and choose the moments to stand aside from their guidance.

The harsh reality of design learning is that it is very hard to initiate and to manage. It requires very special teachers because it throws out the window so many of the comfortable assumptions about what it means to be a teacher. A design learning teacher has to facilitate youngsters' learning without resorting to all the 'normal' practices of teaching. There are no right answers, just multiple possibilities, with some (perhaps) better than others. And the task is to help the learner find good (and safe) ways forward with their work without telling them what they are. They need to see that for themselves. So the teacher spends every minute of the day making decisions about how learners' work *might* move forward, and then every next minute thinking how to raise the issues appropriately so as to enable the learner to see those options and make good decisions. Guiding the work without instructing; steering the work without directing; probing learners' thinking without criticizing; advising without telling. It is an endless and exhausting challenge.

This process might be seen by some as pedagogic, but if it is, then the concept of pedagogy must be infinitely flexible. I argue that design learning – as described here - is closer to views of heutagogy, because at the heart of design learning lies the absolute requirement of learner autonomy.

3.2 *Learner Autonomy*

The opportunity to explore learners' autonomous decision-making arose through a project we in TERU[3] were asked to undertake for the Qualifications and Curriculum Authority (QCA) in 2004. Their interest was in promoting learners' creative performance in design & technology and they asked us to find ways of assessing such performance. So was born the project 'Assessing Design Innovation'.

There are two distinct kinds of learning going on simultaneously in a design task. First there is *procedural* learning; the 'what am I going to do next' kind of learning. This is all about learners managing themselves through the activity. And second there is *content/idea* learning; 'it will be like ... I want it to do ...' This kind of learning centres on the content of learners ideas and how they can be developed into a solution. (Kimbell & Stables, 2007). Procedural learning might be thought to be more important – since it is generaliseable to any task, whereas the specific ideas of a particular project may be irrelevant in the next project. But in both cases it is possible (though very rare) to see learners operating autonomously.

Within the ADI project we were deliberately prioritising creative thinking, so *ideas* and how learners can generate and develop them was central to our interests. Moreover, since we were attempting to develop an examination – with time limits – we chose to take away the procedural, project management functions from learners, leaving them to spend all their energies on creative exploration. Based on a 6 hour activity, we created a teacher script to walk learners through a process, and a folding paper portfolio in which all their work was recorded. But having outlined the task, we left learners completely free to respond to it with ideas of any kind. An example of the procedural framework was that we required learners to take photos of their work approximately every hour. The instruction was: "Please stop what you are doing for a moment. Whatever you are working on at this moment, please take a photo of it now". This happened six times throughout the life of the project and a photo-storyline was pasted (as it emerged) into the portfolios (Figure 21.1).

FIGURE 21.1 Photo storyline

The photo story line was one of a set of pedagogic devices that we used to promote the creation of the portfolios that captured all the learners' designing activities during the 6 hours of the the project took.

The whole A2 portfolio (Figure 21.2) progressively grew through the activity with notes, drawing, photos, and whilst it was entirely the learners' own work that was being captured, the procedural structure of the activity was our construction. We described it as choreography, since we were in effect choreographing learners' behaviour through their designing.

In terms of creative autonomy we described the process in the following terms.

> It's a bit like jazz with a rigid 12 or 16 bar rhythm. Within that tight structure, the most outlandish improvisation can be liberated. So too with our activity booklets and the teacher script. By taking away from students the need to think about how they will organise and present their work they are empowered to concentrate on the ideas that drive their designing. (Kimbell et al., 2004, p. 61; Kimbell & Stables, 2007, pp. 117–127)

I recognise that readers will see this as a strict example of teacher control, and the choreography was unquestionably pedagogic. But the interesting thing here is the effect it had on the other aspect of students' learning. Forgetting for a moment the procedural learning, what about the ideas and the content of learners work. Throughout the whole project we exercised NO control over the ideas that learners created, or on how they developed them. Our structure was there to help them to have and develop ideas, and to ensure that whatever they did, they left behind a good record of it.

We operated the project with (typically) age 15 learners in schools all over the country, from Northumberland to Cornwall and Wales, and we collected their reactions to the whole thing. Concerning their autonomy to respond to

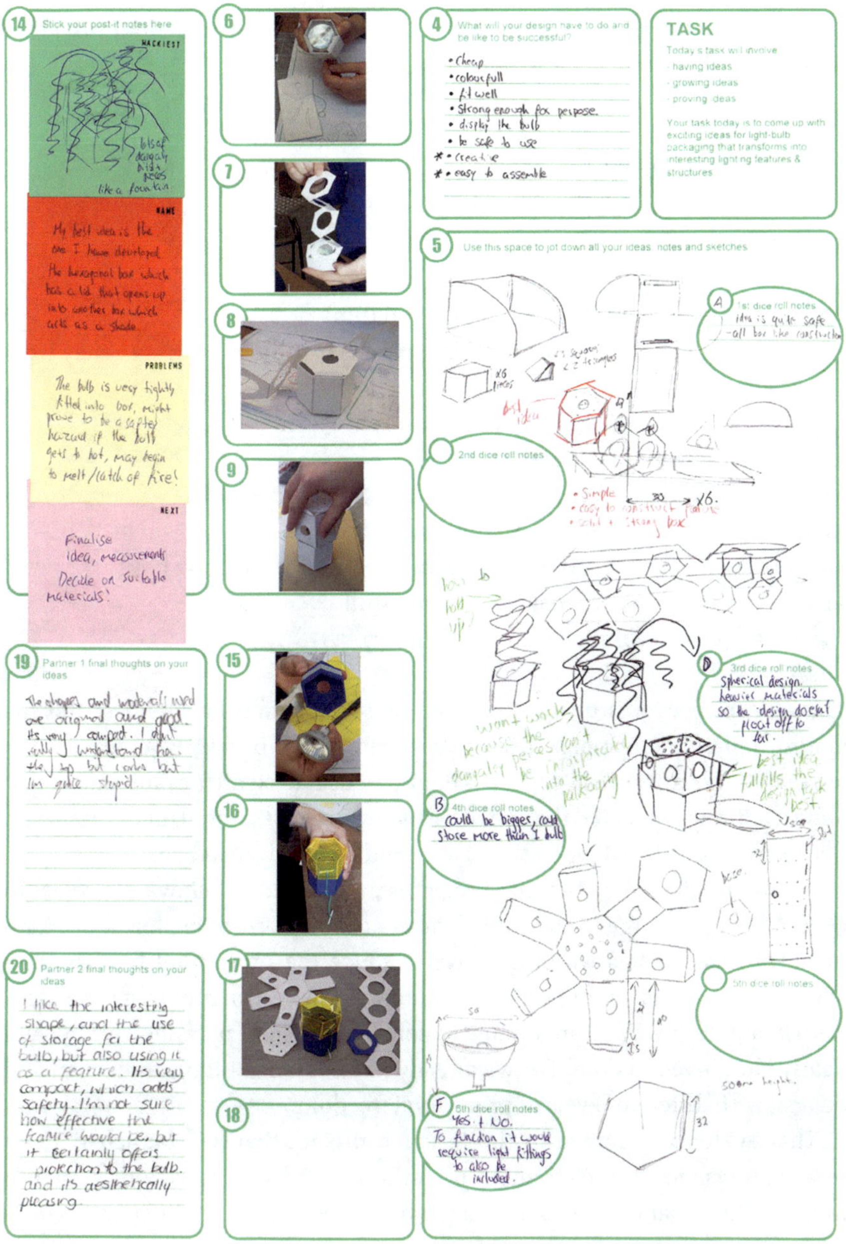

FIGURE 21.2 Student Portfolio

the design challenge in their own ways they were unequivocal about what they liked about their unusual freedom of action.

AB119 - Being able to use my own ideas how I wanted to
AB1117 - Its easier to understand when you try it out for yourself
AB216 - You can make things you didn't expect
AB2110 - Its fun designing
AB228 - You got to show your ability
AB2210 - Starting from scratch to make you think
AB314 - Using your own ideas ; making models instead of drawing - they work better
AB315 - It was fun to design something and actually make it
AB319 - Seeing your idea develop
AB324 - I realised that I could do more than I thought I could
AB411 - I liked getting my imagination go wild
AB4110 - It was a very good way of finding how many skills you have
AB4112 - I learned how to make a lot more stuff
(Learner database Kimbell et al., 2004, pp. 37–38)

This QCA project was being supported by several Awarding Bodies, and the almost immediate response from OCR (Oxford, Cambridge, RSA) was to create their 'Innovation Challenge', a 6 hour practical designing examination that almost entirely replicated the activity structure we had created for our research. The examination is still operational as a module at Advanced level.

At the conclusion of the ADI project, we proposed a 'follow-on' project to QCA in which we argued that the obvious next step was to explore what happens with learners as we progressively reduce the procedural framework – allowing more and more *procedural* autonomy whilst maintaining the *idea/content* autonomy. We were unable to convince them of its value, but I would dearly have loved to do it. For we already had evidence that learners can and do deal with different levels of procedural freedom.

That evidence emerged in the research project that had launched TERU in 1985. It was for the Assessment of Performance Unit (DES) with a brief to monitor the capability of 10,000 (age 15) learners. There is not space here to deal with the details of that project, but suffice to say that we had developed a set of activities with varying structures and with very different levels of procedural autonomy. We assessed all the work and sought to identify reasons for the differential performance of various sub-groups across the different test structures. In the context of learner autonomy we commented …

> There is a clear effect at work here … the tests at the opposite ends of this chart are complete opposites in their structure; the Modelling tests being the most active procedurally and the most loosely structured, while the Evaluating Products test are the most reflective and the most tightly structured. The combined effect on low-ability girls is enormous … the messages are not about the *content* of the tasks but about the ways in which tasks are *structured* … (Kimbell et al., 1991, p. 206) (see also Kimbell & Stables, 2007, Chapter 5)

There was an obvious need to pursue this matter; to explore how different kinds of procedural autonomy affect learners' growth towards capability. But we were unable to convince QCA to launch that project, and it remains research that desperately needs to be done.

4 Conclusion

The thrust of this chapter has been to argue that learner autonomy is completely central to the idea of design learning. All design learning experiences will involve teachers and learners grappling with the challenges of helping learners to make their own good decisions as they pursue their design tasks.

> The ultimate position, towards which both learners and teachers should strive, is learner autonomy. As this point is reached, learners are able to negotiate their tasks independently and pursue them imaginatively and rigorously without external direction, using resources appropriately. (Kimbell & Stables, 2007, p. 32)

If learners are to take responsibility for their own decisions then those learners need to be in a position not only to make those decisions but also to see them through to the point where their wisdom can be judged. Learners making mistakes is part of that process. As is learners, supported by their teachers, looking for ways to rectify the deleterious consequences of such mistakes.

But there is more to it than that, for there is no magic point at which this learner autonomy suddenly appears. To say (as we have above) that the *ultimate* position is to achieve this autonomy, shies away from the equally central truth that this autonomy starts at the outset. Learner autonomy is so central to design learning that as soon as the course of study begins, then so too must begin the process of learners taking responsibility. And it grows as learners take more and more responsibility for their own learning. But inevitably this requires that learners progressively take that decision-making away from teachers. This is a complicated process for learners and especially for teachers to navigate, not least because the very identity of teachers is rooted in their subject expertise. So teachers need to find ways to reassure learners that they understand this subtle handing-over process of decision-making. Because by far the most important feature of teachers' identity is not that they are *subject* experts but that they are *learning* experts.

> Children will only develop the power to think for themselves by thinking for themselves … The function of the teacher in a foundation course in design education is to steer children towards the goal of independent thought and action along a tortuous path of guided or supported freedom. (Kimbell, 1982, pp. 15–16)

At the time I wrote this, it was based more on personal conviction than on research evidence. I am pleased to report, however, that in the subsequent 42 years, the supporting evidence has been accumulating.

Notes

1 Raising of the School Leaving Age [from 15 to 16].
2 Office for Standards in Education: Department of Education.
3 Technology Education Research Unit: Goldsmiths University of London.

References

Appleton, K. (1995). Student teachers' confidence to teach science: Is more science knowledge necessary to improve self confidence? *International Journal of Science Education* 17(3) 357–369. https://doi.org/10.1080/0950069950170307

Association of Advisers in Design & Technical Studies. (1980). *Craft, Design & Technology: a Curriculum Statement.* https://www.stem.org.uk/resources/elibrary/resource/33683/craft-design-and-technology-curriculum-statement

Blaschke, L. M., & Hase, S. (2016). Heutagogy: A Holistic Framework for Creating Twenty-First-Century Self-determined Learners. In B. Gros, Kinshuk, & M. Maina (Eds.), *The Future of Uniquitous Learning* (pp. 25–40). Springer.

Bonawitz, E., Shafto, P., Gweon, H., Goodman, N. D., Spelke, E., & Schulz, L. (2011). The double-edged sword of pedagogy: Instruction limits spontaneous exploration and discovery. *Cognition, 120*, 322–330.

Bronowski, J. (1973). *The Ascent of Man.* London British Broadcasting Corporation.

Delahunty T., & Kimbell, R. (2021, June). (Re)Framing a philosophical and epistemological framework for teaching and learning in STEM: Emerging pedagogies for complexity. *British Educational Research Journal, 47*(3) , 742–769. https://doi.org/10.1002/berj.3706

Department of Education. (2021). *Delivery and Teaching Strategies: Statutory Guidance.* GOV.UK https://www.gov.uk/government/publications/relationships-education-relationships-and-sex-education-rse-and-health-education/delivery-and-teaching-strategiesGOV.UK

Department of Education and Science (DES). (1981). *Understanding design and technology.* Department of Education and Science, HMSO.

Department of Education and Science (DES). (1988). *National Curriculum: Task Group on Assessment and Testing. A report.* Department of Education and Science, HMSO.

Dobson, S. (2012). The Pedagogue as Translator in the Classroom. *Journal of Philosophy of Education, 46*(2), 271–286. doi:10.1111/j.1467-9752.2012.00852.x

Eggleston J, Pemberton A, Taberner D, Brough L, & Hall R. (1974). *You are a designer* Schools Council Design & Craft Education Project Edward Arnold London Schools Council Publications.

Evans, M. (2004). *Killing thinking: The death of the universities.* Continuum.

Foucault, M. (1990). *The will to knowledge* (Vol. 1.). Penguin Books.

Gweon, H., Pelton, H., Konopka, J. A., & Schulz, L. E. (2014). Sins of omission: Children selectively explore when teachers are under-informative. *Cognition, 132*(3), 335–341. doi:10.1016/j.cognition.2014.04.013

Gov.uk. (2020, August). *Creative industries statistics.* https://www.gov.uk/government/statistics/creative-industries-statistics-august-2020

Hicks, G. (1983). "Another step forward for design and technology". *APU Newsletter, 4.* DES.

Kimbell, R. (1982). *Design Education; the foundation years.* London Routledge Kegan Paul.

Kimbell, R. (2010). "Wrong ... but right enough" The Reflection piece in *Design & Technology Education: An International Journal. 15,* 7–9. Design & Technology Association.

Kimbell, R., Stables, K., Wheeler. T., Wozniak, A., & Kelly, V. (1991). *The Assessment of Performance in Design and Technology.* School Examinations and Assessment Council (SEAC) D/010/B/91.

Kimbell, R., Bain, J., Miller, S., Stables, K., Wheeler, T., & Wright. R. (2004). *Assessing Design Innovation*. Goldsmiths University of London.

Kimbell, R., & Stables, K. (2007). *Researching Design Learning* Dordrecht Netherlands Springer

Knowles, M. S. (1972). Innovations in Teaching Styles and Approaches Based upon Adult Learning. *Journal of Education for Social Work, 8*(2), 32–39. doi:10.1080/00220612.1972.10671913

Knowles, M. S., Holton, E. F., & Swanson, R. A. (2005). *The adult learner: the definitive classic in adult education and human resource development* (6th ed.). Burlington, M. A. [etc.]. Elsevier Butterworth Heinemann.

Lee, M. J. W., & McLoughlin, C. (2010). *Learner-Generated Contexts: A Framework to Support the Effective Use of Technology for Learning* (pp. 70–84). IGI Global.

Levitt, R. (2008). Freedom and Empowerment: A Transformative Pedagogy of Educational Reform. *Educational Studies, 44*, 47–61. http://doi.org/10.1080/00131940802225085

Luckin, R. (2010). *Re-designing learning contexts: Technology-rich, learner-centred ecologies. 1. publ.* Routledge.

McAuliffe, M., Hargreaves, D., Winter, A., & Chadwick, G. (2009). Does pedagogy still rule? *Australasian Journal of Engineering Education, 15*(1), 13–18.

Mithen, S. (2001). The Evolution of Imagination: An Archaeological Perspective. *SubStance*, 30(1/2), 28.

Osborne J., and Collins, S. (2000). *Pupils' and parents' views of the school science curriculum.* A study funded by the Welcome Trust. Kings College London.

Pearl J. (2018). Causal and Counterfactual inference In M. Knauff & W. Spohn (Eds.), *The Handbook of Rationality*, Section 7.1, pp. 427–438, The MIT Press, 2021.

Ranciere, J. (1991). *The ignorant schoolmaster: Five lessons in intellectual emancipation.* Stanford University Press.

Sharifi, M., Soleimani, H., & Jafarigohar, M. (2017). E-portfolio evaluation and vocabulary learning: Moving from pedagogy to andragogy. *British Journal of Educational Technology, 48*(6), 1441–1450. doi:10.1111/bjet.12479

Shore, C., & Wright, S. (1999). Audit culture and anthropology: Neo-liberalism in British higher education. *Journal of the Royal Anthropological Institute, 5*, 557–575.

Slaughter, S., & Leslie, L. (1997). *Academic capitalism: Politics, policies and the entrepreneurial university*. The Johns Hopkins University Press.

Wang, T. J., & Huang, K. H. (2018). Pedagogy, philosophy, and the question of creativity. *Teaching in Higher Education, 23*(2), 261–273.

Young, M., & Muller, J. (2013). On the powers of powerful knowledge. *Review of Education 1*(3), 229–250. https://doi.org/10.1002/rev3.3017

CHAPTER 22

Directions for Technology Education in Finland

Sonja Niiranen, Aki Rasinen, Pasi Ikonen, Timo Rissanen and Hanna Kauppinen

1 Introduction

Finland has had a strong tradition of implementing technology education in basic education as part of the crafts, particularly technical craft, curriculum. However, the situation has changed over the last 10 years. Technology education is not an independent subject in basic education; rather, technological topics are decentralised and taught through various subjects (The National Core Curriculum for Basic Education 2014, hereinafter NCCBE 2014). As the role of technology education is still undefined in Finland, strategic planning, and research to develop the necessary procedures and operations to achieve improvements in the future is still needed.

In this article, we will continue reflecting on the findings of an international Delphi study collaboration that aimed at identifying current and future trends and issues in technology education in Finland, Japan, Low Lands (Belgium/Flanders and The Netherlands) and the USA. In this article, we will focus on discussing the findings of the Finnish study. In their Delphi studies, international experts have identified several concerns that leaders need to address as a matter of priority. In the USA these were mainly related to the shortage of technology and engineering teachers, the quality of teacher education and the lack of funding for programs and teachers (Moye et al., 2020). In Flanders and The Netherlands, similar top priority findings have been identified: the shortage of technology education teachers, lack of skills and knowledge in implementing TE, the weak status of technology education and the position of TE in the school curriculum (Ardies & de Vries, 2023). In Japan experts pointed out that following current trends and issues are essential to address when developing technology education: programming education, principles of technology education, teaching hours of technology education, learning contents of technology education, and connection to elementary schools (Yata & Moye, 2021).

© SONJA NIIRANEN ET AL., 2025 | DOI:10.1163/9789004710948_022

2 Development of Craft and Technology Education in Finnish Basic Education

In Finnish general education schools, there has never been a school subject called "technique" or "technology". When observing the five curricula of the past 50 years, one finds the concepts of technique or technology mainly under craft subjects, particularly in "technical work" contents. The 1970 curriculum stated the objectives and contents for different school subjects, and craft education was divided into two sub-areas: technical and textile craft. Technology as a concept is not found in the 1970 curriculum. Consequently, the concept of technique can be found under "technical craft". (POPS, 1970a; POPS, 1970b.)

The concept of "technology" can be found (but is undefined) in the 1985 Framework Curriculum for Comprehensive Schools, only under "Craft, technical work, and textile work". During technical work lessons, pupils should also learn to manage technology. (POPS, 1985.) Technology is clearly stated in the general objectives of the 1994 curriculum. All citizens should have a new kind of readiness to use technical adaptations and to influence the direction of technical development. Students, regardless of their gender, must have the opportunity to acquaint themselves with technology, and critically consider the effects of technology on nature. (POPS, 1994.)

One of seven cross-curricular themes in the National Core Curriculum for Basic Education 2004 (NCCBE, 2004) is "Human being and technology". All cross-curricular themes should have a central emphasis on education and should be included in studies of various subjects. It appeared that much of the technological content of the "Human being and technology" theme was studied during technical craft lessons.

In NCCBE 2004, technology education was focused particularly on the objectives and contents of technical craft, while in the present NCCBE 2014, technology education seems to be emphasised in science studies. This type of development, when technology education is integrated into science education and when having a weak status, might lead to deadly consequences for technology education (de Vries, 2018).

3 Methodological Approaches

To gather empirical data, a three-round Delphi process ascertaining and prioritising the critical trends and issues in technology education was conducted between December 2019 and April 2020 in Finland. During this process, both quantitative and qualitative approaches were utilised in gathering information and analysing the findings. Traditionally, Delphi studies have aimed at reaching expert consensus; however, one application of Delphi, namely Disaggregative Delphi, assumes that expert communication will not lead to consensus but rather to various schools of thought (Tapio et al., 2011). The aim of our study, on the other hand, was to obtain the most reliable consensus of a group through a series of questionnaires but also to examine varying viewpoints concerning technology education.

The first and second phases were distributed by conventional questionnaires and analysed by use of quantitative methods to find out the topics which reached consensus among the panellists. In Round 2, data were analysed by observing means, and standard deviations and using a boxplot analysis for each of the statement. Due to a rather small panel for a quantitative approach, the third round of the Delphi process was organised by use of a qualitative approach to capture the full spectrum of expert opinions.

The following research questions were developed for investigation:

1. What are the trends currently impacting the technology education discipline?
2. What are the trends that will most probably impact the technology education discipline in the future (3–5 years)?
3. What are the issues currently impacting the technology education profession?
4. What are the issues that will most probably impact the technology education profession in the future (3–5 years)?

An invitation was emailed to 52 individuals, and 31 agreed to participate in the study (see Table 22.1). The invitation described the research with its objectives and highlighted the importance, anonymity, and reliability of participation in the research; the voluntary nature of participation in the research; the possibility of refusing at any time; and the opportunity to enquire about the research. Because the success of the Delphi technique relies on the use of informed opinion, random selection was disregarded when selecting the Delphi participants in Finland. The selected participants were considered well-informed and leading authorities in technology education by their position, experience, and research. The criteria used in selecting the participants were based on their involvement in professional associations (teaching, research, developing) representing technology education.

In Round 1, participants were asked to exhaustively identify the trends and issues facing technology education in Finland using four open-ended guiding questions 1–4 between December 11 and 20, 2019. After the first round, researchers formulated 49 key descriptors as statements based on the participant's

TABLE 22.1 Panellists

Panellists	The number of invited panelists and how many joined the study
Educators and/or researchers at the university	33 (21)
Researchers, developers or project workers outside of the university	19 (10)

answers on the guiding questions 1–4. Round 2 was designed to prioritise the identified trends and issues and begin the process of consensus. Panellists were asked to mark their degree of agreement on a Likert scale (1 = I disagree to 5 = I agree). Panellists (n = 28) answered the survey between January 29 and February 19, 2020. After the second round, a consensus was reached with 27 statements by 70 percent of the panellists who agreed or fully agreed/disagreed with the statement.

Round 3 sought to improve the levels of consensus by asking participants to debate those 22 statements on which no consensus could be reached after Round 2. Unlike Round 1 and 2 questionnaires, the last open-ended questionnaire was answered anonymously by the panellists. Hence, the statements were submitted to the online questionnaire as open-ended questions and were supplemented with more specific questions based on the Round 2 analysis. Panellists were asked to comment on and justify either all the allegations or only those they found interesting or necessary. Twenty-six of the panellists responded to the Round 3 questionnaire between April 2 and 23, 2020.

4 Findings from the Second Round

In the following chapter, we will present the main findings from Round 2 of the Delphi study (Niiranen & Rasinen, 2022).

The role of technology education in basic education: Almost all panellists felt that it is important that hands-on and learning-by-doing should remain part of technology education also in the future. They also strongly agreed that all students should have equal opportunities to develop their technological skills and that getting girls more involved in technology is important. Panellists believed that craft subjects and their technological content would continue to inspire young people in the future. However, several of them felt that the current distribution of craft lessons (between hard and soft materials) restricts the learning of technology education. They were also concerned that technology education as a multidisciplinary overarching theme has a risk of being marginalized and that allocated time resources for implementing it are limited in relation to the wide scope of technology education content. They pointed out that the objectives of large-scale technology education are too interpretative, and a technology education curriculum should be developed for basic education soon, covering a wide range of different content areas of technology. In addition, they felt that the cutting down of technology education facilities would increase regional inequalities.

Resources and teacher education: The available resources are in danger of disappearing if the next National core curriculum for basic education and school facilities do not allow for a wide range of hands-on activities and technical subjects. Regarding teacher training in craft, the panellists raised concerns about the subject management of future teachers. As contact teaching has declined, they claimed that universities should invest more in teaching and developing students' professional skills (technical know-how). Panellists also felt that there is an increased need for in-service training for teachers in technology education and that teachers need more collegial professional support.

5 Reflections on the Situation Today

In the next chapter, some of the key themes of the third round of the Delphi study are discussed in more detail (see Niiranen et al., 2022).

5.1 *Definition of Technology Education*

Regarding the status of technology education, half (13/25) of the panellists agreed with the statement: The definition of technology education is missing from the basic education curriculum and should therefore be more closely defined among technical craft. In their arguments, they highlighted how technical craft and its learning contents and operating culture provide a natural and effective learning environment and a framework for technology education. On the other hand, some panellists called for instructions on how technology should be dealt with in various subject areas and that the objectives and contents of technical craft should be defined from the point of view of technology education. Considering the history of craft curriculum during the past 50 years, the statements of the panellists are justified. Moving technology education from craft studies towards science studies, as the current NCCBE 2014 seems to suggest, might jeopardise all the good work that has been done to develop technology education in crafts, particularly in the technical craft context. The threat in this shift would be that active hands-on activities including inventiveness and innovative thinking, may turn to applied science and in the worst case to observing processes only (see Kimbell, 2011; Lebeaume, 2011; Williams, 2011). Of course, there must be strong cooperation between these two subject areas. And not only these two, the broader the cooperation, the more holistic the curriculum becomes.

When reflecting on the statement: The requirements and phenomena of technological literacy have changed and how this change affects the demands of technological literacy, the panellists' views varied. Many (15/24) stated that technological literacy must change because complex technological systems due to digitalisation are part of daily life. Some (4/24) emphasised the importance of "basic technologies" that should remain part of technology education despite the change. With these basic technologies, respondents meant the traditional contents of technical craft, such as tools, materials and machinery. In this context, it is advisable to discuss on one hand the need to be able to learn how to solve everyday technological problems and on the other hand to learn to use modern digital technology devices. However, too often technology education is regarded only as robotics, using 3D machines, laser-cutters and other CNC machines. To be able to learn the "logos", thinking, reasoning, evaluating, and applying of technology, the student needs know-how about technics, including tools and machines and their skilful use. In engineering, both "low-tech" and "high-tech" knowledge and skills are needed. The activities that take place in technology education learning environments are "basic engineering". Innovation and technological problem-solving can also be explored through basic tools – particularly in primary education. The more versatile the tools and machines (including the latest digital devices), the wider understanding of the field of technology the student will gain.

Regarding what technology education should be in basic education, half (11/22) of the panellists thought that: Working with new and easy materials should be a stronger part of technology education in the future. However, they suggested that new and traditional materials should be used in parallel, complementing each other and making full use of the opportunities they offer. Also, almost half (9/20) of the panellists agreed that Digitalisation is an opportunity for technology education to be an integrator of learning. Some of them pointed out how digitalisation is fundamental in daily life. Others argued that digital tools and applications offer versatile opportunities to integrate learning. Another half (8/20) thought that digitalisation does not affect technology education either way. Moreover, regarding digitalisation, half (11/21) of the panellists agreed that digitalisation has decreased children's technological know-how and narrowed their understanding of technology education by reflecting that technology and technology education are mostly understood as ICT.

5.2 *Role of Technology Education in Finnish Multi-Material Craft Subject and Teacher Education*

Regarding the allocated time and resources for implementing multi-material craft and technology education most (19/21) respondents agreed with the statement: *The merging of craft content (textile and technical) and the resulting cuts to teaching threaten to undermine teachers' motivation and job satisfaction.* Concerns about the adequacy of resources and time to deliver high-quality multi-material craft education emerged from the responses, as the time available is not perceived to be sufficient to teach all content. The long tradition of Finnish school crafts has brought many materials, techniques, and tools to school crafts over time. However, it is often difficult to abandon and question these, which has created challenges in combining the extensive content with the time available. On the other hand, the panellists' responses also revealed that it is not the multi-material craft itself that weakens teachers' motivation, but factors related to the organization of teaching, such as the distribution of teaching hours or technical issues related to the timetable, that cause a decrease in motivation. Teacher's motivation and job satisfaction were also seen to be related to the teacher's competence and the smooth cooperation between teachers. Thus, the professional identity of the teacher can also be seen influencing the motivation and work-life balance of a teacher teaching multi-material craft. On the other hand, the situation was seen as a positive opportunity to learn new skills and develop one's competencies. However, to develop competence, teachers hoped to get more in-service training.

In Finnish crafts education, there is a strong tradition of dividing the teaching of crafts into technical and textile work, which has determined the implementation of crafts education (Porko-Hudd et al., 2018). Along with this tradition, technology education has been implemented as part of the crafts curriculum, especially technical crafts (Niiranen et al., 2021; Rasinen et al., 2006), but the activities in school and the development of teaching have relied on the personal interest of individual teachers. The limited and unclear role of technology education has been influenced by the lack of clarity of concepts related to craft and technology education. The tradition of dividing crafts in grades 5–7 has created a situation whereby girls (mainly) who studied textile craft in these grades did not participate in technology-related activities that are part and parcel of technical craft studies. Ten years ago, multi-material craft offered a way out of these gendered practices, as every pupil is studying the same craft content during their basic education. However, the NCCBE 2014 did not precisely define the concept of multi-material crafts, which has led to a wide range of interpretations of the concept and confusion among teachers. Thus, also the role of technology education in multi-material crafts is unclear and not defined. In this ambiguous situation, the development of technology education has taken a back seat. Kröger (2016) presents that one way of looking at multi-material craft is to divide school craft into three craft orientations: design-oriented, expression-oriented and technology-oriented craft. These orientations make it possible to address a wide range of school craft content without dividing craft into technical and textile content areas, as projects are based on appropriate choices and design (Kröger, 2016). These orientations also highlight technology education as an area of its own. Since technology education is not inherently material-based, technology education can be treated as a broad concept and is not limited to specific materials or working methods.

The majority (23/25) of the panellists were fairly unanimous in their assertion that the role of technology education as part of multi-material crafts is unclear and challenging. Their arguments were linked to the change and turbulence in the craft discipline. The rationale given for this view is the breadth of the content in multi-material craft, which is seen as having little room for technology education content. In addition, some respondents argued that the concept of technology education is not further defined. They also saw the lack of a curriculum for technology education as a problem. The same challenges identified by the panellists can be identified in teacher education. Teacher qualification studies include the subjects mentioned in the NCCBE 2014 and thus technology education is not included in the teacher education curriculum as a separate course or topic. A few panellists raised the question of the dominance of science in STEAM. STEAM as a concept has not gained an important role in discussions of Finnish school pedagogy. However, the holistic planning and implementation of cross-curricular teaching, which is part of the STEAM concept, is very familiar to Finnish teachers and teacher education. The idea of subject integration and cooperation has been discussed and implemented under slightly different names (holistic teaching, multidisciplinary learning, integration of cross-curricular themes, aiming for transversal competence) since the 1980s. Particularly, at the Finnish primary school level (grades 1–6), where one teacher usually teaches all or almost all subjects, offers good opportunities for cross-curricular teaching, which is very well suited to technology education. Teachers have a lot of pedagogical leeway to build cross-curricular units so that technology education can be integrated into teaching according to the teacher's interest and competence.

The problematic relationship between Finnish teacher education and positioning technology education in teacher education curricula is very similar to the situation in basic education. In some teacher education institutions, the objectives and contents of technology education have been included in the studies of craft pedagogy since the 1990s, and in some universities less attention has been paid to them, or technology education has been interpreted mainly as ICT technology. The teacher qualification requirements in Finnish legislation may have contributed to making it difficult to include technology education in teacher education. Studies leading to the qualification of primary school classroom teachers have long been interpreted as being linked to certain traditional school subjects so there has been no room for technology education in these studies. Multidisciplinary learning modules are included in the present NCCBE 2014, and these courses can also be found in the curricula of teacher education institutions. Depending on the teacher's interests and competencies, the learning modules can be based on STEM or technology education. Thus, the situation is better and universities offering teacher education courses have greater scope to include technology education content in their curricula.

6 Future Directions for Technology Education in Finland

In recent years, there has been increasing discussion in Finland about the direction in which craft education should be developed. We can only guess when the next national curriculum work will start, but the current NCCBE 2014 will probably only be valid for a few more years. Some educators, teachers and developers think that when considering multi-material craft subjects, a new subject called "Technical Craft and Technology" (TCT) should be developed parallel to Textile craft. This became evident already five years ago in our Delphi study when panellists reflected on the statement: *A new Technical Craft and Technology (TCT) subject would bring concreteness to technology education.* More than half (12/22) of those panellists who responded considered the idea of having a new subject of TCT appropriate. They argued that this would clarify and boost the status of technology education in Finnish basic education. On the other hand, almost half (10/22) of the panellists were against this idea because it would restrict the concept of technology education and divide the new multi-material craft subject.

The main tasks of TCT would be to develop students' technical-logical thinking and abilities, to inspire them to become interested in technology, to develop basic skills in technical fields for both secondary and tertiary and higher education engineering studies, to make visible to the pupil the links between technology, mathematics and science in the field of engineering and to make the content of science and mathematics more concrete (integrating teaching). TCT would also support gender equality and women in engineering, as it would be a subject for all pupils in basic education. One of the roles of TCT would be to highlight the links between engineering, science, and mathematics in the context of hands-on technical learning. TCT and other STEM subjects deal with much the same issues but from their perspectives. TCT would be a subject that focuses on the concrete study of technology, and the links that emerge are addressed from a practical point of view. However, now, this proposal is only at the stage where it is TAO's (Association of Technology Teachers) declaration for the future development of the subject. The strongly divided views regarding the future directions of technology education reveal the unclear role of technology education and underline the need for clarifying it in the next national core curriculum for basic education. If stakeholders have a disharmonious understanding of the role of technology education or who is responsible for it, this will not promote the development. Defining technology education in Finland, but also internationally, has proved to be very challenging. As the international history of technology education is relatively short and its position in curricula varies, debates continue in many countries (de Vries, 2018).

In Finland and internationally, "craft" is often understood only as traditional crafts or textile crafts. To clarify the situation, it would therefore be worth considering a new title for the subject. Whether technology education will in future be part of the craft subject or a subject on its own, the words 'technology education' or 'technology' should appear in the subject title in the curriculum. Hopefully, technology education will be firmly established in the future curriculum.

References

Ardies, J., & de Vries, M. (2023). A Delphi study on the future of technology education. *Australasian Journal of Technology Education, 9.* https://ajte.org/index.php/AJTE/article/view/95/53

de Vries, M. (2018). Technology education: An international history. In M. de Vries (Ed.), *Handbook of technology education* (pp. 73–84). Springer.

Kimbell, R. (2011). Handle with care... *Design and Technology Education: An International Journal, 16*(1), 7–8. https://openjournals.ljmu.ac.uk/DATE/article/view/1743/909

Kröger, T. (2016). Diverse orientations in craft education: Student teachers' conceptions and perceptions. *TECHNE SERIES: Research in sloyd education and craft science, 23*(1), 1–14. https://journals.hioa.no/index.php/techneA/article/view/1445

Lebeaume, J. (2011). Integration of Science, Technology, Engineering and Mathematics: Is this curricular revolution really possible in France? *Design and Technology Education: An International Journal, 16*(1), 47–52. https://openjournals.ljmu.ac.uk/DATE/article/view/1744/910

Moye, J. J., Reed, P. A., Wu-Rorrer, R., & Lecorchick, D. (2020). Current and future trends and issues facing technology and engineering education in the United States. *Journal of Technology Education, 32*(1), 35–49. https://scholar.lib.vt.edu/ejournals/JTE/v32n1/pdf/moye.pdf

NCCBE. (2004). The National Core Curriculum for Basic Education 2004. Finnish National Board of Education. Vammalan kirjapaino.

NCCBE. (2014). The National Core Curriculum for Basic Education 2014. Finnish National Board of Education. Next Print Oy.

Niiranen, S., & Rasinen, A. (2022). Teknologiakasvatuksen tulevaisuus suomalaisessa perusopetuksessa: Käsityön juurilta kohti uutta. *Kasvatus, 53*(1), 33–45.

Niiranen, S., Rasinen, A., Rissanen, T., & Ikonen, P. (2021). Identifying past and current trends in technology education in Finland. *TECHNE SERIES: Research in sloyd education and craft science, 28*(2), 17–24.

Niiranen, S., Rasinen, A., Ikonen, P., & Rissanen, T. (2022). Current and future trends and issues facing technology education in Finland – Taking part in an international Delphi study. *Australasian Journal of Technology Education, 8.* https://ajte.org/index.php/AJTE/article/view/88/45

POPS. (1970a). Peruskoulun opetussuunnitelmakomitean mietintö I. Opetussuunnitelman perusteet. Komiteamietintö 1970: A 4. [Memorandum. Foundations of the comprehensive school curriculum]. Helsinki: Valtion painatuskeskus.

POPS. (1970b). Peruskoulun opetussuunnitelmakomitean mietintö II. Oppiaineiden opetussuunnitelmat. Komiteamietintö 1970: A 5. [Memorandum. Curricula of Various Subjects]. Helsinki: Valtion painatuskeskus.

POPS. (1985). Peruskoulun opetussuunnitelman perusteet 1985. [Framework Curriculum for Comprehensive Schools]. Kouluhallitus. Valtion painatuskeskus.

POPS. (1994). Peruskoulun opetussuunnitelman perusteet 1994. [Framework Curriculum for Comprehensive Schools]. Opetushallitus. Painatuskeskus.

Porko-Hudd, M., Pöllänen, S. & Lindfors, E. (2018). *Common and holistic crafts education in Finland. TECHNE SERIES: Research in sloyd education and craft science, 25*(3), 26–38. https://journals.oslomet.no/index.php/techneA/article/view/3025/2943

Rasinen, A., Ikonen, P., & Rissanen, T. (2006). Are girls equal in technology education? In M. J. de Vries, & I. Mottier (Eds.), *International handbook of technology education: Reviewing the past twenty years* (pp. 449–461). Rotterdam: Sense Publishers.

TAO ry. (Association of Technology Teachers). https://www.tekninenopettaja.net

Tapio, P., Paloniemi, R., Varho, V., & Vinnari, M. (2011). The unholy marriage? Integrating qualitative and quantitative information in Delphi processes. *Technological Forecasting & Social Change, 78*(9), 1616–1628.

Williams, P. (2011). STEM Education: Proceed with caution. *Design and Technology Education: An International Journal, 16*(1), 26–35. https://openjournals.ljmu.ac.uk/DATE/article/view/1748/914

Yata,C., & Moye, J. J. (2021). Identifying current and future trends and issues facing technology education in Japan. *Proceeding International Conference on Technology Eduvation in the Asia-Pacific Region* (pp. 139–145).

CHAPTER 23

Talking Technology into the Future

Wendy Fox-Turnbull

1 Introduction

> Kia whakatōmuri te haere whakamua.
> I walk backwards into the future with my eyes fixed on my past.

Technology first appeared as a part of the New Zealand curriculum when it was included in The New Zealand Curriculum Framework (Ministry of Education, 1993a) as one of seven essential learning areas (Jones, 2006). That same year the draft technology curriculum was published (Ministry of Education, 1993b) and trialled through 1994. After significant consultation with relevant sectors, Technology in the New Zealand Curriculum (Ministry of Education, 1995) was published for the first time in 1995. Learning was organised into eight levels of achievement from Years 1–13 and incorporated three strands: Technological Capability, Technological Knowledge, and Technology and Society.

Full implementation occurred in February 1999, when technology became part of the compulsory curriculum for students in years 1–10 (5–14-year-olds). In 2001, a curriculum stocktake was undertaken to review current curricula, teachers' experiences with it, and international trends. This resulted in the development of a revised and refined curriculum. The launch of the New Zealand Curriculum (NZC) (Ministry of Education, 2007) saw significant changes in technology and included the introduction of two new strands (Ministry of Education, 1995, 2007) and the modification of one. The new strands became Technological Practice, Technological Knowledge and the Nature of Technology (Ministry of Education, 2007). When the 1995 technology curriculum was written, it had never been taught in schools, however before the development of the 2007 technology learning area considerable research was undertaken led by Dr Vicki Compton. In 2007, technology was redefined as "intervention by design, the use of practical and intellectual resources to develop products and systems [technological outcomes]" (Ministry of Education, 2007, p. 32). The new strands allowed for the considered and efficient development of culturally and environmentally situated technological practices and outcomes. The general aim of technology education in NZC is to develop technological literacy through understanding across all three strands

© WENDY FOX-TURNBULL, 2025 | DOI:10.1163/9789004710948_023

(Ministry of Education, 2007, p. 32). NZC (Ministry of Education, 2007) also suggests technological literacy enables students to develop a broader, deeper, and more critical understanding of technology (Compton & Harwood, 2007). The 2007 technology curriculum incorporated five technological areas: biotechnology, control, food technology, information and communication technology, and structural technology (Ministry of Education, 2007) and identified the concepts of the transformation, manipulation, storage, transportation, or control of either materials, energy, or information (Compton & France, 2006).

The New Zealand Minister of Education announced in July 2016 that digital technologies needed to become more explicit in the New Zealand curriculum (Ministry of Education, 2017a; Stuff, 2016). Because digital technologies aim to design and development of digital technologies it was situated within the technology curriculum (Ministry of Education, 2017b) to ensure learning is built on a holistic constructivist model of learning (Fox-Turnbull, 2017). A revised version of Technology in the New Zealand Curriculum was published in 2017 (Ministry of Education, 2017b). The extended *Digital Technologies* covered aspects of computer science, programming and computational thinking (Bell et al., 2016). The technological areas were redeveloped as outlined in Figure 23.1 and included:

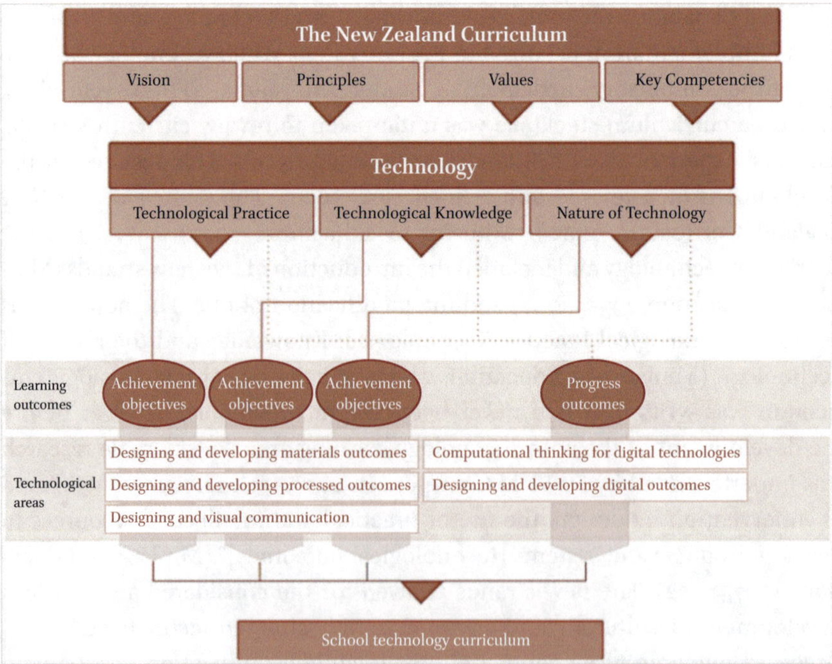

FIGURE 23.1 The 2017 revised New Zealand technology curriculum

- Designing and developing materials outcomes
- Designing and developing processed outcomes
- Design and visual communication
- Computational thinking for digital technologies
- Designing and developing digital outcomes.
  (Ministry of Education, 2017b)

The goal of this change was to ensure students learned about digital technologies, designed and developed digital technologies, and understood the decisions people make when they create and use them.

2 SWOT Analysis of Technology Education in the New Zealand Context

Technology in New Zealand is internationally well-regarded; however, it is not perfect. This section identifies and discusses three to four strengths and weaknesses of technology in The New Zealand national curriculum. In addition, it also discusses three to four potential threats to and the opportunities for the current curriculum and the refreshed New Zealand Curriculum Te Mataiaho currently under development.

2.1 *Strengths*

The key strengths identified and discussed in this section include the opportunity for authentic learning giving students a sense of empowerment, engagement in 21st-century future-focused skills, knowledge and competencies, opportunities for students to deploy their funds of knowledge in a meaningful context and implementation of digital technologies and developing students to be problem solvers and critical thinkers who collaborate and celebrate failure.

As students progress through their technology education journey in schools their skill development, abstract and critical thinking develops. Studying technology through to the end of secondary school presents students with opportunities to design and develop technologies that can make a real difference in the lives of others, thus enabling a sense of empowerment and voice. This was illustrated a number of years ago when a colleague teaching Design and Visual Communication (DVC) shared her student's work with me. The student had designed and developed a New Zealand version of 'Life of a Cow' for her senior secondary assessment. Although the message is not one I totally agree with I was impressed with the sense of empowerment the process of designing and developing such an outcome must have facilitated. In technology, our young people have opportunities to intervene in the made world in meaningful ways.

A goal of the New Zealand education system is to have students experience a rich curriculum that engages and challenges them, is forward-looking, and is inclusive (Ministry of Education, 2007). The 21st century has changed the type of skills, knowledge, and competencies that are needed for success in modern society. Learning through the solving of problems by designing and developing potential solutions in technology offers students authentic opportunities to develop skills and competencies as they are needed. This approach is both engaging and motivating as students understand the role their learning plays in their current lives and facilitate link-making to their own and others' futures.

Another strength of technology education in New Zealand is that it offers opportunities to use their cultural funds of knowledge (FOK). Funds of Knowledge describe the developed bodies of knowledge and skills that are accumulated by a group, for example, a family or the individuals within, to ensure that they can function appropriately within their social and community contexts (Lopez, 2010). In New Zealand Māori and Pacifica people are underrepresented in STEM-related careers. One way to challenge and disrupt this is to highlight the strengths and resources of Māori and Pacifica communities through a focus on their funds of knowledge (Hunter, 2021). In addition, students who are considered by their peers and themselves to be 'less academic' have opportunities to excel in technology as they bring to technology their practical funds of knowledge and problem-solving skills from their extended family thus providing opportunities to develop mana (prestige) and to be seen by their peers in a different light.

Recent additions to the current technology curriculum in New Zealand are the introduction of the technological areas Designing and Developing Digital Outcomes and Computation Thinking for Digital Technologies (Ministry of Education, 2017c), which enable students to know how to use and create digital technologies. Computer science concepts and thinking computationally are being recognised as important for everyone (Bell et al., 2014). Aspects of computation thinking are easily integrated across the curriculum. In a recent research project, a colleague was teaching year 4–6 students to write and debug a small programme on an App developed by the research team. One of the boys indicated that the debugging activity was cool and likened the process to editing a story. He thought 'debugging' his story was 'way cooler' than editing it. In the same research project students learned sequencing, pattern recognition, abstraction, decomposition and debugging as aspects of computational thinking. All skills are applied to both literacy and numeracy thinking.

The final strength that I would like to discuss is that technology education requires students to become collaborative problem solvers and critical thinkers. Undertaking successful technology practice requires collaboration with peers and others, when working in groups success requires seeing ideas from a variety of lenses, compromising, and celebrating their own and others' failures as much as the successes. Failures and mistakes require evaluation, reflection, critical thinking, further investigation and having another informed go. Students who may not experience much success with written work can and do succeed in technology.

2.2 *Weaknesses*

In this section, I outline the key weaknesses of Technology Education as it is currently situated in New Zealand. These weaknesses inform the subsequent threats and opportunities sections below. Key weaknesses identified include a lack of New Zealand-trained technology teachers, the training of technology teachers as well as aspects of our senior secondary school assessment practices.

New Zealand currently has a considerable lack of technology teachers. "The Education Gazette | Tukutuku Kōrero is produced by the Ministry of Education for teachers, leaders and other education professionals working in New Zealand. It's the place to advertise job vacancies, notices and announcements, and to read articles that explore and celebrate the education system" (Ministry of Education, 2024). At the time of writing, I undertook a search for technology vacancies in the Aotearoa New Zealand Education Gazette Tukutuku Kōrero three weeks after the beginning of the academic year on 21 February 2024 when one would expect that vacancies would be at their lowest. The search identified 24 vacancies for technology teachers across all the technological areas. To put this in context in 2023 there were 374 secondary schools in New Zealand.[1] The inability of schools to find technology teachers has several potential implications. If teachers are unavailable schools may cease to offer the subject, non-qualified or 'out-of-field' teachers may be asked to fill the gaps or the vacancy may be filled by foreign teachers whose training and underpinning understanding and teaching philosophy is likely to be different to what is practised in New Zealand.

Having said that there are several challenges associated with the education of technology teachers in New Zealand. Secondary teacher education in New Zealand requires at least a bachelor's degree in their field of teaching. Students subsequently undertake a graduate diploma or post-graduate diploma in teaching to gain provisional certification to teach in New Zealand. Graduate diploma or post-graduate diploma courses typically run over the calendar year.

At the University of Waikato for example, the programme starts mid-January and finishes towards the end of November. The programme is very intense and condensed. To succeed students, need to be very disciplined and time-smart-qualities also required of teachers, however, the challenge is that most student teachers also need to undertake part-time work in order to survive. Many in the initial teacher education sector would like to see this programme expanded to at least 18 months or even two years. This would ease the pressure on students and universities and enable students the time to think, reflect, critique, and evaluate their learning and future practice as teachers.

Another issue associated with the education of secondary technology teachers is the need for a bachelor's degree in their teaching subject. In the past, some of the technological areas have not drawn from people who have degrees. Technology requires and values practical as well as academic skills and knowledge. People from trades and non-degreed fields, who are skilful and experienced have difficulty finding a pathway into technology teaching. Several universities have worked on programmes to mitigate this by facilitating either recognition of prior learning or develop alternative pathways into the profession. For these students working in an academic programme can throw up challenges and some struggle to complete the programme unless they are very well supported by their universities.

Most students in New Zealand undertake their National Certificate of Educational Achievement (NCEA) at three levels in their final three years of secondary school. NCEA Level 1 in Year 11, Level 2 in Year 12 and L3 in Year 13. The qualification is recognised by employers, universities and other tertiary education providers and is managed and maintained by the New Zealand Qualifications Authority (NZQA) (New Zealand Qualifications Authority, 2024). Each year students and teachers select and submit evidence to meet a range of externally and internally assessed achievement standards (AS). In theory, this could be a wonderful opportunity for students to undertake their technology practice and evidence their process, learning and understanding within a technology portfolio and subsequently select appropriate AS as the best evidence of their learning. However, the reality is much more complex. NZQA requires schools and teachers to register the AS they are using for assessment throughout the school year in the first term of the year. This, of course, is understandable, as NZQA need to prepare examinations and summative assessments as well as ensuring rigorous moderation practices are undertaken, however, it does decrease the flexibility for teachers and students to respond to students' individual learning needs and practices.

Another weakness of the NCEA system occurs when students who want to specialise in more than one technological area. For example, a student who undertakes DVC may also want to study resistant materials or textiles. However, each AS can only be assessed once, which means students are limited in the selection of AS for assessment depending on what standards are assessed in their other technological area. Unfortunately, this has led to a movement away from generic technology practice, skills and knowledge standards to more specific and fragmented standards in each of the technological areas.

The final weakness of the NCEA systems that I believe is worth mentioning is assessment-driven practice. I wish I had a dollar for every time I have heard a technology teacher say “I am teaching AS *xy* standard this term”. In my view, teachers should not be teaching achievement standards. They should be teaching the curriculum through carefully designed programmes of work that involve students developing their understanding of technology and undertaking technology practice. The AS can then be used to assess student's work against the standards at levels: not achieved, achieved, merit or excellence.

2.3 *Opportunities*

In this section, I discuss the one most exciting opportunity for technology education in New Zealand at present. That is the Te Mātaiaho/Curriculum Refresh project with the technology refresh currently underway. The Ministry of Education website states

Te Mātaiaho is the refreshed NZ curriculum is designed to give practical effect to Te Tiriti o Waitangi and be inclusive, clear, and easy to use. It has been developed over the last three years in collaboration with the education sector, students, whānau, communities, and a wide range of interested organisations and communities. The updated draft of Te Mātaiaho was released in Term 4, 2023, and included the three refreshed learning areas released so far: te ao tangata | social sciences, English, mathematics and statistics. The other five learning areas are still being refreshed and will be released in 2024 and 2025. Schools and kura have until 2027 to begin using the complete refreshed curriculum (The Ministry of Education, 2024).

Te Tiriti o Waitangi is New Zealand's founding document between Māori and the British Crown and was signed in 1840. Amongst other things, the main purpose of the treaty was to protect the Māori from European lawlessness and as Māori believed to protect their land, language and tikanga (cultural treasures), however in reality for many years the treaty was not honoured, with Māori losing their language and much of their land. The New Zealand Education Act of 2020 stated that education in New Zealand should honour Te Tiriti and support Māori-Crown partnerships (Parliamentary Counsel Office, 2024). At the end of 2022, I was appointed one of two ‘subject matter experts’ (SME) in technology as a part of the refresh process by the New Zealand Deans of

Education. I was the only SME in English medium; the other was in Hangarau (Māori medium technology). With a wide team of practitioners and MOE advisors the draft Technology curriculum was developed in 2023 and was due for release in October that year, however, due to a change in government this release is yet to occur. Although I am unable to share the proposed changes in technology at this time, I can say that I am excited by the opportunities the direction of the refresh may bring. The proposed curriculum is centred on Te Ao Māori (Māori worldview). The released learning areas and the whakapapa (genealogy) of the new curriculum framework celebrate Māori ways of seeing the world and learning. Learning is holistic. In technology, this could mean that generic technology practice is centred around people, practice, and place. In a Māori world view the environment is a living being which is as highly valued as people. This holistic approach will identify Māori as New Zealand's first technologists, it will place high value on the environment, considering both short- and long-term impacts of technological development and how and why materials and information are used. This will mean that students will learn that technology practice is not isolated from our environment and people but an integrated piece of it, hopefully leading to a better world for all.

Another exciting opportunity to evolve recently in New Zealand technology is the increased visibility and developed understanding of designing and developing digital technologies. The technological areas of Designing and Developing Digital Outcomes (DDDO) and Computational Thinking (CT) in the 2017 version of the technology curriculum has heightened awareness of the digital realm but also made explicit what young people learn and do to position themselves as innovative designers and creators of digital solutions and moving beyond solely being users and consumers of digital technologies. This combined with the new approach of te Māitaiaho offers opportunities for students to critically assess the impact of existing and new technologies on society and the environment as well as design for a future safe and inclusive world.

2.4 *Threats*

The key threats to the future of technology education in New Zealand I have identified include a lack of understanding and status of technology, teacher history background and teaching approach, lobby groups pushing their interests at the expense of the integrity of the discipline, change of government, and teacher professional learning and development (PLD).

For many years technology has suffered from a continued lack of status within the school system and its often misunderstood dual academic/ practical nature of the discipline by senior school management, principals, parents, and politicians. The implications of this include less time allocated for technology in a school timetable, students either pushed into technology (because they are 'not academic') or turned away from or discouraged from taking technology (because they are too academic) as a school subject even though technology significantly contributes to understanding and practice in a range of professions such as engineering and architecture.

Another threat is the power of lobby groups to influence curriculum decision-making to ensure their interests are met, while not necessarily understanding the wider implications for students or specific learning areas. This was clearly illustrated in New Zealand with the strengthening of digital technologies. Originally the New Zealand Computer Society called for digital technologies to be a learning area within the New Zealand curriculum, rather than be situated with technology. This issue is partly related to the one above, as those from the New Zealand Computer Society did not have a thorough understanding of technology their assertions were based on outdated practice. Another digital technologies issue also related to the lack of understanding of the technology curriculum was the plethora of organisations that applied to the MOE to deliver professional learning to teachers in digital technologies. I witnessed several examples where digital technologies professional development involved teachers and children engaging in a series of skills-based fun activities related to coding, and programming with few links to the technology curriculum or authentic technology practice. The implications for this include students not able to see the relevance of what they are learning and a lack of connections to their current or future lives, thus they lose motivation and become turned off technology.

The final threat I would like to identify comes in the form of teacher PLD concerning the implementation of the Māitaiaho / Refreshed curriculum in New Zealand. The proposed curriculum has the potential to facilitate huge shifts in education towards a more inclusive and equitable curriculum for all New Zealanders. At the point of writing, there is no certainty that the proposed shift will occur, however, if it does New Zealand is well-positioned to become world-leading in positioning our indigenous thinking, knowledge, and skills within a national curriculum. However, this giant shift in thinking and practice can only occur with very significant support for consistent PLD of high quality for teachers provided by the Ministry of Education. Given the current government's current fiscal rhetoric about our vulnerable financial position, this seems unlikely. Implications of insufficient support for teacher PLD include failure of the curriculum, incomplete shifts in practice and a very confused teaching body, not to mention the students in their classrooms.

3 Conclusion

Technology Education in New Zealand is currently sitting on a precipice, however threats to its future are not insurmountable. I conclude this chapter feeling excited for the release of the draft refreshed technology curriculum. Given the hundreds of hours of work that have gone into the development of the draft refreshed curriculum, I find it hard to believe that any responsible government would drop such a world-leading curriculum. Once released in total much of the required shift will have to occur at a whole school level and will be the responsibility of the school's senior management to ensure it is implemented fully with the support (hopefully significant) of the Ministry of Education. Time and time again I have seen the empowering nature of technology on students. It is well situated to deliver an education within which our students become critical thinking reflective technologists who understand the world they live in and know how to care for it. "Space junk and pollution begone!!"

Note

1 https://www.educationcounts.govt.nz/statistics/number-of-schools

References

Bell, T., Duncan, C., & Atlas, J. (2016). Teacher feedback on delivering computational thinking in primary school. *Proceedings of the 11th Workshop in Primary and Secondary Computing Education.*

Fox-Turnbull, W. (2017). *TENZ Response to Digital Curriculum.*

Hunter, J. (2021). Disrupting deficit discourses in mathematics education: Documenting the funds of knowledge of young diverse learners. *Excellence in Mathematics Education: Foundations & Pathways*, 431.

Lopez, J. K. (2010). Funds of Knowlege. *Learn NC*. Retrieved 9 March 2010, from www.learnnc.org

Ministry of Education. (2007). *The New Zealand curriculum.* Learning Media.

Ministry of Education. (2017a). *Digital technologies-Hangarau matihiki, draft for consultation.* Ministry of Education https://education.govt.nz/assets/Documents/Ministry/consultations/DT-consultation/DTCP1701-Digital-Technologies-Hangarau-Matihiko-ENG.pdf

Ministry of Education. (2017b). *Digital Technologies and the national curriculum – what's it all about?* Retrieved 5 December 2017 from http://elearning.tki.org.nz/Teaching/Curriculum-areas/Digital-Technologies-in-the-curriculum

Ministry of Education. (2017c). *Strengthening digital technologies hangarau matihiko in the curriculum*. Retrieved 3 November from https://assets.education.govt.nz/public/Documents/Ministry/consultations/DT-consultation/Digital-Technologies-Hangarau-Matihiko-ENG-July.pdf

Ministry of Education. (2024). *The Education Gazette | Tukutuku Kōrero* Retrieved 20 February from https://gazette.education.govt.nz/about-this-site/

New Zealand Qualifications Authority. (2024). *New Zealand Qualifications Authority (NZQA)* Retrieved 20 February from https://www2.nzqa.govt.nz/

Parlimentary Counsel Office. (2024). *New Zealand legislation: Education and training act 2020*. Retrieved 21 February from https://www.legislation.govt.nz/act/public/2020/0038/latest/LMS170680.html?search=sw_096be8ed81df65ae_Te+Tiriti+o+Waitangi_25_se&p=1&sr=2

Stuff. (2016). Digital Technologies coming to NZ curriulum. (7 July 2016). http://www.stuff.co.nz/national/education/81763452/Digital-technology-coming-to-the-NZ-Curriculum

The Ministry of Education. (2024). *Te Mātaiaho | the refreshed NZ curriculum*. Ministry of Education. Retrieved 20 February from https://curriculumrefresh.education.govt.nz/te-mataiaho

CHAPTER 24

A Strategy for Moral Formation of Professionals

Henk Jochemsen

1 The Challenge[1]

The strong and growing pluralism regarding worldviews in European societies challenges all professionals to be able to deal with opposing worldviews, as well as a plurality of faiths. Students from all worldviews and religious backgrounds who want to enter the professional world will face this potent pluralism. One of the strongest challenges facing students, whatever their view of life, is to be spiritually authentic not only in their school setting but also in their profession. And at the same time be a good professional and collaborate with colleagues of very different backgrounds and world views. These young professionals will enter a world where they are urged to "sell themselves on sight", meaning they are enticed to choose a profile of a type of person who is all flashy portfolio and slick résumé. What prevents them from adopting a profile commonly referred to as the "empty suit" or the "designer personality"? Even in highly professional vocations, young people may feel compelled to engage in impression management in order to survive. How do educators prepare students to face these onslaughts on the integrity of their character, once they are employed as professionals?

Rediscovering our calling as professionals is more than knowing our psychological identity. It is also more than knowing our worldviews. It is the ability to combine the right normativity to be identified in each specific context, with a determined focus to be moral. This is what we call having a moral profile. It is being moral in a concerted way because we access normative sources such as our worldview, our spirituality, and our ethics, and we apply these in our individual lives.

In this contribution, we will outline briefly a kind of educational strategy for moral formation.

Moral formation is the process of gaining the morality required by one's spirituality and ethics as manifestations of one's worldview and fitting the professional role one fulfils. In this contribution, it is treated as development that activates ethical knowledge, skills, and attitudes to become moral in practice, based on personal spirituality. Here we understand spirituality as the functioning in everyday life of one's worldview or faith. So we take a functional

© HENK JOCHEMSEN, 2025 DOI:10.1163/9789004710948_024

approach to spirituality as an anthropological structure of the human being. This implies that every person, religious or not, has a spirituality.

The steep rise of new educational developments, such as competency-based learning, constructivist learning practices, student mentoring and supervision involving extensive reflection, portfolio learning, and digital learning, creates new demands for moral formation.

Over against the modern emphasis on the individual as autonomous, we place the young professional as an individual within the normative confines of practices. We argue that their choice of a virtuous profile in a profession must reflect the normative basis of the profession in a way consistent with their worldviews. In short, we urge them to seek a proper spiritual and moral formation in the vocational context.

Surely, moral formation is not just left to the devices of students. We believe that educators must be involved to the point where they have a significant role in coaching young professionals. They must stimulate the student to make a committed choice to adopt a calling and aim for a livelihood in a certain professional practice.

2 A Proposal for a Strategy for Moral Formation

Based on our research, we contend that four dimensions or windows hold for an educational strategy regarding moral formation in higher education in which we use worldview training, ethics, and reflection. We use the metaphor of windows to explain how these four conditions shed light on this type of moral formation.

The four windows can be indicated with the acronym PISA[2]

P: Being Practice minded.
I: Being Integral
S: Being Spiritual
A: Being Answerable

The acronym PISA outlines our strategy of moral formation. Below, we introduce the four elements of our PISA strategy as four complementary ways of looking at moral formation in higher education. These four perspectives are also useful for inquiring about the perspectives that hold for moral formation.

2.1 *P: Being Practice Minded*

The window P stands for knowing the characteristics of social practices and subsequently being able to position one's profile within the ethics of professional practice. We open this window first of all to grasp the normativity inherent in social practices. This requires an analysis of the practice. In our view, a fruitful start for such an analysis is Macintyre's definition of a practice. This famous definition runs:

> By a 'practice' I am going to mean any coherent and complex form of socially established cooperative human activity through which goods internal to that form of activity are realised in the course of trying to achieve those standards of excellence which are appropriate to, and partially definitive of, that form of activity, with the result that human powers to achieve excellence, and human conceptions of the ends and goods involved, are systematically extended. (MacIntyre, 1981, p. 187)

The fundamental presupposition of this concept of social practice is that in reality, normative structures are operative that the professional should take into account in his professional work (de Vries & Jochemsen, 2019). I will briefly explain the central concepts of this definition. Professional practices have an internal good, a core value whose realisation is pursued by the practice. I often use the word telos in this context (Jochemsen, 2006, p. 71). Physicians pursue the health interests of their patients, lawyers pursue justice, and engineers pursue the shaping of meaningful devices,[3] to just mention a few practices. This is also evident from professional education curricula that embody what it is to be, e.g., a good journalist, church worker, physician, or lawyer.

The telos of a practice can be realised by observing a constellation of principles and rules; MacIntyre speaks of standards of excellence. In the performance of their practice professionals like physicians and technical engineers observe a whole constellation of norms to realise the specific value the practice stands for. The ICT programmer who develops new software for the analysis of big data should not only know the workings of computers, take into account the wishes of the client, and be able to deal with the client in an ethical way, and take into account relevant legislation, but also be aware of the risks of unwanted prejudices in the algorithms. And 'unwanted' always gets its meaning from ethical positions.

To identify these principles we have used the theory of modal aspects of Herman Dooyeweerd (de Vries & Jochemsen, 2019). The rules derived from the principles often concern a form of 'know-how'; for instance a professor who notices the nonverbal signs of her students as a matter of habit, referring to the psychic and lingual aspects. This is why in this approach competent performance requires the virtues in which moral behavior has become a habit. Being oriented in relation to P means that the student is able to understand how the practical virtues of his moral profile fit the demands made by the professional practice. An important virtue in this context of the classical virtue of phronèsis, a practical wisdom. The person having the quality of phronèsis will use his cognitive faculties to consider the type of action needed to bring about a certain change. Phronèsis involves the ability to determine through reflection how to achieve the telos of a certain practice (Hegeman et al., 2011, p. 127).

In addition to the constitutive side, consisting of the constellation of principles and rules, in our approach, we distinguish a regulative side to practices. Distinguishing this side emphasises that any performance of practices involves a specific interpretation and actualisation of the rules (cf. the interpretation of a piece of music in a particular performance). Such an interpretation departs from a wider interpretative framework concerning the meaning of that practice for human life and for society and, hence, on the direction performances of that practice should have. In other words, the regulative side of practices pertains to motivations and beliefs about human life in the world, the reason for existence and the role and meaning of the practice for human life and society. These fundamental attitudes, beliefs and motivations reflect the worldview and, if they are religious, the religious beliefs of the people involved. So, any performance of a practice is regulated by those worldview beliefs and religious beliefs. We come back to this in the presentation of the window Being Spiritual.

2.2 *I: Being Integral*

The window I of PISA, Being Integral, stands for being in touch with all of the facets of one's life, and being able to incorporate them into one's choice of a moral profile as a professional. It expands on what is sometimes referred to as the relationships between "head, heart, and hands". This window can be related to the three aspects of competencies, knowledge (head), skills (hands) and attitude (heart). Sometimes the iceberg is used as an image of competencies, in which the knowledge and skills are largely in the 10% above the water surface and the attitude (the heart), below the surface. In professional education and in professional practice we often see that students and professionals have difficulty in connecting their deep personal worldview or religious convictions with their everyday practice. Or they do so in an unreflected and poorly argued way. Being integral is closely connected to integrity, the quality of being honest and having strong moral principles and living in accordance with them.[4] So from the perspective of window I, moral formation must go beyond learning competencies as a set of standardised requirements for an individual to properly perform in a specific vocation. Only if moral formation encompasses a combination of knowledge, skills, personal value beliefs and behaviour utilised to improve performance, will there be 'learning'. Thus, more than the ability to perform a specific role, it is a person knowing authentically if his personal profile fits the competency profile of his particular practice

(Hegman, 2011, p. 57). An educational method to further this Being Integral is reflection in the context of supervision and intervision (peer learning). I will come back to this below.

Frits Korthagen, a Dutch professor of education, has put forward a good model of behaviour we are to reflect on in supervision, the so-called Onion model consisting of a number of concentric circles. These represent, from outside in: (1) environment, (2) behavior (performance), (3) competencies, (4) beliefs, (5) identity, and (6) mission. At the very centre of the model, Korthagen puts 'core qualities', whereas we would propose seeing it as the self of the person. This is the seat of a person's spirituality (see next window). According to Korthagen

> a harmonious alignment between the layers is the key to effective functioning. Once the connection with the inner layers has been established, personal core qualities in the practitioner are triggered, such as enthusiasm, curiosity, courage, steadfastness, decisiveness, openness, flexibility, and so forth. [...].Core reflection is aimed at promoting awareness of these core qualities as well as personal reflection on the relation between the various layers within oneself.[5]

We have found that amongst our students in nursing and social work, supposing this also applies to other careers, there can be a barrier between the inner layers, notably the layers of beliefs and competencies. We name this the problem of weak externalisation, of not being able to put into one's practice what lives within ourselves as persons. That means that the virtues a person may possess are not utilised by that person in professional practice. These problems very much need treatment in supervision. Each layer of the onion represents an aspect of our behaviour that is supposed to be touched on in reflection on our practice (Hegman, 2011, pp. 231–234). This will be discussed below.

2.3 *S: Being Spiritual*

This window departs from the view that human beings actively relate to their living conditions and experiences, and always attempt to interpret and clarify them in the light of a set of symbols, motifs and views, in short, a world view, even if this is not always made explicit. All people, and not just followers of an organised religion, are concerned with what they see as 'essential and fundamental' in existence and take this into account in shaping their lives and circumstances. Here we follow a functional approach to spirituality, which emphasises that all persons function with spirituality. As indicated above we understand spirituality as 'the religious and/or existential mode of functioning of a person including experiences and question of meaning and purpose' (Hegman, 2011, p. 156).

So, S is the window open to understanding how spirituality feeds morality. Common sense tells us that no student will be morally formed if there is not an explicit choice, an innate and intrinsic desire of that person to do so. It looks at the basic motives that guide one's calling and reflection. By allowing individuals to explore their spirituality, the student can realise which fundamental sources of meaning and quests for transcendence aid them in choosing their calling in life (Hunink et al., 2009).[6] Focusing on this condition allows us to see what students are actually doing when they try to relate their spirituality to their learning, particularly in practice (Hegman, 2011, p. 50).

This brings us to consider the process whereby students and educators use religiously based virtues to guide their moral behaviours. An important question is how this process can be integrated into moral formation. Once again we come across the challenge to connect the inner layers of the human person with one's actual behaviour in professional practices, showing the required virtues to realise the telos of the practice. And again we point to a form of reflection in the context of peer learning.

2.4 *A: Being Answerable*

Window A refers to our ability to be answerable, or responsive to normativity. Its core is the willingness to apply the principles, rules and values of which one knows they count. Answerability also indicates the willingness and ability to account for one's choices concerning one's moral identity. Through this lens, we see how a student reflects on her own behaviour while engaging in her education. We observe in particular how she partakes in supervision, intervision, and reflection exercises to discover who she is within professional practice. This window is closely connected to the virtue of authenticity. Without being authentic in the reflection practices these will not contribute to moral formation.

Furthermore, the window of Being Answerable can sharpen the focus on the formation of moral identity in order to become responsible as a professional. Answerability is disclosed through analysing certain basic assumptions about identity.

Clearly this treatment of the condition Being Answerable converges on the importance of reflection, not just on one's actual performance in a specific case but also on the normative motives that guide that performance and how those motives relate to the practitioner's concept of a competent professional. Without proper reflection, no serious attempt at becoming answerable can be expected.

This leads us to discuss the role of reflection in moral formation.

3 Deep Reflection

From the presentation of the four windows of the PISA strategy of moral formation it will be clear that in our view reflection practices address the integral sense of oneself so that the supervisee can draw on his spirituality to meet conditions "I" and "A" in order to observe all the relevant normativity in the performance of his practice. The practitioner's perception of her moral identity is therefore a needed element of supervision. Without this deeper look into her identity, reflection in forms of peer learning could easily remain superficial.

Let us have a closer look at the common reflection in professional mentoring sessions.

Reflection sessions in the context of professional training refer to meetings in which the students, either guided by a supervisor or in a peer group, reflect on their practice through observations and feedback from colleagues. Mostly one of the participants brings in a case from his own experience in which it was, at least at first-hand, not clear what the professionally adequate way of acting was. The supervisor coaches the supervisees so that they, and in particular the one who presented the case, can integrate their work experience in their professional formation based on reflection and personal choice. Reflection in such sessions involves the distancing of the participants from the specific experience or event to analyse it from different perspectives. The first main question is: were the professional actions performed correctly, i.e. according to the professional standards? Often this question dominates the supervision sessions. This means that the reflection concentrates on professional-technical, psychological and social aspects of the practice.

A second reflection question is: did the professional do the right things? So, not so much the technically correct performance of the actions is reflected upon but whether the actions carried out were adequate in that situation or whether a different way of acting should have been chosen. An example of agriculture could be the management of nitrogen (N) in livestock. A reduction of N-containing substances could be achieved by changing housing conditions in which manure and urine are maintained separated to avoid the release of ammonia. The question of whether the livestock expert is doing the right thing may lead to concentrating more on feed for the cattle that contains less N.[7]

We argue that for moral formation in the PISA Strategy, a third step is necessary that will lead the reflection to what we call deep reflection in the two steps of common reflection mentioned above the basic values and beliefs that function as standards are presupposed and accepted silently or explicitly. In the example mentioned above these regard the presuppositions of our modern large-scale technically supported industrial agriculture (Jochemsen & Rademaker, 2019). In the third step of deep reflection, those common standards are evaluated in the light of personal beliefs and aims (one's moral profile) that themselves should be expressed and discussed with respect to the consequences of the practice. In this reflection on the performance of professional practices the windows of Integrality, Spirituality and Answerability should be opened to give space to the deepest normative attitudes of the professionals involved in the reflection and strengthen the virtues required to realise the telos of the practice. Important virtues in this context are courage, integrity, authenticity and also the cardinal virtue of phronèsis (see under P). For the example of agriculture, this could lead to raising the question of whether in certain contexts the paradigm of industrialised agriculture is sustainable and justifiable. There are groups of farmers who themselves have answered this question negatively and seek to practice according to a different paradigm.[8]

4 Final Remarks

It will be clear that in the presentation of the strategy for moral formation indicated with the acronym PISA, I have chosen a functional approach, describing the concepts and processes without giving them content from a specific worldview of religious belief. I think this strategy can work using different worldviews and belief systems. At the same time, it holds an understanding of professions and professional practices that depart from the view that morality is completely a matter of historical construction, as a full-swing postmodern view would have it. Morality in my view is also always a response to normativity discovered in reality that asks for a positivisation in everyday life to make life flourish as much as possible in a world in which evil and brokenness also is an indelible element. This is precisely why we need morally well-formed professionals. The strategy PISA presented in this contribution is put forward in light of that need.

Notes

1 To a substantial degree this contribution is based on: Johan Hegeman, Margaret Edgell & Henk Jochemsen, *Practice and profile*. Eugene, OR: Wipf & Stock 2011.

2 In educational terms there is another widely used PISA acronym that stands for Programme for International Student Assessment.

3 This is rough short formulation for the sake of this paper that could use some elaboration. On this point Marc de Vries is an expert who has emphasised the importance of meaning in reality and also in technique.

4 See https://dictionary.cambridge.org/dictionary/english/integrity opened on February 22, 2024.

5 See https://korthagen.nl/en/focus-areas/core-reflection-multi-level-learning/opened February 22, 2024

6 Gert Hunink, René van Leeuwen, Michel Jansen & Henk Jochemsen, Moral Issues in Mentoring Sessions, *Nursing Ethics* 2009; 16 (4), 487–498.

7 Of course this is a simple presentation of a complex issue and the two approaches does not exclude one another. It intends to illustrate the differences and complementarity of the two reflection questions.

8 See for example https://caringfarmers.nl/

References

Gerrit Glas, G. (2019). Roots of the Normative Practice Approach: The Philosophy of Herman Dooyeweerd. In M.J. de Vries & H. Jochemsen (Eds.), *The Normative Nature of Social Practices and Ethics in Professional Environments* (pp. 15–30). IGI Global.

Hegeman, J., Edgell M., & Jochemsen, H. (2011). Practice and profile. Wipf & Stock 2011.

Hunink, G., van Leeuwen, R., Jansen, M., & Jochemsen, H. (2009). Moral Issues in Mentoring Sessions. *Nursing Ethics*, 487–498.

Jochemsen, H. (2006). Normative Practices as an Intermediate Between Theoretical Ethics and Morality. *Philosophia Reformata*, 96–112.

Jochemsen, H., & Rademaker, C. J. (2019). Food Systems: How can the Normative Practice Approach Help Toward a Just and Sustainable Food System? In M. J. de Vries & H. Jochemsen (Eds.), *The Normative Nature of Social Practices and Ethics in Professional Environments* (pp. 134–163). IGI Global.

MacIntyre, A. (1981). *After Virtue. A Study in Moral Theory.* Duckworth.

de Vries, M. J., & Jochemsen, H. (2019). *The Normative Nature of Social Practices and Ethics in Professional Environments.* IGI Global.

CHAPTER 25

Family Prospects and the Oblate Spheroid

Steve Keirl

Oblate spheroids are quite rare
and should be treated with great care;
this one, 8,000 miles across,
is crucial to our lives, because
another one like this would take
four and a half billion years to make;
I've had a look; there's none in store;
so don't break this one; there's no more.

BARKER (n.d./2022)

1 Epilogue

A group of city school teachers engaged in a mountain leadership training course were loaded into a minibus and driven to a remote spot in the mid-Wales mountains. After a couple of miles of steady walking up to higher ground and into dense mist and light rain, each participant was given a map and compass. Acting alone, each was to take a turn at leading the others in a navigation exercise across undulating terrain littered with rocky outcrops and extensive ponds of water. The instructor's direction to each lead participant was the same each time: "From where you think you are now, take us to (map reference) X". No comment was made at the end of each leg of the exercise as to whether the referenced point had actually been accurately reached.

One by one, the teachers navigated in the very poor conditions from 'where they thought they were' to 'where they thought they should be'. The requirement, of course, was total concentration not only on one's own judgment of where you were and where you were going but also on the accuracy of each prior lead colleague's judgement. Terrain, map and compass had to be read in harmony.

© STEVE KEIRL, 2025 | DOI:10.1163/9789004710948_025

The experience was salutary; there were surprises, humour, annoyance, errors, dismay, difficulties, and successes. Key to the learning were such concepts as *back bearings* (a kind of projecting backwards in order to go forward) and *cumulative error* (which occurs if one, or more, of the participants made wrong decisions on top of a prior participant's wrong decisions). All such outcomes mattered if these teachers were eventually to take school children into similar conditions where, the *prospect* of getting lost is not an option to be entertained.

I offer this story as an allegory. If there is a genuine need to consider (to be considerate towards) our oblate spheroid, what kind of assessment might we make in order to answer the question: *Where do we think we are now?* Such an assessment of the current global situation is a pre-requisite to answering the question: *Where do we want to get to?*

(*Visiting aliens may well be stunned that such a dominant and creative species could have arrived at a situation where asking such questions was even necessary. Perhaps the visitors might explore what the other millions of species make of it all.*)

However, when we look at our prospects of getting from whatever is the answer to the first question to whatever is the answer to the second, we might recognise some of the places we *don't want* to go. Thus, even if we can say *where we want to be,* we may have to address some serious obstacles en route. Looking back in order to look forward can be helpful. Understanding our cumulative errors to date might perhaps inform how we mitigate future-oriented erroneous decision-making. This is to say that knowing where we want to be is not the same as knowing how to get there. Regardless, the first thing to assess is just where we have ended up – temporally, physically, psychologically, politically, economically, and more.

2 Conclusion/s?

In sketching the following overview of where we are now, different kinds of *conclusion* offer themselves. A selection of global, research-based summaries, all combine to give a picture of 'where we're at' or ... where we seem to have ended up.

(*Visiting aliens may well be intrigued by how we live by the technology of the* TLA *– the three-letter acronym. There never seem to be enough* TLA*s...*)

Setting aside older familiars such as materialist Standard of Living (SOL) and its value-richer comparator Quality of Life (QOL – see, e.g., WHO for one definition), we can compare approaches of, say, WWF with WEF. The World Wildlife Fund (Johnson et al., 2020) and Ranger et al., (2023) of the ECI (Environmental Change Institute) point to the trillions of dollars that loss of global nature costs and how our global problems are as much economic as environmental. Meanwhile, the World Economic Forum explores ways to sustain growth-driven (capitalist) economic models (WEF, 2024). Compare, too, the constrained and embedded GDP (Gross Domestic Product) with the rather richer GPI (Genuine Progress Indicator, n.d.). There is a plethora of governmental, institutional and academic reports (TLA and otherwise) which offer perspectives on the sorry state of where we are at and what our collective prospects amount to. Before presenting four fuller perspectives to illustrate the current state of play, a word on a word …

Globally, we now witness crisis upon crisis – some quick (pandemic, war) some slow (climate change, loss of biodiversity, ocean degradation) – and, when multiple crises overlap, coincide or interplay they have the capacity to trigger further crises. We are now living the *polycrisis*. Writing in the Financial Times, Tooze (2022) describes the polycrisis in the context of economics (he attributes the term's origin to French complexity theorist Edgar Morin). As Tooze puts it: 'A problem becomes a crisis when it challenges our ability to cope and thus threatens our identity. In the polycrisis the shocks are disparate, but they interact so that the whole is even more overwhelming than the sum of the parts.' (Tooze, 2022). For a critique of the term 'polycrisis' see Lähde (2023).

Three of the following perspectives address the polycrisis. The fourth takes a look at three powerful technologies.

2.1 *The SDGs (Sustainable Development Goals)*

In 2015, guided by their principle of 'Leave no one behind', the United Nations presented their 17 Sustainable Development Goals as part of a 'shared blueprint for peace and prosperity for people and the planet, now and into the future.' describing the goals as 'an urgent call for action by all countries – developed and developing – in a global partnership. (United Nations, n.d.). The goals were targeted towards 2030 and, while none mentions technology overtly, there is (arguably) a case to be made for technology and technology education to play a role in all 17 goals.

Subsequently, annual progress reports have been published with the most recent (United Nations, 2023b) marking halfway to the 'deadline' of 2030. The report makes bleak reading noting that progress on over half of the targets is 'weak and insufficient' and that, on 30 per cent, it has 'stalled or gone into reverse' (United Nations, 2023b, p. 2) and cites the UN's Secretary-General thus:

'Unless we act now, the 2030 Agenda will become an epitaph for a world that might have been.' (United Nations, 2023b, p. 78).

In parallel, the United Nations commissioned quadrennial reviews by an independent group of scientists. Fortunately (for this author) the constructions of 'science' and 'scientists' in the report allow for a richly ecumenical approach regarding knowledge, and the transformational, and synergistic practices necessary globally for quickly achieving the SDGs. 'Science' reads as 'knowledge' rather than any particular discipline and there is no mention of STEM (perhaps a TLA that co-opts a vowel to create a soundbite).

The scientists' report (United Nations, 2023a) echoes the 2023 Progress Report noting that 'the world is far off track' regarding the SDGs and that even greater efforts are needed across many fronts and across disciplines. (United Nations, 2023a:XVII–XVIII). In the report, 36 indicators, each ranked by 5 descriptors (1 = Very far from target; to 5 = Target met or almost met) are used to assess the progress towards the 17 SDGs. Only three of the 36 are reported to have trended forwards in the 2020–2023 period. The report argues for science that is 'multidisciplinary, equitably and inclusively produced, openly shared, widely trusted and embraced, and "socially robust" and relevant to society'. (United Nations, 2023a, p. XXVII). Such richness is invigoratingly refreshing given dominant institutional silo organisation and positivist constructs such as STEM.

2.2 *The 26 GTPs (Global Tipping Points)*

For a different take on where we're at, Lenton et al. (2023) have compiled a substantial analysis with proposals for action regarding 26 Global Tipping Points.

A tipping point is where a small change can make a big difference to a system. Associated terminology talks of tipping points as *thresholds* often abruptly crossed when *triggered* by a small change of conditions. The resulting state will often be *self-perpetuating* and difficult or *impossible to reverse*. The researchers suggest that there are five major tipping points on the verge of being crossed while three others threaten for the 2030s. Once a tipping point has been reached the full resulting damage will be far greater than any initial impact – with a resulting *cascading* effect.

While, globally, the principal concerns and perceptions are (accurately) of negative tipping points, the review is written with some optimism toward positive ones. These, however, require concerted and rapid action. 'Currently, there is no adequate global governance at the scale of the threats posed by negative tipping points. The world is on a disastrous trajectory. (and the authors call for ...) ... a new global governance framework.' (Lenton et al., 2023, p. 10)

The report cites six tipping points in the cryosphere, 16 in the biosphere and four in the oceans and atmosphere. Some of these are no longer high-impact, low-likelihood events but are 'rapidly becoming high-impact, high-likelihood events. (Also) Earth's tipping systems can interact in ways that destabilise one another, making tipping "cascades" possible.' (Lenton et al., 2023, p. 20). Across the report, it is noteworthy that 'technology' and 'science' are accorded their due respect and recognition, each in their own right, the culturally embedded inappropriate compound of 'science and technology' being eschewed. It is the report's contention that BAU (business-as-usual) is no longer tenable, that is, that BAU is unsustainable.

2.3 *The 14 Anthropocene Traps*

The term *Anthropocene* is used to depict the current geological time period which, it is argued, is that period which is being shaped significantly by human action – via our exploitation of resources, capacities to alter environments, impact the climate, and so on. Whilst the term is gaining traction as a classification, it remains contentious with geologists.

Søgaard Jørgensen et al. (2023) have contributed one (of 18) articles for an impressive theme issue of Royal Society Philosophical Transactions on evolution, sustainability and the Anthropocene. The authors discuss whether the accelerating change and complexity of current global challenges (the polycrisis) constitute a form of trap for humanity. They develop the biological concept of evolutionary traps in the context of humans and, thus, discuss *anthropogenic traps*. They present 14 traps subdivided as five global, five technology and four structural and suggest that 12 of these could be well advanced and, as with tipping points, very hard to reverse. Such traps can also interact to reinforce each other in ways that strengthen the severity of their impact. Rarely do they serve to lessen negative impacts.

While this work is largely theoretical and is advocating new approaches for collaborations across evolutionary science and sustainability science, it is nonetheless a welcome addition to a collection of lenses to help us see 'where we are at' and to help us consider where we could be. Equally welcome is the refined application of 'technology' (rather than the adjectival 'technological') as one of the three framing categories.

2.4 *Three Brief Takes on the Artificial (i.e. That Which Is Not Natural; Contrived; Made by Humans)*

First, it seems to me that the terms *artificial intelligence* (AI) and *machine learning* are, linguistically, quite problematic. Nevertheless, these phenomena, taken alongside *robotics* and other data-based technologies (including the ill-named 'social' media), have brought us to a condition that is both beyond the comprehension of most people and is powerfully reshaping, even undermining, life and society as we know them. Second, using gentle nomenclature from the world of nature, is *The Cloud* with all of its associated computational kit. With a greater carbon footprint than the global airline industry, a soft and fuzzy cloud it is not (Gonzalez Monserrate, 2022). Third, we are now entering the age of the *post-human*. This is no simple concept or unitary philosophical field but, as a so-called intelligent species, we are hopelessly ill-educated and under-prepared for what is currently manifesting. I share a few indicative sources about what 'we' are 'making':

Articulating *technological singularity*, Vinge contended: 'Within thirty years, we will have the technological means to create superhuman intelligence. Shortly after, the human era will be ended.' (Vinge, 1993). See Kurzweil (e.g. 1999, 2005) over the same decades on the *technological posthuman* and Bostrom (2003, 2009, 2019) on *transhumanism*. More recently, calling for a pause in all AI development, over 1000 AI experts wrote:

> AI systems with human-competitive intelligence can pose profound risks to society and humanity … (and they report on the) … out-of-control race to develop and deploy ever more powerful digital minds that no one – not even their creators – can understand, predict, or reliably control. (Bengio et al., 2023)

The circumstances that underpin the four perspectives, above, portray 'where we're at'. Put another way, they portray *the world we have created*. Yet, as I have argued:

> I, for one, deny having created, or having had any part in creating, the technological world. It must have been somebody else – not me. I also refute the implication in the 'we'. I didn't join with others in creating this world. And so, we distance ourselves from, absolve ourselves of, any responsibility for technologies. The 'we', of course, is comfortably ambiguous - all-embracing yet anonymity-guaranteeing. It's the species, not me. 'Include me out' of any part in the decision making. (Keirl, 2006, p. 83)

Perhaps others might concur. So to other conclusions …

I begin by taking a back bearing to the atomic bomb era of the 1940s. (I record, here, my gratitude to the film-makers for the timely delivery of Oppenheimer for this chapter!). The ethical, existential and political reverberations following the use of the bomb were global and last to this day. The great philosopher of technology, Lewis Mumford, reflecting on the gravity of the situation facing humanity, concluded in his chapter 'Program for Survival' that: 'Mankind (sic) … cannot give itself time to idle at the oars, while deliberating at length over the changes that must be made in our educational and political institutions'. He stressed, using a rare application of italics: *'This means that there is no part of our modern world that we must not be ready to scrap, if the need to scrap it is the price of mankind's* (sic) *safety and continued development.'* (Mumford, 1946/1956:229. Original italics). Food for thought, indeed, for considering where we're at for our future prospects. One thing does seem sure: BAU is no longer tenable (My penultimate TLA).

While 'apocalyptic' is the word usually associated with weapons of mass destruction we are broadly aware of other forms of ending. Thus, one conclusion is that we are simply living our natural or (pre-)determined (but not chosen, singly or collectively) path towards post-humanity: 'Technology is evolution by other means' said Kurzweil (1999, p. 14). We could shape this conclusion otherwise and discuss the range embraced by the term *existential threat.* Each of the tipping point thresholds, Anthropocene traps, and artificial creations can be considered separately or in any combinatory form as threatening existence/s. They occasionally appear in the foreground but, certainly in combinatory form, creep forwards with subtlety in the background.

Whither our prospects? Or, to engage the wordplay ... w(h)ither our prospects? Which cards do we play: optimism, pessimism, chance, faith, fate, hope – none of which seem too scientific or practical, although I wouldn't doubt the debates here? As we turn to such questions, we join the PATT family.

> I shall mention no further TLAs other than perhaps MdV to whom this festschrift is dedicated. Although a potential move from acronyms to anagrams does invite speculation on whether Marc de Vries drives a Merc.

3 Family Fortunes

(The name of a long-running television quiz show – formerly Family Feud in the US)

Marc has often talked of the 'PATT family' and there is no doubt of that as a reasonable descriptor - an extended family for sure but a welcoming one across the decades. (Many are the academic gatherings one can attend and feel estranged or isolated.) For sure, this family has its stalwarts, cliques and individuals. Distant relatives (by technological blood or marriage) drop in now and again. The family is particularly good at hosting family gatherings around the globe where all are embraced. It has its introverts and extroverts; its conservatives and its boundary-pushers; its theoreticians and its practitioners.

Taking a curriculum perspective, I'm not sure that technology education has ever really known its proper home so I'm also not sure we can talk of a diaspora in the sense of being dispersed from the homeland but scattered we surely are – geographically and educationally. I won't elaborate but little has changed since Layton (1994) documented his six 'principal actors' in Technology Education's curriculum drama. Such are the multiple scenes in our own family drama that I often recall the view of a dear friend, a former Design and Technology teacher, who had witnessed yet another curriculum configuration (in England of course). He suggested the subject be dismantled and cast to the three winds of the practical to Art; the sociological to Social Studies; and the theoretical to Science. Meanwhile, it's quite common for family members to join the STEM (whatever that is) club. Perhaps a PATT family group could form a cell.

Such is the PATT family: bringing together, under a technology umbrella, a largely dedicated bunch of comfortable companions with a broad range of knowledges, philosophies, practices and passions shared in not-too-disputatious ways.

3.1 *Introduction/s?*

For reasons that should now be apparent, the Mumford text I have drawn on for both title and some of the inspiration for this chapter is *The Human Prospect.* In response to the brief that this book's authors received, I'd now like to introduce a group of six *considerations* for how we might strive for a better state of affairs.

First, and paramount, wherever we are and whatever our roles, I suggest we take time to *consider just what it means to be human.* Doing this calls for temporary suspension of the everyday stuff of our lives. This exercise should be a never-ceasing part of universal education.

Such consideration must inevitably extend to all other species, the planet and our technologies – three entities without whom/which we cannot exist (that is, be human). Hopefully, our considerations would embrace concepts such as humility, empathy, peace, genuine sustainability, and remembering that the true measure of 'progress' (the jester in technology's court) is how we look after the vulnerable (regardless of species or planet).

> Until man's (sic) personal disciplines and social controls are capable of mastering the immense forces he now commands and putting them to wise use, he must desist from aggravating the danger by fostering further effort along the lines that have brought on the present crisis. (Mumford, 1946/1956, p. 250)

Second, there's the puzzle of what to do about the Big C (formerly cancer, now capitalism). To challenge C invariably brings boos, sighs, guffaws and moans (to use the academic parlance) along with the usual charges of naivety, or 'there's no alternative?'. Notwithstanding, I have yet to be shown that this pernicious technology (for C is just another technology) of competition, profit, ownership, exploitation, inequity and much more, is either logical or ethically defensible or sustainable. While the debate continues around the *Anthropocene*, the case arises for the *Capitalocene* (see Arons, 2023)

Third, I'd call for *healthy* education to help counter the unhealth of Big C. Currently, most education 'systems' serve economies first and democracy second. Testing regimes, intra- and between-country, are not only competitive but also shape curricula. The subjects being taught and tested are privileged at the expense of others. Curriculum is increasingly state-controlled and being narrowed. While the Arts and Humanities are being intentionally marginalised by STEM curricula across the planet today, we can reiterate Mumford's call to 'multiply the number of anthropologists, sociologists, and psychologists, (and to) diminish their opposite numbers in the physical sciences: (thus) endow schools of humanities and *curtail schools of technology*.' (Mumford, 1946/1956, p. 251. My italics). Significantly, 'education' is largely uncritical. Learning to question, to critique and to debate – in social, personal, political and philosophical senses – is an unwelcome threat to the status quo but a prerequisite of healthy education and democracy.

Fourth, despite humankind's intimate co-existence with technologies and their seeming invisibility to us (we are the fish, they are the water), our proper understanding, and critique, of them is shallow. It is time to interrogate this co-existence which we've taken for granted for millennia. Matters are compounding as we realise that the balance of our Human-Technology co-existence tips towards technological dominance. Thus, a *particular* technology education is a prerequisite for the aforementioned healthy education.

In turn, the fifth consideration invites Technology education to pay respect to the full extent of technology's philosophical richness (witness, e.g. Scharff & Dusek, 2003; Kaplan, 2004). Herein lie the foundations for a re-conceptualisation of the field – away from the current orthodoxies of the economy, applied science, skilling, kit, professions, and more. Since the technological-philosophical range is extensive I'd nominate ethics as a *locating field* to lead from – being concerned at its core about right motives, right actions and what constitutes good. Also, I'd probably take Hickman's (2001) productive pragmatism for the public sphere as a useful *philosophical tool*. However, I'd maintain in the foreground a collection of characteristics that apply to all technologies which could be applied to frame potential healthy education.

Thus, sixth, considering some of technologies' common characteristics offers a quite different educational picture than any from the status quo. We might consider how:

- technologies constitute a key to what it means *to be human* – personally and as a species;
- all technologies (allowing for those created by other species) come to be as a result of human *intentional acts* of changing one state of affairs into another i.e. by *design*. I include, here, those generated by programmes which had their genesis in prior technologies;
- all have *five 'phases' to their being* (lightly articulated in Keirl, 2009) namely: *conception, design, realisation, use*, and *consequences*. Applying, for example, critique and ethical interrogation to each phase is educationally far more rewarding than doing so with the whole (which usually presents as a thing or seemingly classifiable entity);
- technologies are usually developed *without full forethought or ethical consideration*. Codes of practice and legislation come after problems become apparent. The Precautionary Principle has little voice in the rush to market;
- no singular body of knowledge is drawn upon when technologies are created – nor when they are engaged with. Technologies in their fullness are *multi-disciplinary*;
- they are *multistable*, that is, they can 'be' in different ways in different circumstances (Ihde, 2002);
- they are *polyvalent* (Sclove, 1995), that is, they afford, and can be afforded, multiple values. Thus, no technology is universally good. All technologies have *contested values* and are non-neutral. Most have unintended consequences;
- technologies embody politics and powers that are largely *undemocratically created and distributed*. (Feenberg, 1999, 2002). (Name one technology whose design you were invited to contribute towards).

To educate for the above as a basis for improving our collective planetary prospects, no school 'subject' is enough. Cross-disciplinary 'learning areas' do offer richer learning opportunities however, for (design and) technology education, STEM is a model of worst-practice while STEAM merely strains the acronym into colonisation. Ultimately, it is essential that a *rich technological literacy* becomes the norm (see, e.g. Dakers, 2006, 2014). That is, a technological literacy that permeates the whole curriculum just as language literacy or numeracy must do. Here, every teacher has a role and truly integrated curricula, as witnessed in many primary years settings, offer invaluable models.

Polycrises are arguably a function of technological determinism – the kind of underpinning philosophy of those who uncritically say that technology has something to do with 'progress' and that it can't be stopped so that's why we are 'where we're at'. However, as Warnock (1998) says, ethics is anathema to determinists because to act ethically – to consider right motives and right actions – is to act towards the future, towards where we want (and don't want) to be.

'Where we're at' can be seen as the result of our cumulative errors. It such a situation, it is possible to take time to re-calculate, perhaps with several back-bearings and a few breaks, (witness Bengio et al, above) in order to get to where we want to be. The prospects for all – our own species, millions of other species, and our planet – call for us to read our metaphorical terrain, map and compass in harmony. The situation its less about getting to map reference X than always knowing all of: where we've come from, where we're at, and where we are heading but, of the three, it's knowing where we're at (at every step) that is key to the prospects of the children we're leading.

4 Prologue

A long time ago (for the matter under discussion), Stephen Hawking said:

> 'Our future is a race between the growing power of our technology and
> the wisdom with which we use it.
> Let's make sure that wisdom wins.'
> (Hawking, 2018:196).

> Oblate spheroids are quite rare.
> Barker (n.d/2022)

References

Arons, W. (2023). 'We Should Be Talking about the Capitalocene', *TDR: The Drama Review, 67*(1), 35–40. Accessed February 8, 2024, from https://doi.org/10.1017/S1054204322000697

Barker, L. (n.d.). *Oblate Spheroids*, Personal communication 3rd December 2022.

Bengio, Y., Russell, S., Musk, E., Wozniak, S., ... & Hogarth, I.,. (2023). *Pause Giant AI Experiments Open Letter*, Future of Life Institute, Pub. 22 March 2023. Accessed December 4, 2023, from https://futureoflife.org/open-letter/pause-giant-ai-experiments/

Bostrom, N. (2003). 'The Transhumanist FAQ: A general introduction', World Transhumanist Association. https://nickbostrom.com/views/transhumanist.pdf

Bostrom, N. (2009). The Future of Humanity. In J. K. B. Olsen, E. Selinger, & S. Riis, (Eds.), *New Waves in Philosophy of Technology,* pp. 186–215. Palgrave Macmillan.

Bostrom, N. (2019). The Vulnerable World Hypothesis. *Global Policy, 10*(4), 455–476.

Dakers, J. R., (Ed.). (2006). *Defining Technological Literacy: Towards an epistemological framework*. Palgrave Macmillan.

Dakers, J. R., (Ed.). (2014). *Defining Technological Literacy: Towards an epistemological framework* (2nd ed.). Palgrave Macmillan.

Feenberg, A. (1999). *Questioning Technology*. Routledge.

Feenberg, A. (2002). *Transforming Technology: A critical theory revisited.* Oxford University Press.

Genuine Progress Indicator (GPI). (n.d). Wikipedia. Accessed February 8, 2024, from https://en.wikipedia.org/wiki/Genuine_progress_indicator

Gonzalez Monserrate, S. (2022). *The Staggering Ecological Impacts of Computation and the Cloud.* MIT Press. Accessed December, 9, 2023, from https://thereader.mitpress.mit.edu/the-staggering-ecological-impacts-of-computation-and-the-cloud/

Hawking, S. (2018). *Brief Answers to the Big Questions*, John Murray.

Hickman, L. A. (2001). *Philosophical Tools for Technological Culture: Putting Pragmatism to Work.* Indiana University Press.

Ihde, D. (2002). *Bodies in Technology*. University of Minnesota Press.

Johnson, J. A., Baldos, U., Hertel, T., and others. (2020). *Global Futures: modelling the global economic impacts of environmental change to support policy-making. Technical Report*, January 2020. https://www.wwf.org.uk/globalfutures

Kaplan, D. M. (Ed.). (2004). *Readings in the Philosophy of Technology*, Rowman & Littlefield.

Keirl, S. (2006). Ethical technological literacy as democratic curriculum keystone. In J. R. Dakers (Ed.), *Defining Technological Literacy: Towards an epistemological framework* (pp. 81–102). Palgrave Macmillan.

Keirl, S. (2009). Seeing Technology Through Five Phases: a theoretical framing to articulate holism, ethics and critique in, and for, technological literacy. *Design and Technology Education: An International Journal, 14*(3), 37–46. https://openjournals.ljmu.ac.uk/DATE/article/view/2274

Kurzweil, R. (1999), *The Age of Spiritual Machines: When computers exceed human intelligence.* Allen & Unwin.

Kurzweil, R. (2005). *The Singularity is Near: When humans transcend biology*, Penguin.

Layton, D. (1994), A School Subject in the Making? The Search For Fundamentals. In D. Layton (Ed.), *Innovations in Science and Technology Education* (Vol. V, pp. 11–28). UNESCO.

Lähde, V. (2023). The case for polycrisis as a keyword of our interconnected times', in *Aeon*, 17th August 2023, Melbourne: Aeon Media. Accessed February 8, 2024, from URL: https://aeon.co/essays/the-case-for-polycrisis-as-a-keyword-of-our-interconnected-times

Lenton, T. M., Armstrong, D.I., McKay, S. & others (Eds.). (2023). *The Global Tipping Points Report 2023*. University of Exeter, Exeter, UK.

Mumford, L. (1946/1956). *The Human Prospect,* London: Secker & Warburg.

Ranger, N., Alvarez J., Freeman, A., & others, (2023). *The Green Scorpion: the Macro-Criticality of Nature for Finance – Foundations for scenario-based analysis of complex and cascading physical nature-related risks*. Oxford: Environmental Change Institute, University of Oxford. https://www.eci.ox.ac.uk/sites/default/files/2023-12/INCAF-MacroCriticality_of_Nature-December2023.pdf

Scharff, R.C., & Dusek, V. (Eds.). (2003). *Philosophy of Technology: The technological condition – an anthology*. Blackwell Publishing.

Sclove, R. E. (1995). *Democracy and Technology*. The Guilford Press.

Søgaard Jørgensen, P., Jansen, R. E. V., Avila Ortega, D. I., & others. (2023, November). Evolution of the polycrisis: Anthropocene traps that challenge global sustainability. *Philosophical Transacti*ons of the Royal Society B: Biological Sciences, *379*(1893). Accessed December 9, 2023, from https://royalsocietypublishing.org/doi/10.1098/rstb.2022.0261

Tooze, A. (2022). Welcome to the world of the polycrisis, 28th October 2022, London, Financial Times. Accessed February 8, 2024, from https://www.ft.com/content/498398e7-11b1-494b-9cd3-6d669dc3de33

United Nations Department of Economic and Social Affairs (UNDESA). (n.d). *The 17 Sustainable Development Goals*. Accessed February 8, 2024, from https://sdgs.un.org/goals

United Nations. (2023a). *Independent Group of Scientists appointed by the Secretary-General, Global Sustainable Development Report 2023: Times of crisis, times of change: Science for accelerating transformations to sustainable development,* United Nations, New York, 2023. Accessed February 8, 2024, from https://sdgs.un.org/sites/default/files/2023-09/FINAL%20GSDR%202023-Digital%20-110923_1.pdf

United Nations. (2023b). *The Sustainable Development Goals Report: Special edition. Towards a rescue plan for people and planet,* New York: United Nations. Accessed February 8, 2024, from https://unstats.un.org/sdgs/report/2023/The-Sustainable-Development-Goals-Report-2023.pdf

Vinge, V. (1993). *What is The Singularity?* Accessed November 26, 2023, from http://mindstalk.net/vinge/vinge-sing.html

Warnock, M. (1998). *An Intelligent Person's Guide to Ethics,* Duckworth.

World Economic Forum (WEF). (2024). The Future of Growth Report 2024, Geneva, World Economic Forum. Accessed February 8, 2024, from https://www3.weforum.org/docs/WEF_Future_of_Growth_Report_2024.pdf

CHAPTER 26

Technology Ethics at Work

Innovation and Education

Maarten J. Verkerk

1 Introduction

I have worked with Marc de Vries for many years. Everything indicates that he is an engineer: he thinks like an engineer, he behaves like an engineer and manages his own life like an engineer. He is also a passionate teacher: he is fully committed to education and the innovation of education. Particularly in the field of technology ethics. Finally, Marc is a philosopher. Specifically, a philosopher of technology. Through his thinking, he wants to contribute to the education of engineers and thereby to the flourishing of society.

This chapter is about engineering. About developing new innovations and the values that need to be upheld. It also concerns renewing the engineering practice itself and therefore about how engineers think about technology and how they develop it. Philosophy is needed for this. Philosophers can precisely indicate which types of values are at play and should be translated into design specifications. This chapter ultimately focuses on education. How do we teach students to innovate and use innovations in a responsible way? Is it possible to translate philosophical insights and ethical considerations into models, methodologies, and frameworks?

In this chapter, I will address two important developments in science and technology. Firstly, the focus is on social or organisational practices (Schön, 1991; Sandberg & Tsoukas, 2011; Nicolini, 2012; Verkerk, 2014; de Vries and Jochemsen, 2019). This involves how engineers work, what they say, and how they think. Secondly, the focus on value-sensitive design (Flanagan et al., 2008; Van den Hoven et al., 2015; Friedman & Hendry, 2019). This concerns the question of 'what is valuable' in technological products and systems for humans and society. I will show that these two developments are crucial to consider when thinking about the future of ethics education for students.

© MAARTEN J. VERKERK, 2025 | DOI:10.1163/9789004710948_026

2 Three Stories

Through my network, I connected with Fred Holtkamp. Fred was an orthopaedic engineer. He was at his wits' end. He was researching the quality of orthopaedic devices and told me that many patients are dissatisfied with their devices. Sometimes they are so dissatisfied that they put their (expensive) device in the closet because it causes more pain and discomfort than it alleviates complaints. He couldn't put his finger on this problem and asked me for help. Especially because he wanted to improve education – he taught orthopaedic technology at a college. We have worked closely together for several years. This cooperation has resulted in a couple of publications (e.g. Holtkamp et al., 2016; Verkerk et al., 2017; Van Dijk et al., 2022).

Jan Harmsen is an old acquaintance of mine. Occasionally we would run into each other. Jan is an engineer at heart. At that time, he was working as a consultant in the process industry. At one point, I told Jan that I had developed a new model with engineers to ask the right ethical questions. Jan was sceptical. At one point he challenged me. He said, 'I doubt if ethics and philosophy can help me. I am soon going to write a book on process intensification. But go ahead and show me what you have to say.' Two years later, our book *Process Intensification. Breakthrough in Design, Industrial Innovation Practices, and Education* (Harmsen & Verkerk, 2020) was published.

I am a member of a group of philosophers and engineers who regularly discuss texts with each other. Recently, the question was raised: Shouldn't we think more together about developing and using AI? A first inventory showed that the field is developing faster than ethical reflection. It turns out that there is even a pandemonium. There are no ethical frameworks embraced by engineers that are firmly grounded in philosophical ethics. This collaboration between engineers, ethicists, and philosophers led to the development of an ethical framework that aligns with the way engineers think and work and is firmly grounded in sociology, ethics, and philosophy (Verkerk & Voordijk, to be published).

Three stories. Stories that teach us how complex innovation is, how challenging the work of engineers is, and how intensive collaboration between engineers, ethicists, and philosophers can lead to solving practical problems and developing ethical frameworks. A key to this seems to lie in the concept of social or organisational practices.

3 Technology as a Social Practice

The notion of social practices was originally introduced by the philosopher MacIntyre (1984). Later, this concept was further developed by sociologists, ethicists, and philosophers (Nicolini, 2012; Verkerk et al., 2016; de Vries and Jochemsen, 2019). What all these approaches have in common is that (1) they take everyday practice as a starting point, (2) they recognise that there are different types of practices, (3) they emphasise that each practice is characterised by specific objectives, quality standards, virtues, and values, and (4) they acknowledge the complexity of practices and their interactions with the outside world.

This approach is also increasingly being used in the field of technology (Sandberg and Tsoukas, 2011; Verkerk, 2014; Rosenberger and Verbeek, 2015; de Vries and Jochemsen, 2019). The concept of social practices forces us to investigate what exactly happens in those practices. Who does what? Who talks to whom about what? Who decides on what? Which virtues, values, and norms come to the fore in daily practice? What are the power dynamics? The concept of social practices unlocks a rich complexity of phenomena that largely determine the actions, behaviours, and outcomes of social practices. Intuitively, we feel that there are great differences between technological practices. For example, the practice of developing electric cars is incomparable to that of developing an AI system to combat fraud. But there are also similarities. Practice models help us map out differences and similarities.

Figure 26.1 provides a three-layer representation of the practice approach, the Triple S model (Verkerk & Voordijk, to be published). The first 'S' – the inner layer – is about the social practice itself. It concerns its actors, values, and methods. The second 'S' – the middle layer – refers to its stakeholders. These are parties that are at a greater distance from the practice, have a certain interest, and want to exert a specific influence. The third 'S' – the outer layer – is about society: citizens, networks, and organisations. This layer also includes the collection of opinions, ideals, and motives that 'resonate' in society – the spirit of the time.

The three-layer model invites us to ask various questions. It starts with the question: What kind of social practice is it about? Is it about fundamental research? Is it about technological development? Or is it about the professional use of technology? The next question is: Which sector does it concern? Agriculture, food industry, oil industry, healthcare, education, sports and leisure, or arts? The layers of stakeholders and society also lead to various specific questions. Other questions relate to the dynamics within the three layers themselves. For example, we can ask questions such as: (1) who are the key actors? (2) what values do they consider important, and (3) how can we translate those values into design requirements? Finally, it is also exciting to look at how the different layers influence each other. Among other things because the participants in social practices are also citizens of society.

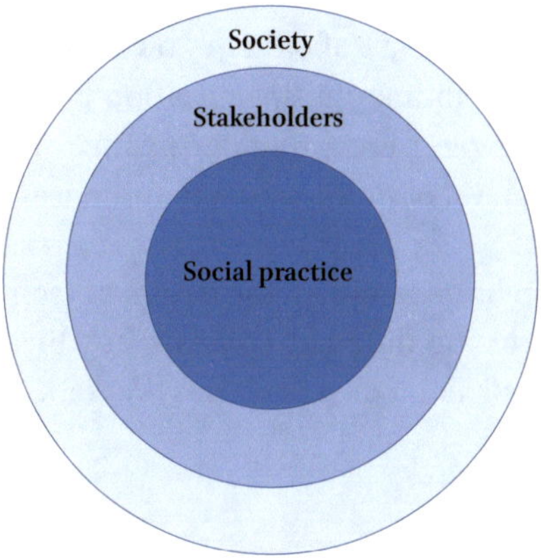

FIGURE 26.1
Triple S model

4 Understanding User Practices

Fred Holtkamp is an orthopaedic engineer. One of his research interests is the quality of ankle-foot orthoses. This orthosis is externally applied and intended to control the position and motion of the ankle. Reasons for the use of this aid vary from congenital malformations to diseases and traumas such as poliomyelitis, multiple sclerosis, and cerebrovascular accident (stroke), resulting in an aberrant gait. The main quality problem of ankle-foot orthoses is non-use and dissatisfaction. Even satisfied and very satisfied users have serious complaints concerning the quality of the device.

I had several conversations with Fred about this problem. We focused on the logistic chain and the different practices within it. A breakthrough in our thinking occurred when we highlighted the question: What does the life of patients look like? In what contexts do they use this device? Does the orthopaedic engineer – who measures for orthoses – really know these contexts well enough? Shortly after, we coined the term 'user practices' for the different contexts in which patients use their orthosis.

Using various methods – e.g. interviews, audio recordings, and video recordings – we conducted research on the key user practices in patients' lives and to what extent the orthopaedic engineer discusses those practices with the patient and translates them into design requirements. From this research, it was found that the most important user practices are daily life at home, transportation, work, and leisure (Holtkamp et al., 2016). To determine the design requirements for an orthosis, it is necessary to inventory which movements and activities are done in each of these practices for each patient. What did our research show? During intake, orthopaedic engineers address only 25% to 50% of the movements and activities that patients do in their daily lives while using the orthosis. Negatively formulated, 50% to 75% of patients' movements and activities are not considered when determining the specification of the device (Holtkamp et al., 2016; Van Dijk et al., 2022). From that perspective, it is not surprising that there is a relatively high level of non-use and that there are many complaints about the quality.

The lesson of the foot-ankle orthoses can be generalised: the design of technology requires a detailed understanding of the different contexts in which that technology is used. In other words, it's all about user practices (Verkerk et al., 2017).

5 Developing Sustainable Industrial Practices

The chemical process industry is facing the challenge of innovating its processes to contribute to the Sustainable Development Goals of the United Nations. In particular, the transition from fossil energy to renewable energy is a significant task. One way to achieve this is through process intensification. The core of this approach is integrating multiple process steps into one step in a chemical process. The promise is that the renewed process consumes less energy, has lower capital costs, and has lower operating costs.

Jan Harmsen's objective was to write a book in which successful process intensifications would be described and lessons for the future could be learned. I told Jan about the Triple S model. I explained that in the inner layer, the values of engineers are central, that in the middle layer attention is focused on the (justified) interests of various stakeholders, and in the outer layer, societal ideals or beliefs are discussed.

I told Jan that I was convinced that the reason why there are so few successful examples of process intensification can be found in the three layers. Jan looked at me incredulously. He said, 'I have read all the articles, and no one talks about values, interests of stakeholders, and ideals and beliefs in society.' I replied to him, 'I believe you won't find these considerations in the various articles. But that's because engineers don't think in these terms. Go interview them and ask the right questions. Ask about their values, the stakeholders they collaborate with, and the pressure from society.'

A few weeks later, he called me excitedly. 'I interviewed a researcher who was at the forefront of one of the biggest successes in the field of process intensification. He really enjoyed that I asked him all kinds of questions about values, stakeholders, and ideals. He kept talking non-stop. It was as if he always wanted to tell this story and now finally had the opportunity.'

In total, we have spoken with the inventors of five successful process intensification projects. We have learned the following. First, the values of engineers play a crucial role in success. Second, R&D, M&S, and Operations – which can be described as different practices – collaborated closely without any organisational or cultural barriers between them. Third, there was close collaboration with relevant stakeholders: potential customers, research institutes, equipment suppliers, etc. In other words, in the course of time innovation networks were established. Finally, the pressures, interests, and values of the local context played a significant role.

The case of process intensification clearly shows that the practice model provides a powerful methodology for researching technical practices. It offers insight into the nature of different practices (R&D, M&S, Operations), the interaction between those practices, and the networks that stakeholders form with each other. It also provides insight into success factors: people's motivation, the values that play an important role, collaboration between practices and the formation of innovation networks (Harmsen & Verkerk, 2020).

6 Ethical Framework for AI

Recently, I had a meeting with a group of engineers, ethicists, and philosophers about artificial intelligence. The question was: Can we develop an ethical framework for engineers developing AI, professionals using AI, and policymakers developing policies related to AI? This question is particularly relevant. First, because the field of AI is developing at a rapid pace. Can ethics keep up with this speed? Is it possible to develop a framework that easily adapts to developments in the field? Secondly, because the current situation in the field is not encouraging. There is pandemonium. There are many codes and guidelines developed by engineers who have little basis in ethics or philosophy. In addition, there are publications by philosophers and ethicists that have a good academic level but do not sufficiently connect with the world of engineers, professionals, and policymakers. This situation is highly unsatisfactory. Especially when technology is developing rapidly and the urgency for an ethical framework is very high considering the expected societal effects. Our initial question was: Could the practice model in combination with Dooyeweerd's philosophy provide a solution?

In this group, engineers worked together with ethicists and philosophers to develop a practical ethical framework. This framework is based on the layered model discussed earlier. Figure 26.2 provides a summary. We start with a pre-question about the practice that is being discussed. Then we focus on the three layers. For each layer, we ask three questions: (1) Which actors are relevant? (2) Which values do the different actors consider important? and (3) How to translate abstract values in concrete design requirements. We conclude with

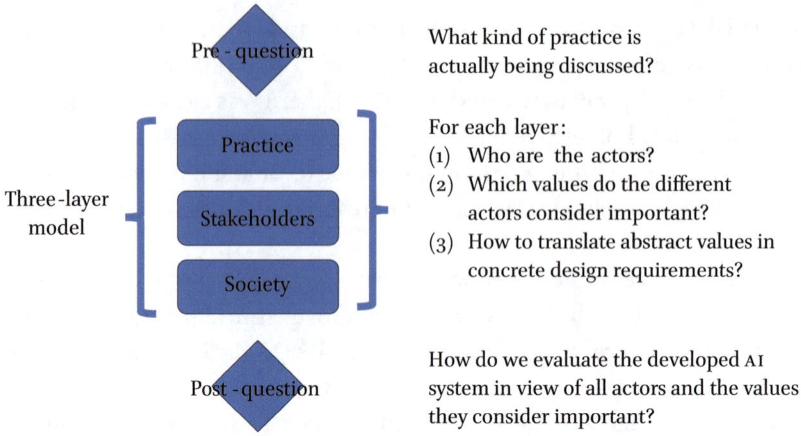

FIGURE 26.2 Ethical framework for AI development, professional use of AI, and AI policy

a post-question: How do we evaluate the developed AI system in view of all actors and the values they consider important?

The pre-question is very important. What kind of practice is it? Is it a technological practice where AI is being developed? Is it a professional practice where AI is being used? Or is it a policy practice where concrete policies regarding AI are being developed? Answering this question is relevant. After all, each of these three practices are different: both in terms of the social practice itself, the stakeholders, and the influence of society.

The next step is the analysis of the three layers based on the previously mentioned questions. This analysis must be done separately for each layer. We start with the first layer. The first question is about the actors in practice. These can be both internal and external actors. For example, the doctor in a medical practice is an internal actor (he or she belongs to the organisation) but the patient is an external actor (he or she comes from outside). The second question is about the values that are important for each actor. We make a distinction between core values and supporting values. Core values refer to the essence of a practice, e.g. loving care, empathy, and respect for a medical practice. Supporting values refer to values that are conditional, e.g. economic, and juridical values in a medical practice. Finally, the third question is about translating values into specifications. For this, we use a modified model by Van de Poel (2013), see Figure 26.3.

An intriguing and extremely important question is how we can identify the relevant core values and supporting values. The philosophy of Dooyeweerd provides a framework for this (Verkerk et al., 2016). First, based on Dooyeweerd's systematics, it is possible to present an overview of all types of values that come into play, see Figure 26.4. This overview can be used as a checklist or

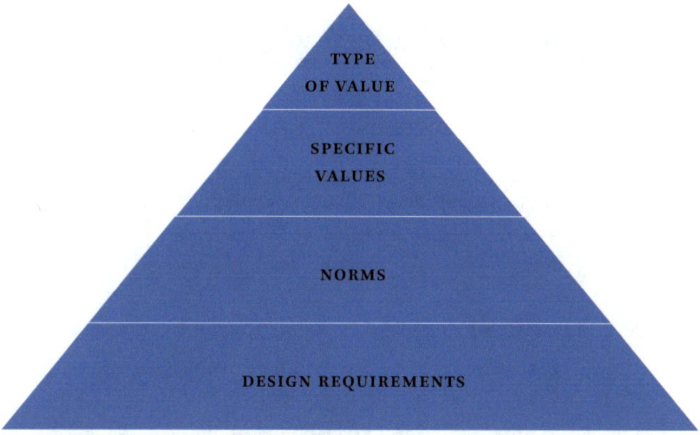

FIGURE 26.3 Translation of abstract values in design requirements

Type of values	Specific values
Faith values	Belief, existential certainty, spirituality, trust, compassion
Moral values	Love, care, healing, respect, empathy, integrity
Juridical values	Constitutionality, legality, human dignity, justice, privacy, retribution, equality
Aesthetic values	Beauty, harmony, joyful, balance
Economic values	Profitability, productivity, stewardship
Social values	Solidarity, autonomy, intercourse, communion, interconnectedness, respect
Lingual values	Symbolic meaning, signification, clarity
Formative values	Serving needs, design competence, controlling power, responsible design, creative design
Analytical values	Rationality, logical,analytical approach

FIGURE 26.4 Different types of values and a first elaboration thereof

tick-off list. Second, it is possible to create a table showing the core values and their specific values for different practices or organisations, see Figure 26.5. We would like to note that an analysis has to be conducted for each specific AI development, application, and policy to identify the types of values involved and inventorying the relevant specific values for each type.

7 Teaching Ethics

How to teach ethics to engineers developing AI, to professionals applying AI, and to policymakers developing policies? From experience, it appears that the well-known philosophical ethics (virtue ethics, duty ethics, and utilitarian

Practices/ organisations	Type	Specific values
Churches, mosques	Faith values	Spirituality, respecting God, transcendence
Healthcare, charities	Moral values	Loving care, closeness, empathy, respect
Authorities, courts, parliaments	Juridical values	Public justice, flourishing of society, security
Museums, orchestras, music bands, art expositions	Aesthetic values	Beauty, harmony, creativity, playful
Business, banks, financial agencies	Economic values	Profitability, reliability, quality
Friends, clubs, social networks	Social values	Coziness, support, respect, connectedness
Publishers, reading clubs, libraries	Lingual values	Meaning, symbolic meaning, richness, aesthetic, creativity
Engineering companies, design agencies, factories	Formative values	Serving needs, creativity, responsible design, power and control, robustness
Research institutions, policy agencies	Analytical values	Rationality, logical, analytical approach

FIGURE 26.5 Practices, organisations, types of values, and specific values

ethics) do not provide sufficient guidance. Firstly, because these ethical theories still need a significant translation into the various AI practices. But also, because engineers, professionals, and policymakers are accustomed to thinking in terms of methodologies and frameworks that provide them with sufficient guidance.

Marc de Vries writes: 'We need to work on a technology education practice that convinces governments, school administrators and school boards, teachers, parents, and pupils that teaching and learning about technology is not only fun but also a necessary component of every (future) citizen's general education (de Vries, 2006, p. 9). The challenge that De Vries formulates for education in technology applies even more specifically to education in the ethics of technology. The following must be added. In Western society, neoliberal values set the tone. Everything revolves around the individual and the realisation of financial or economic values. I do not want to diminish the importance of these values. But I do want to address their one-sidedness. It is not just about individuals but also about communities and society. It is not just about economic values but also, for example, formative, social, artistic, juridical, moral, and religious values.

Educational practices, technological innovation and education are closely interconnected. The three cases show that engineers should not be trained as 'specialists' but as participants in a technological practice. Engineers must learn to understand the nature of their own practice, conduct an analysis of relevant values, and translate them into design specifications. In addition, engineers need to map out the entire field within their organisation and identify the different networks that are relevant. Ultimately, it all comes down to actors, values, and design specifications again.

Most students start a technical study with the goal of mastering their chosen field. Whether it be civil engineering, medical technology, artificial intelligence, or any other subject, they are passionate about making an impact in their respective industries. In my experience, when speaking with these students, they not only want to contribute but also want to make a meaningful difference. Students, with few exceptions, recognise that technology can enhance the well-being and prosperity of individuals and society as a whole. They are aware that simply relying on technology is not enough, but they lack the vocabulary to articulate this understanding. They also lack the language, methodology, and guidelines necessary for addressing these broader considerations. Ethics is frequently viewed as a challenging obstacle rather than an integral part of design processes.

The cases presented in this chapter are real-life examples, providing insight into various fields of expertise and the challenges they are currently facing. Each case introduces specific vocabulary and methodology that can be utilised to integrate ethical questions into the design process. In Figure 26.6, we outline a methodology or framework for incorporating ethical considerations into design practices. The are two pre-questions: In what kind of practice am I working? Which other practices do I need to know very well and analyse further? After that, an analysis of the three layers of the relevant practices follows. The methodology closes with the post-question: How do we evaluate the developed system considering all stakeholders and the values they find important?

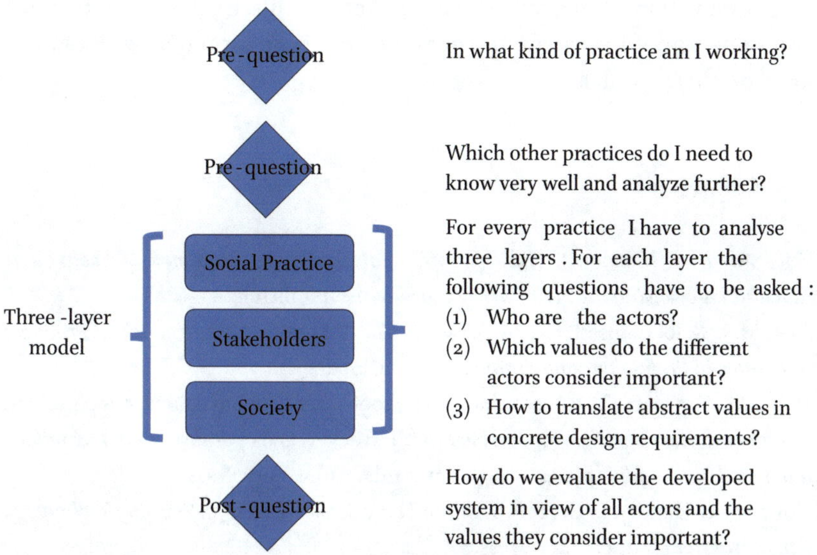

FIGURE 26.6 Methodology or framework to integrate ethics in developing technology, using technology, and developing technology policy

8 Conclusions

This chapter is about ethics at work. It is about how ethics can be integrated into education in such a way that technology students see ethics as an integral part of the design process. I would like to draw a couple of conclusions.

The first conclusion is that the combination of a practice approach, value-sensitive design, and the philosophy of Dooyeweerd offers a fruitful basis for developing both a language and methodology for integrating ethical considerations into the design process.

The second conclusion is that the practice approach provides insight into technological practices. This approach challenges engineers to reflect on their own practice and explicitly address the question of which other practices they need to understand to do their work well.

The third conclusion is that the Triple S model provides a tool for engineers to map all relevant values of their own practice, relevant stakeholders, and society, and how to translate them into design requirements.

The fourth conclusion is that Dooyeweerd's philosophy insight provides an overview of all types of core values and supporting values that could be relevant to practice.

The conclusion is that the three case studies, the practice approach, and the values-sensitive design collectively lead to a language and methodology or framework for teaching ethics at a technical university or college.

I would like to end with an invitation. I invite Marc to delve deeper into the topics of innovation, ethics, and education that were discussed in this chapter – as a culmination of his career. It would be a privilege for me if we could collaborate on this together.

References

de Vries, M. J., & Mottier, I. (Eds.). (2006). *International Handbook of Technology Education: Reviewing the Past Twenty Years*. Sense Publishers.

de Vries, M. J., & Jochemsen, H. (Eds.). (2019). *The Normative Nature of Social Practices and Ethics in Professional Environments*. IGI Global.

Flanagan, M., Howe, D. C., & Nissenbaum, H. (2008). Embodying values in technology. Theory and practise. In J. Van den Hoven & J. Weckert (Eds.), *Information technology and moral philosophy* (pp. 322–353). Cambridge University Press.

Friedman, B., & Hendry, D. G. (2019). *Author Value Sensitive Design: Shaping Technology with Moral Imagination*. MIT Press.

Harmsen, J., & Verkerk, M. J. (2020). *Process Intensification. Breakthrough in Design, Industrial Innovation Practices, and Education.* Berlin: De Gruyter.

Holtkamp, F. C., Verkerk, M. J., Van Hoof, J., & Wouters, E. J. M. (2016). Mapping user activities and user environments during the client intake and examination phase: An exploratory study from the perspective of ankle foot orthosis users. *Technology and Disability 28*, 145–157.

MacIntyre, A. (1984). *After Virtue. A Study in Moral Theory* (2nd ed.). University of Notre Dame Press.

Nicolini, D. (2012). *Practice Theory, Work, and Organization: An Introduction.* Oxford University Press.

Rosenberger, R., & Verbeek, P.P.C.C. (2015). A field guide to postphenomenology, In R. Rosenberger & P. P. C. C. Verbeek (Eds.), *Postphenomenological investigations: Essays on human-technology relations* (pp. 9–42). Lexington Books.

Sandberg, J., & Tsoukas, H. (2011). Grasping the logic of practice: Theorizing through practical rationality. *Academy of management review*, 36(2), 338–360.

Schön, A.A. (1991). *The Reflective Practitioner. How Professionals Think in Action.* Farnham: Ashgate.

Van de Poel, I. (2013). Translating Values into Design Requirements. In D. MichelFelder, N. McCarthy, & D. Goldberg (Eds.), *Philosophy and Engineering: Reflections on Practice, Principles and Process.* Vol. 15 of Philosophy of Engineering and Technology. Dordrecht: Springer (253–266).

Van den Hoven, J., Vermaas, P.E., & Van de Poel, I. (2015). *Handbook of Ethics, Values, and Technological Design: Sources, Theory, Values and Application Domains.* Springer.

Van Dijk, L. S., Holtkamp, F. C., Verkerk, M. J., & Wouters, E. J. M. (2022). Better understanding the life of the user to improve the design of orthopedic footwear. *Technology and Disability* 34, 271–277.

Verkerk, M. J. (2014). A philosophy-based 'toolbox' for designing technology: The conceptual power of Dooyeweerdian philosophy. *Koers – Bulletin for Christian Scholarship* 79 (3), 1–7.

Verkerk, M.J., Hoogland, J., Van der Stoep, J., & de Vries, M. (2016). *Philosophy of Technology. An Introduction for Technology and Business Students.* Routledge: Abingdon on Thames, UK.

Verkerk, M.J., Holtkamp, F.C., Wouters, E.J.M., & Van Hoof, J. (2017). Professional practices and user practices. An explorative study in health care. *Philosophia Reformata, 82*, 167–191.

Verkerk, M.J., & Voordijk, J.T. (to be published). Ethical Framework. Three-layer Actor Values Model. In E. Araujo, C. Van Burken, M. J. Verkerk. (Eds.). *Ethical framework for Artificial Intelligence. Layers, actors, and values* (working title).

CHAPTER 27

Technology and Tradition

Roel Kuiper

Marc de Vries is a professor in the area of technology and a man of tradition. Reflecting on the impact of the latest technological innovations he will never forget his background, as a man of faith and learning. This combination of technology and tradition might seem rare in a high-tech environment where students are trained to look forward and develop technological solutions for the future. In philosophy, this combination is not rare, especially not in critical thought. The future is not a technological project, even when people can still have their utopian dreams of control, progress and human autonomy. There are strong voices arguing that we should become more humble and modest, less progressive and more cautious, rather than learning from history and then rejecting it, in order to avoid technological catastrophes. There are philosophers of technology, like Marc de Vries, distancing themselves from the *hybris* of man in a 'technological society'. Here is my question: what is the importance of the wisdom of traditions with regard to modern expectations of technology?

In this short contribution, I will make a few remarks about this notion of tradition in relation to technology in reformational philosophy. As Hendrik van Riessen (the first representative to be mentioned here because he taught in Delft) has demonstrated in his dissertation *Filosofie en Techniek* (1949), technological design and form-giving are a matter of culture.[1] Human beings do have the capacity to mould and develop things, instruments, and objects, in short: they create culture. Therefore, he stated, there is always something historical in technology. When Dooyeweerd and Vollenhoven, the two 'fathers' of this philosophy, discussed this 'historical aspect', Vollenhoven wrote that this had to do with the element of 'power, including tradition and technè'.[2] Interestingly, Vollenhoven combines the notion of 'technè', which refers to making designs, instruments and projects, with the notion of 'tradition'. Tradition means that we never start from scratch, we get ideas, manners, observations, rules, and standards, handed over from previous generations.

One would expect that reformational philosophy, in its criticism of human autonomy and modernistic expectations of technology, should have paid more attention to this notion of tradition. That is not the case. When we consider the main publications since Van Riessen it is hard to find this notion at all. Van Riessen himself made some brief remarks about the historical embeddedness

© ROEL KUIPER, 2025 | DOI:10.1163/9789004710948_027

of technology. In technological projects, there is always an idea of form-giving, which is partly determined by historical patterns. These remarks about the historical dimensions of technology reflect perhaps the influence of Vollenhoven on his early work, I would suggest. Van Riessen's focus however, and also the focus of reformational philosophers of technology that came after him, was elsewhere: the enclosure ('ontplooiing') of new technological objects and culture in general.[3] Technological development in a positive sense was seen as a differentiation of cultural possibilities, always moving away from the past.

My impression is that these philosophers of technology followed Dooyeweerd in his account of the historical-formative aspect of cultural activities. Dooyeweerd differed with Vollenhoven on this point and dropped the notion of tradition deliberately.[4] He acknowledges that tradition indeed is a 'meaning-moment' of that aspect, but rejected it at the same time as a notion to work with.[5] Traditions, he admitted, can guarantee continuity with the past, but they don't have a normative voice in the formation of culture. The cultural mode of form-giving was defined by Dooyeweerd as 'a controlling manner of moulding after a free project'.[6] To be free from history was important for him. To be bound by traditions would hinder a culture and human communities from moving forward. We should not be held captive by 'traditionalism and conservatism', wrote Dooyeweerd, and not forget 'that the fulfilment of the cultural task of mankind demands a continuous striving forward'.[7]

These are problematic expressions. They uncover a modern idea about moving forward to a future with endless extended possibilities and progressive solutions. Today we are impressed by the darker sides of progress and especially by technological progress that causes large-scale devastations of natural and cultural environments. Some scientists have become *AI doomers*, believing that the world will be destroyed by computers. Dystopias become more realistic in late-modern minds than utopias. How then can we endorse this 'continuous striving forward' in a technological society without reservations? In order not to derail we should find more value in traditions, I suggest. Philosophically, we could return to the combination of ideas that Vollenhoven thought to be complementary: tradition and technè, technology embedded in patterns of historical wisdom. This movement to what can be valued in traditions seems to be a key lesson to be learned today. Isn't this a task for philosophers of technology, especially when they feel inspired by this school of reformational thought?

What does this suggestion of keeping technology and tradition together entail? The notion of tradition is essential, first of all, because all human practices are built on knowledge and skills of the past. We are always culturally bound to what is handed over to us. The notion of tradition is also essential in a larger sense, to collect the wisdom of the ages before us in order to see how we can build on seasoned insights. A culture is also a conversation between the generations. Modernism did not invest in such a conversation. One of the reformational philosophers at Delft, Ad Vlot, once remarked that in technological projects the 'sense for time, traditions and the coherence of history is eroded'.[8] It is 'technoscientific thought', as he phrased it, that stood in the way of this sense of tradition. Vlot is arguing for a 're-integration' of technology in a broader historical framework. He argued even for spiritual renewal, another understanding of time and wisdom. The philosopher Egbert Schuurman, also one of Marc's predecessors in Delft, often emphasised the need for a 'reformation of our technological culture'.[9]

As a man of faith and learning Marc de Vries touches in his writings at the very same point, understanding with Vlot that technology always turns to the future. He is very critical of the utopian representation of an ideal world shaped by technology. In one of his many articles Marc de Vries points critically to the technological dreams about a world without limitation and with endless possibilities. But this belief in 'makeability' causes 'wicked problems', he argues.[10] The world that should be controlled is running out of control. Wicked problems in a technological world are the antinomies and paradoxes that arise from a one-sided technological approach, lacking (ethical and other) standards in dealing with the consequences of modern technology and its applications. In understanding the nature of these wicked problems Marc de Vries talks about a more 'coherent framework' that we need, including our broader worldviews.

Although he did not mention the role of traditions, I assume that Marc de Vries is open to a more Vollenhovian idea of doing technology. My point is that one cannot reform the dominating attitude in Western culture, this progressive idea of continuously striving forward, without a thorough reorientation with regard to what should be learned from the past, and having a respectful conversation between the generations. I cannot elaborate on this notion of tradition at length here, but will conclude with some observations.

The idea of tradition revives in several ways today, and I will address two of them, very briefly, suggesting they might direct the way ahead. Firstly I will discuss the growing interest in professional practices, including the tradition of professions. Secondly, I will discuss the importance of education and learning to introduce broader pathways of wisdom for our time.

1. In 2019 Marc de Vries edited *The Normative Nature of Social Practices and Ethics in Professional Environments.* This book was written against the background of what he called 'the crisis in professionalism'. A functionalist and technical phase in the development of professions has been followed by a postmodern phase leaving professionals feeling dehumanised

and alienated from their work.[11] Professions, however, are important settings for life, requiring more competencies than just the ability to use a set of skills. They are connected with 'communities of practices' (Wenger) that include broader meaningful cultural and social contexts of life. By the end of the twentieth century, new ideas about professional practices became prominent and Marc de Vries and others launched the idea of the Normative Practice Approach (NPA). The NPA is inspired by Alasdair MacIntyre who defined a practice as a socially embedded set of professional activities according to standards.[12] In MacIntyre's thought, there is a traditional side: the history of a profession is part of its embeddedness. In the NPA this returned as a normative element of every practice, however not strongly articulated in this model. But it is an element at the 'constitutive side' of practices, where practices start and have their historic foundation. This element of tradition guarantees continuity, as it should, and gives a practice a sense of its roots and limits. My point is that this element of professionalism should play a bigger role in our reflection on practices, because it hands over the knowledge of practitioners and their work in the past, their ethical standards and their failures. The crisis of professionalism is that it reduced professions as something just technical, whereas thinking of profession as part of practice brings them back home to the everyday reality of society and history.

2. We can reappraise traditions in relation to technology in the context of professional practices, we can also do that in the context of society at large and education. The criticism of reformational philosophers has been that science and technology combined with cultural ideas about human autonomy and modern individualism has created a one-sided and reductionist 'technological society'. We have become instrumentalists, refusing to look back and learn from history, living without limitations. This attitude contributes to the crises we experience today in society in our care for one another, in our stewardship of planetary systems, in health care, in public governance and in safety policies. There is a sense of darkness and disorientation in all the broken promises of modernism, a profound lack of perspective. Some still believe that technoscientific solutions will set us free, others are convinced that the technoscientific approach brought us where we are now and therefore cannot serve as a remedy. In the precarious world of AI technology, there are *slow-downers* and *accelerationists*. At this critical point in history, it is important to be aware of the fact that our responsibilities do have historical dimensions. We should learn from the traditions that guided humankind in the past facing the challenges of their times. This is why MacIntyre was talking

about the virtues and their rehabilitation for today's world. We need to learn from the conversation between the generations to overcome our limited views and moral ignorance, we need to learn from religious and ethical traditions and what they hand over to us from the wisdom of former generations. Here is a task for schools, and educators, for 'civic education'. We need a new awareness of our historical responsibilities, our connection with past and future generations, and the way traditions can inform us how to make our technological choices. For Marc de Vries Biblical wisdom is part of this education, it is at his heart. Hopefully, he will get new opportunities in the years to come to hand over that wisdom as an essential part of the traditions that we need in a technological society.

Notes

1 H. van Riessen, *Filosofie en techniek* (Kampen: Kok, 1949), 636–637.
2 D.Th. Vollenhoven, *Isagôgè Philosophiae* (VU Uitgeverij, 2010), 109.
3 Egbert Schuurman, for instance, criticised modern ideas of human autonomy behind modern technology and at the same time explained that human beings that obey the laws of God can see the future as progression. *Techniek en Toekomst* (Assen: Van Gorcum, 1972), 353, 358.
4 Vollenhoven maintained it, while Dooyeweerd kept making his objections.
5 Herman Dooyeweerd, *A New Critique of Theoretical Thought* (Amsterdam. Philadelphia, 1953), II, 230. Abbreviated as *NC*.
6 Dooyeweerd, *NC*, II, 231.
7 Dooyeweerd, *NC*, II, 242.
8 Ad Vlot, 'Cultural Diversity and Technological Development' in: *Philosophia Reformata* 66 (2001), no. 1, 58–59.
9 For instance in his 'The technological culture between the times', in: Sander Griffioen and Bert Balk (editors), *Christian Philosophy at the Close of the Twnetieth Century. Assessment and Perspective* (Kampen: Kok, 1995), 197.
10 Marc de Vries, 'Wicked Problems in a Technological World', in: *Philosophia Reformata* 85 (2020), no. 2, 125–127.
11 Marc de Vries and Henk Jochemsen, *The Normative Nature of Social Practices and Ethics in Professional Environments* (IGI Global, 2019).
12 Zee: Alasdair MacIntyre, *After Virtue. A Study in Moral Theory* (London: Duckworth, 1985), 187. See for his argument that practices do always have histories, 220–222.

CHAPTER 28

Whose Idea Is It Anyway?

Designing with Generative AI

Alice Schut and Christian Detweiler

1 Introduction

Design is all around us, and the practice of design has always taken a key position within our world. The importance of design is widely recognised and accepted, which is why it is a crucial component within most technology education curricula (McDyer et al., 2023). Designing is an inherently creative activity (Barlex, 2007; Goldschmidt, 2016). This is due to the open-ended nature of design, with ill-structured problems and undefined solutions (Dorst, 2003). When practising design, students cannot fall back on any existing formula with a 'right answer. Instead, they need to engage in creative processes to generate and develop possible solutions. Even though creativity and design go hand in hand, this does not mean it is necessarily easy for design students to behave creatively (Schut et al., 2019).

In this chapter, we zoom in on creativity in design through the dependence and interplay that exists between designers and the use of tools. Creative design processes do not take place within a cognitive vacuum. The material (and social) world in which designers operate plays a substantial role in how designers address the problems they are working on. Within design education, especially competency-based design education, technology education seems to implicitly adopt the view of design tools as instruments that passively follow the designer's will. Here, we want to offset that view by arguing that designers might not be fully autonomous and that creativity in design also relies on the interplay between designer and tool. In other words, the designer and their tools are co-constitutive of creative practice.

While this view already has its implications when considering 'traditional' design tools, we specifically aim to address this in relation to the rise of a 'new' category of tools utilising Generative Artificial Intelligence (GAI). At this moment, various GAI tools[1] have made their way into numerous domains, and design practices and education are no exception. When looking at creativity in design as a co-construction of a designer and their tool, these new GAI tools

© ALICE SCHUT AND CHRISTIAN DETWEILER, 2025 | DOI:10.1163/9789004710948_028

bring about a whole set of different implications than the preceding tools used by designers.

In the upcoming sections, we will outline a revaluation of the position of tools within a creative design process, specifically in relation to the current rise of GAI tools, and discuss the potential implications on creative design practices. In conclusion, we advocate for the integration of reflective tool practices within design education as an essential component of design creativity.

2 The Active Role of Design Tools

Tools play a fundamental role in creative practices. At the very least, they are a means through which materials can be manipulated to produce creative works (e.g., using a pencil to create marks on paper using a chisel to shape a piece of wood). It is tempting to see tools as merely a means in the creative process, objects (pencil, chisel) that a designer uses to manipulate some material (paper, wood) into a form that matches some idea already formed in the creative practitioner's imagination. In this view, creativity takes place in the mind, and tools are instrumental in realising creative ideas. A contrasting view of tools in creative practices is that they determine or bring about certain design outcomes. Using design tools often involves performing a sequence of tasks to produce a certain output (e.g., select text box, drag text box onto onscreen "canvas", select font, type). In a sense, the design tool prescribes a set of methodological steps, and through this tool, its use determines an aspect of design practice.

Another perspective on the relationship between tools and creativity attributes creative influence to both humans and tools and holds that tools are inseparable from creativity. Following this perspective, a work of creative expression emerges from the relationship between the person creating, the material in the tools, or the *artist-tool-material system* (Baber, 2023). Tool use in creative practice does not only involve offloading some aspects of (creative) cognition to the tool, for example, quickly sketching a design for later recall; external representations do not only externalise aspects of the problem space that the practitioner would otherwise have to retain in short-term memory. Rather, they are physical things "that allow epistemic actions to be performed on them" (Baber, 2023). In this way, tools constrain not only the processes involved in creating the work (e.g., using a Bezier curve tool versus using a pencil tool in image editing software) but also constrain the cognitive processes involved in the conceptualisation of the work (Baber, 2023). So, the materials of design - both the tools and the materials being manipulated to form the eventual design or prototype - are co-constitutive of creative practice.

Jane Bennett's notion of "vibrant materiality" takes this view a step further, arguing that we need to come to terms with the "capacity of things [...] not only to impede or block the will and designs of humans but also to act as quasi agents or forces with trajectories, propensities, or tendencies of their own" (Bennett, 2020, p. VIII). In the context of creative practice, tools then have their agency–they have the power to shape the creative process and its outcomes. They can be seen as active participants that contribute to the creative flow. Creative acts are then neither fully an outcome of a designer's will nor fully determined by the designer's tools. Rather, creative action is an effect of some hybrid of designer-tool-material, where the boundary between subject and object is not so clear-cut, and both humans and non-human tools have a form of agency.

From a design cognition perspective, the integrated perspective on tools and the designer is supported by the Extended Design Cognition Theory. This theory gives equal attention to the internal and external factors involved and emphasises an integrated model of design cognition (Blom, 2022; Blom & Bogaers, 2020; Haupt, 2015). Thinking does not only reside in a designer's mind and body but extends beyond into the social and physical environment. This notion is connected to the work of Menary (2012), who identifies enculturated cognitive actions – interactions with external resources like tools and representations. Blom (2023) goes on to explain how these are not just extensions of thought but *active components* shaping designers' cognition during their design activities. This reinforces the idea that tools are not merely passive extensions but active participants in the design process. However, this perspective is not currently prevalent within design education, where a view of tools as being subordinate to a designer's will is dominant.

Competency-based design education focuses on design skill acquisition. In this type of education, curricula tend to prioritise mastery of specific tools (e.g., prototyping software) over a deeper understanding of tools' constitutive roles in creativity and the historical context in which specific design tools emerged. This approach implicitly adopts the view of design tools as instruments that passively follow the designer's will, as opposed to that of an active participant. Within this type of design education, design students develop skills that may or may not involve tool use. For example, a course might have a learning objective that students are able to use given interaction design techniques to create and demonstrate an interaction design. Students in such a course might learn about various interaction design techniques and apply those techniques in assignments to ultimately (further) develop the skill of designing interaction. Part of learning about those techniques could involve learning how to use tools that support the techniques, such as a prototyping tool to create wireframes. In such cases, learning tools often involve following instructions on which steps to take in the interface to produce a certain result (e.g., adding a representation of a button to a wireframe) and internalising those steps through repetition.

This approach to tool use in design education presents potential drawbacks when following an integrated view of designers and their tools. By following a tool's prescribed workflow, students are more likely to adopt the tool's "inscribed vision" of a (part of a) design process. This vision dictates which steps are involved in creating a low-fidelity prototype and in which order. This approach does little to encourage critical engagement with the tools designers use, which can lead to a limited view of their implications. In turn, this can lead to uncritical acceptance of tools' constraints and biases.

Generative AI tools are increasingly used in design processes, and it seems that our analysis of the relationship between designers, tools, and creativity also applies to these newer tools. However, these tools have several characteristics that complicate the analysis, which we will turn to in the upcoming section.

3 Generative Artificial Intelligence as a Creative Design Tool

Advances in deep neural networks have led to the widespread availability of Generative Artificial Intelligence (GAI) systems that can generate text, images, audio, and video in response to text prompts and other modalities. GAI systems synthesise images or other data by analysing training examples in vast datasets (of creative works) and learning patterns (e.g., Jovanovic & Campbell, 2022). Many of the currently popular tools were not explicitly or exclusively created to support creative processes. For example, ChatGPT "was trained to predict the next word in a document, using publicly available data (such as internet data)".[2] While others, such as Google's Magenta Studio, were created to explore "the role of machine learning in the creative process".[3] Nonetheless, these new GAI tools are used by many designers in different stages of their creative design process with the intent to support creative actions like idea generation or cheap and fast prototyping iterations.

So, designers have adopted these new GAI tools in their creative processes. However, the nature of these tools raises new questions about the relationships between designers, tools, and creativity. Some of these questions stem from characteristics of GAI systems—the way they work "under the hood": how datasets are built, how models are trained, and how systems are further refined. Other questions stem from how these systems and tools are presented to users, the kinds of interaction paradigms they use, and the kinds of expectations users form based on this. These characteristics have implications for the relationships between designers, tools, and creativity, which we will discuss now.

4 Artificial Creativity

GAI tools rely on vast datasets of what is essentially creative work (text, images, audio). Generative AI models learn patterns and structures from the data they are trained on, learning underlying features, including features of the dataset's creativity. These tools generate new content by replicating patterns and structures distilled from the training data. This can create the impression that the generated work is, in a sense, creative by replicating features of the training set, though some work argues it should not be seen as such and even that it should be labelled "artificial creativity" (Runco, 2023). Indeed, analysis of creative works using machine learning is limited because "we only hear what we listen for, and the machines will be conditioned by what we instruct them to do and thus potentially marginalising crucial elements of [...] expression which might go ignored if they are not registered for processing in the system" (Magnusson, 2019. p. 161). In an overview of creativity and machine learning Franceschelli & Musolesi (2021) describe the different ways in which creativity is aimed to be integrated into various machine learning models. They highlight the shortcomings of current models in terms of creativity in how they learn their space of solutions and how the observations are generated from them, as well as the difficulty and inadequacy of the creative metrics and methods proposed till now to evaluate the creativity of the artefacts generated by artificial agents.

Though such outputs arguably should not be considered creative, generative techniques often do have pronounced aesthetics; for example, some tools are built to produce images that are as "realistic" as possible (Epstein et al., 2023). Furthermore, these tools can reproduce biases present in their training sets; in other words, "the aesthetic and cultural norms and biases embedded in the training data of generative-AI models affect their outputs" (*ibid.*, p. 5). What's more, in training models, (algorithmic) decisions are made about which types of outputs to reward, for example, "common" versus "unique" inputs, which can decrease aesthetic diversity (*ibid.*).

These characteristics of GAI tools potentially affect creative practices. If a creative practitioner views a system's output as "creative" and uses it as such, then that person is incorporating machine-learned and replicated features in their creative process. Furthermore, if generated outputs reproduce aesthetic and cultural norms and biases present in the training data, then a user who uncritically uses the output may unintentionally include those norms and biases in his or her work. Cultural and aesthetic influences of a work can become obscured.

5 Black-Boxing Creativity

We touched on how generative AI models are trained to aim to 'include creativity' and the criticisms surrounding it. However, this does not yet touch on the opaque nature of the computational process itself. The interaction with, generative AI tools like ChatGPT, Dall-E, and Stability AI's Dream Studio, as of this writing, do not show any intermediate steps involved in generating the output—the process involved in assembling an image, an arrangement of pixels, that is the "best" match for a string of characters given as input. This process is not represented in any way, save for a "loading" indicator. For the user, there is no insight into the types of computation taking place, the model being used, the features represented in that model, how features are selected, how the output is synthesised, or how the output relates specifically to the prompt or the training set. There is no possibility to gain an understanding of how a potentially creative output is produced—"algorithms [...] decide *what is to be* in the single output signal" (Amoore, 2019, p. 155). This opaqueness obscures the 'creative' labour that went into producing the works present in the training set and stands in the way of giving credit where it is due. Furthermore, little insight is given into the materiality and spatiality of the infrastructures and resources involved in producing output. The output is "finished", high resolution, possibly "realistic", and seemingly detached from the materials (creative works, physical infrastructure, and resources) that make it possible in the first place.

Regarding the relationship between designers, GAI tools, and creativity, this "black-boxing creativity" can complicate the ability to locate human agency, raising questions about the extent to which a GAI system's output is based on a user's or somebody else's choices. These concerns have also been identified by teachers in craft, design and technology education, who recognised the difficulties in locating human agency in such black-box systems and the power that might give to its creators (Vartiainen & Tedre, 2023).

6 Dependence on the Designer's Initiative

Besides obscuring the process of output generation, GAI systems afford and structure specific types of interactions that contribute to expectations users form and how they integrate these systems as tools in their creative processes. The systems are often described in ways that suggest they are creative partners or collaborators of some kind. For example, part of OpenAI's product description for ChatGPT states that it can "generate, edit, and iterate with users on creative and technical writing tasks, such as composing songs, writing screenplays, or learning a user's writing style".[4]

Several researchers have experimented with the collaboration between GAI systems and designers to enhance and or stimulate creative processes and outputs. Examples of such co-creative systems are the Drawing Apprentice (Davis et al., 2015), The Sentient Sketchbook (Liapis et al., 2013) and the Collaborative Ideation Partner (Kim et al., 2021). The Drawing Apprentice is a co-creative drawing system that generates drawing 'answers' to users' drawing actions (Davis et al., 2015). Think for example of completing the rest of an object, drawing the same object in a different style, suggesting an object that matches the picture, or colourising. Through the interaction, the system is intended to spark alternative ways of completing the drawing than would be the case when working alone. The Sentient Sketchbook is intended to support game developers in the development of gaming levels (Liapis et al., 2013). While a designer edits a gaming map, the system tests it for playability and, from that, generates new and unexpected map alternatives to the users' current design. Through these suggestions, the tool intends to challenge the user's current design focus and provide unforeseen alternatives to achieving the designer's goals. The Collaborative Ideation Partner leverages generated images based on conceptual proximity to the design challenge at hand to stimulate the variety, quantity, and novelty of ideas generated by the designer (Kim et al., 2021). Through what is meant to be an iterative process, the designer can initiate the generation of inspiring images with a conceptual relevance to the design challenge, which can be used as a means to inspire and stimulate their idea generation.

In the examples described above, the 'collaboration' between a designer and GAI system relies on interaction paradigms that revolve around turn-taking, for instance, by directly modifying a shared artefact or by providing suggestions to the user rather than direct modification. In those systems, the turn-taking process and 'answers' that are generated are catered towards the creative stimulation of the user. For example, the 'answers' that are generated by the Collaborative Ideation Partner in the form of images with conceptual closeness and relevance to the design challenge intently utilise the strategy of Design-by-Analogy as a means to stimulate novelty, variety, and quantity of ideas (Kim et al., 2021; Moreno et al., 2014).

ChatGPT also utilises a turn-taking paradigm, in which the user prompts the system, which in turn takes this prompt to generate a response. However, there are no specific creative strategies behind the content generated to answer the prompt. This means that, when taking its turn, the 'answers' to the prompt will vary in their ability to stimulate creativity. However, unlike the previously described GAI systems, ChatGPT is not inherently designed to stimulate users' creativity in a co-creative manner for a specific context based on fitting creativity stimulation strategies. Instead, it is a system fitting for all types of use purposes. This means that the initiative of getting a more creatively stimulating 'answer' lies with the user prompting the system.

ChatGPT and similar systems' range of possible uses arguably make them highly adaptable for different creative tasks. However, the need for users to take the initiative introduces potential challenges concerning creativity. Such systems only produce output if prompted by a user, and the quality of the output depends, in part, on the quality of the prompt. People using such systems in a creative process need to be able to understand which type of output might help support them creatively. This understanding involves not only a reflection on their own creative needs but also knowledge of creative 'good practices', as most people are not aware of what actually might help stimulate their creativity. For example, to help kickstart an ideation activity for a design challenge, a user might prompt the system for a set of initial ideas to use as a starting point and move past the 'blank canvas' in front of them. Although the designer might experience this set of ideas as a creative stimulant, working from examples is a well-known cause of design fixation, which often has a negative impact on the creative process and output (Jansson & Smith, 1991). Novice designers are known to be more at risk of fixation on example stimuli (McLellan & Nicholl, 2009; Nicholl & McLellan, 2007), which gives reasons for concern when design students embrace GAI tools without proper guidance.

A more effective strategy would be to stimulate the ideation activity based on 'good practices' from research, such as Design-by-Analogy. Yet the user needs to, firstly, be aware of these strategies and, secondly, be able to prompt the system to correctly create output fitting to that strategy. Furthermore, utilising such a strategy is likely to be (perceived as) more labour-intensive and a less direct route toward a solution to the design challenge. Even if the designers can prompt the system correctly, they might opt for the path of least resistance instead. In short, the creative quality of the 'collaboration' between such multipurpose GAI systems and a designer is therefore highly dependent on the ability and initiative of the designer interacting with it. Using these tools in the context of design education, where novice designers are still developing these needed abilities, therefore creates a need for guided and reflective practices.

7 Shifting Creative Agency

Another factor that could influence this quality of use within a creative design context is related to the previously touched-on turn-taking paradigm that most GAI systems and tools utilise. It can be argued that this type of interaction further feeds into the separated view of tool and designer, which was already argued to be prevalent in the realm of design. What can be argued to strengthen this separation even further is the use of anthropomorphism. Various popular GAI systems are often described in anthropomorphising terms, such as "ChatGPT can see, hear, and speak" and "you can now use voice to engage in a back-and-forth conversation with ChatGPT"[5] (cf. Lipton & Steinhardt, 2019). They also offer certain forms of interaction that further contribute to their anthropomorphisation. For example, ChatGPT's interface is visually similar to a messaging or chat application. The text input field even says "Message ChatGPT". When ChatGPT produces output, text appears on the screen letter-by-letter, simulating words being typed, to give the impression of conversational, human-like interaction. It reinforces this impression by using the "I" pronoun in its output, for example, "The more details you provide, the better I can assist you".

This approach of externalising the tool to the designer, thereby almost making it another 'entity', can strengthen the agency that is ascribed to it. Most educators are likely not unfamiliar with the phrase "But it knows so much, and says it so convincingly". when students refer to an AI language system like ChatGPT. This perspective of students could harm a more critical consideration of the effects such a tool can have on their creative design processes.

The characteristics of GAI systems discussed above contribute to a situation in which, if they are used uncritically, they potentially diminish designers' creative agency. We have argued that tools used in creative practices already, in a sense, have agency in how they shape creative thought, perception, and action as part of a designer-tool-creativity system. GAI tools potentially shift the balance of agency in the designer-tool-creativity system towards the tool. They introduce relations to learned patterns, content, and creative labour that goes into the dataset—all decisions made "elsewhere", i.e., not by the designer—simultaneously obscuring these relations through limitations of the technology (it is currently almost impossible to know which specific elements of a training set contributed to a given output) and through how they structure interaction with users (e.g., by only presenting "finished" outputs). What's more, the way these tools are designed (e.g., the use of anthropomorphisation) potentially *invites* uncritical use. In turn, uncritical use can diminish designers' agency by relying on aesthetic judgments, cultural norms, and patterns of creativity produced "elsewhere" (i.e., present in the training set or produced in training the model).

8 A Need for Reflective Practices

As we have argued, competency-based design education's treatment of design tools does not adequately account for the entanglements between designers, tools, and creativity. This relationship is complicated by emerging generative AI tools, which potentially diminish designers' agency. This calls for a new approach to teaching tool use in design education. Such an approach should create space to examine the relationships between designers, their tools and creativity. It should explore how different tools influence creative processes differently and how each tool was developed in a specific context and reflects certain ideas about creativity and creative processes. Making the relationships between designers, tools and creativity explicit can help designers in training adopt a more critical-reflective approach to tool use in creative practice. In light of newer GAI tools, such an approach can help novice designers consider how these tools influence (enhance and constrain) their creative agency and think about the ethical implications (e.g., ecological impact and effects on creative labour) of using these tools. This kind of reflective tool use can help designers develop practices that balance agency between designers and tools in creativity.

Notes

1 We use the terms GAI systems and GAI tools interchangeably in this chapter. That said, when we use systems it generally refers to the larger collection of technologies (training data, computational infrastructure) that makes the tool that a user interacts with possible in the first place.
2 https://openai.com/research/gpt-4
3 https://magenta.tensorflow.org/
4 https://openai.com/product
5 https://openai.com/chatgpt

References

Amoore, L. (2019). Cloud ethics : algorithms and the attributes of ourselves and others.

Baber, C. (2023). Tool use and creativity. *The Routledge International Handbook of Creative Cognition* (pp. 570–584). Routledge.

Barlex, D. (2007). Creativity in school design & technology in England: a discussion of influences. *International Journal of Technology and Design Education, 17*(2), 149–162. https://doi.org/10.1007/s10798-006-0006-x

Blom, N. (2022). Design cognition in design and technology classrooms. In A. Hardy (Ed.), Debates in Design and Technology Education (2nd ed.). Routledge. https://doi.org/10.4324/9781003166689

Blom, N., & Bogaers, A. (2020). Using Linkography to investigate students' thinking and information use during a STEM task. *International Journal of Technology and Design Education, 30*(1), 1–20. https://doi.org/10.1007/s10798-018-9489-5

Davis, N., Hsiao, C.-Pi., Singh, K. Y., Li, L., Moningi, S., & Magerko, B. (2015). Drawing Apprentice. Proceedings of the 2015 ACM SIGCHI Conference on Creativity and Cognition, 185–186. https://doi.org/10.1145/2757226.2764555

Dorst, K. (2003). *Understanding Design.* BIS Publishers.

Epstein, Z., Hertzmann, A., Herman, L., Mahari, R., Frank, M. R., Groh, M., Schroeder, H., Smith, A., Akten, M., Fjeld, J., Farid, H., Leach, N., Pentland, A., & Russakovsky, O. (2023). *Art and the science of generative AI: A deeper dive. Science.* 380(6650), 1110–1111. https://doi.org/10.1126/science.adh4451

Franceschelli, G., & Musolesi, M. (2021). *Creativity and Machine Learning: A Survey.* http://arxiv.org/abs/2104.02726

Goldschmidt, G. (2016). Linkographic Evidence for Concurrent Divergent and Convergent Thinking in Creative Design. *Creativity Research Journal*, 28(2), 115–122. https://doi.org/10.1080/10400419.2016.1162497

Haupt, G. (2015). Learning from experts: fostering extended thinking in the early phases of the design process. *International Journal of Technology and Design Education*, 25(4), 483–520. https://doi.org/10.1007/s10798-014-9295-7

Jansson, D. G., & Smith, S. M. (1991). Design Fixation. *Design Studies, 12*(1), 3–11.

Jovanovic, M., & Campbell, M. (2022). *Generative Artificial Intelligence: Trends and Prospects. Computer*, 55(10), 107–112. https://doi.org/10.1109/MC.2022.3192720

Kim, J., Maher, M. Lou, & Siddiqui, S. (2021). Studying the Impact of AI-based Inspiration on Human Ideation in a Co-Creative Design System. Proceedings of the ACM IUI 2021 Workshops.

Liapis, A., Yannakakis, G. N., & Togelius, J. (2013). Sentient Sketchbook: Computer-Aided Game Level Authoring. International Conference on Foundations of Digital Games.

Lipton, Z. C., & Steinhardt, J. (2019). Troubling Trends in Machine Learning Scholarship. Queue, 17(1). https://doi.org/10.1145/3317287.3328534

Magnusson, T. (2019). Sonic Writing. Bloomsbury Publishing.

McDyer, M., Buckley, J., Dunbar, R., Blom, N., & Seery, N. (2023). Problematising and unpacking the uncertainty of design within technology education. PATT40: The 40th International Pupils' Attitudes Towards Technology Conference Proceedings. https://openjournals.ljmu.ac.uk/PATT40/article/view/1531

McLellan, R., & Nicholl, B. (2009). "If i was going to design a chair, the last thing i would look at is a chair": Product analysis and the causes of fixation in students' design work 11–16 years. *International Journal of Technology and Design Education, 21*(1), 71–92. https://doi.org/10.1007/s10798-009-9107-7

Menary, R. (2012). Cognitive practices and cognitive character. Philosophical Explorations, 15(2), 147–164. https://doi.org/10.1080/13869795.2012.677851

Moreno, D. P., Hernández, A. A., Yang, M. C., Otto, K. N., Hölttä-Otto, K., Linsey, J. S., Wood, K. L., & Linden, A. (2014). Fundamental studies in Design-by-Analogy: A focus on domain-knowledge experts and applications to transactional design problems. *Design Studies*, 35(3), 232–272. https://doi.org/10.1016/j.destud.2013.11.002

Nicholl, B., & McLellan, R. (2007). "Oh yeah, yeah you get a lot of love hearts. The year 9s are notorious for love hearts. Everything is love hearts". Fixation in pupils' design and technology work (11–16 years). *Design and Technology Education: An International Journal, 12*(1), 34–44.

Runco, M. A. (2023). AI can only produce artificial creativity. *Journal of Creativity, 33*(3). https://doi.org/10.1016/j.yjoc.2023.100063

Schut, A., Klapwijk, R., Gielen, M., van Doorn, F., & de Vries, M. (2019). Uncovering early indicators of fixation during the concept development stage of children's design processes. *International Journal of Technology and Design Education.* https://doi.org/10.1007/s10798-019-09528-2

Vartiainen, H., & Tedre, M. (2023). Using artificial intelligence in craft education: crafting with text-to-image generative models. *Digital Creativity, 34*(1), 1–21. https://doi.org/10.1080/14626268.2023.2174557

CHAPTER 29

Adapting the Design and Technology Curriculum

A Framework for Evaluation and Evolution

Alison Hardy

1 Introduction

The design and technology curriculum is not nor should be static. The professional and research aspects of design and technology create new knowledge. The nature of design and technology is that it evolves, and the subject content itself grows. For example, as new technological processes products and artefacts are developed (like artificial intelligence and autonomous vehicles) educators question whether and how they should be taught in schools. External political pressures also influence the subject's curriculum content. For example, STEM and STEAM came from political and economic ideas and challenges, that education was then expected to help address through changes to curriculum content and design. Although these agendas can have some positive contribution, primarily through funding and status (e.g. TVEI in the UK, the Perkins Act in America and Taiwan's focus on technology as a superpower), ultimately, they can lead to a lack of clarity about the nature of the subject. This leads to educators asking, debating or insisting that this new knowledge should be taught in schools. My view is when this happens it highlights a lack of consensus, dialogue and sensemaking for the subject and consequently, the planned curriculum is at risk of being overloaded with content. I liken the current style of curriculum evolution to the method described by the fictional Professor Peddiwell in *The Saber-Tooth Curriculum* (Benjamin 2004). Peddiwell recounts to Benjamin a curious tale about how the creation of a palaeolithic curriculum led to debate and tensions about the content. In England the design and technology content for one exam runs to over 60 pages, I think because we do not talk about what should be excluded as well as included.

In this chapter, rather than reimagining a new curriculum, I present an approach for evaluating proposed new content or direction for design and technology. Following this proposed approach would lead to an agreed recommendation for the new content or direction to be added to the curriculum, or rejected. Rather than viewing any changes to the curriculum as an endpoint, I think we should see each development as a resolution. Each curriculum

© ALISON HARDY, 2025 | DOI:10.1163/9789004710948_029

development and iteration is viewed as a design resolution, answering: what is relevant and appropriate for the design and technology curriculum now?

As with the development of a product, the curriculum needs to adapt, reduce, and maybe reject some longstanding features. The alternative is an awkward addition, a part that might not fit, potentially making historic content 'grates' or sit uncomfortably in the new resolution when it should have been removed.

The approach involves two parts. First, who makes the decision and how? Second, I propose seven criteria for evaluating changes to the curriculum content. I end with a brief case study demonstrating how the criteria might apply to some content design and technology educators and researchers are currently testing.

2 Community Decision

2.1 *Who Decides?*

In the land of the saber-tooth tiger (Benjamin, 2004), two groups tried to influence the curriculum – the "thoughtfuls" and the "radicals", but only one group controlled the school – the wise old men. When D&T was first mooted as a national curriculum subject, a forum (officially a "Working Party") was created to propose its content, purpose and aims (Department for Education and Science and the Welsh Office, 1988). McCormick's (1990) analysis of the group shows that members were design education advisors and consultants, education officers chief executive officers from non-government organisations and industry, a computer science university lecturer, and a professor of science education, plus two design and technology school department leaders. Totalling thirteen with a chair, the group comprised eleven men and three women. Through their membership, each had a stake in the development of D&T, but McCormick's analysis shows that there was an imbalance in the expertise and legitimacy of the group, he seems to ask – did the group have salience? In other words, were they important to or connected with the development of a D&T curriculum? Williams (2007) explores this when he uses Mitchell, Agle and Wood's (1997) typology of stakeholders to analyse the attributes different stakeholders have concerning a given D&T curriculum situation, such as implementing a new curriculum. I have developed the use of Mitchell, Agle and Wood's typology in my research on the value of design and technology, investigating the changing salience and values of stakeholders at different levels of curriculum influence and implementation (Hardy, 2017 unpublished thesis).

Modifying Mitchell, Agle and Wood's (1997) business-focused definitions with a D&T focus the three attributes are defined as:

- Power: A stakeholder 'has power to the extent it has or can gain access to coercive, utilitarian, or normative means, to impose its will' (p. 865) on D&T's nature, content or status.
- Legitimacy: a stakeholder exhibits their legitimacy through their desirable, proper or appropriate actions towards D&T. Their actions are deemed to be legitimate because they are constructed within a system of socially accepted norms, values and beliefs (Suchman, (1995) in Mitchell et al., (1997)).
- Urgency: 'the degree to which a stakeholder's claims [on D&T] calls for immediate attention' (ibid, p. 867). Urgency is a dynamic attribute that 'exists when two conditions are met: when a relationship or claim is time-sensitive and when that relationship or claim is important or critical to the stakeholder' (Mitchell et al., p. 867).

Mitchell, Agle and Wood identified seven classes of stakeholders each with a descriptive name and description arguing that each class has low, moderate or high salience defined as latent, expectant and definitive (Figure 29.1). Classes 1, 2 and 3 are *latent* stakeholders with only one attribute. Classes 4, 5 and 6 are *expectant* stakeholders with two attributes. The *definitive* classification (7) is the only group to have all three attributes and therefore the highest salience, that is they have the most prominence and are the most notable. Those possessing none of the attributes are non-stakeholders, or 'potential stakeholders' (Mitchell et al. 1997, p. 873). However, this model is dynamic and stakeholders from other groups can acquire other attributes by a change of situation or circumstance, such as coalition building or broadening the urgency to include other stakeholders (see Williams, 2007 for examples). But I think Mitchell, Agle and Wood's theory does not sufficiently explore the dynamic influence of the 'levels of these attributes' on stakeholder's salience. What I mean is, although Mitchell, Agle and Wood exemplify how a stakeholder can acquire attributes and therefore move between the seven classes they do not consider whether a stakeholder has more than one position in relation to the business, or if there are different levels within each class of stakeholder, such as national or local level. My view is that these different positions and levels are also dynamic, so using three of Van den Akker's (2003) four levels of curriculum, the committee's membership also needs to include stakeholders from each of the following levels:

- system/ society/ nation/ state (or macro).
- school/ institution (or meso).
- classroom (or micro).

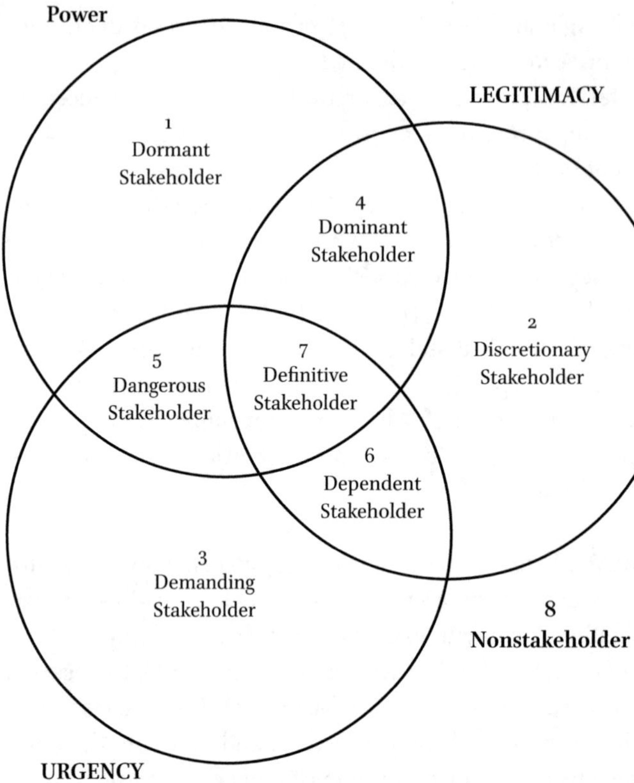

FIGURE 29.1 Stakeholder Typology (Mitchell, Agle and Wood 1997, p. 874)

Using this typology as a way of forming the "committee", I suggest focussing on the attributes of the members is more useful than on the who. Consequently, the group's attributes and salience would be consistent even when members left because their replacement's attributes would be the focus rather than their personality.

I suggest the committee includes members from these ten groups.

1. Curriculum policy writers because they know the policy and its interpretation.
2. Design and/or technology professionals who create new knowledge and will employ or train young people who come through with this new, or dysfunctional, knowledge.
3. D&T teacher educators and researchers ("D&T academics") who have expertise in D&T pedagogy and educational settings.
4. An ethical advisor who would ensure the panel and curriculum had integrity, transparency, and a commitment to promoting ethical education

practices education. The assumption too often is that the purpose of education is to feed a nation's economy; this member would be there, in part, to counter this viewpoint.

5. Exam boards that manage external qualifications.
6. Higher education (HE) design academics who work in one of the many fields involving design (e.g. product, fashion, architecture) and, like the design and/or technology professionals, create new knowledge and require qualified applicants for their courses.
7. Primary and secondary school D&T teachers as they know the limitations and potentials of curriculum changes.
8. School inspectorate representative: In England, this would be Ofsted.
9. School leaders who allocate resources to the subject.
10. Special Interest Groups (SIGs) organisations, such as subject associations and others might create professional development who work with individual teachers and national influencers. These groups have a legitimate and, at times, urgent interest in D&T.

This categorisation will cause some debate among readers; I do not expect agreement with my placement of the stakeholders, and some of the stakeholders' position is fluid and their influence could change if they pair up with another stakeholder group (Williams, 2007).

TABLE 29.1 Categorisation of each stakeholder at each level using the typology from figure 1

Stakeholder	Macro	Meso	Micro
Curriculum policy writers	7	1	8
Design and/or technology professionals	2	8	6
D&T academics	2	8	6
Ethical advisor	2	2	2
Exam boards	6	8	3
Higher education design academics	2	8	6
Primary and secondary school D&T teachers	6	6	7
School inspectorate	8	3	4
School leaders	8	7	1
Special Interest Groups (SIGs)	6	3	6

2.2 *How will They Decide?*

The group needs to talk but it is how they talk that is crucial. One approach would be for a democracy, where the members vote and if the new content has more "yes" than "no" then it's in. But this breeds discontent and disharmony. Instead, I propose the group engage in a dialogue to make sense of how the proposed curriculum change could work or not. For this, I am drawing on the work of Bushe and Marshak's (2015) dialogic organisational development. This means the group focus on the ideas and variation in opinion about the content rather than on the people. To do this, individuals come with a dialogic mindset that involves accepting that the group is making meaning of the proposed changes, the decision emerges from the group's dialogue and is negotiated (Bushe & Marshak, 2015).

By taking a dialogic approach the implicit idea is that curriculum development is about continual development, not a start and end point. The group realise there will not be a perfect curriculum. Instead, they see the curriculum as evolving, talking about how it can be improved, and whether and which issues outside the subject should affect it. Therefore, the group would need to talk on a regular basis.

With each proposed change the group needs to have unanimous agreement before moving on. Initially, this would be a time-consuming process but with familiarisation of the process and criteria, this would change. What would emerge from the dialogue would be stories that create a space for a shared culture. Differing opinions within the group would be seen as positive and part of the wider puzzle, meaning everyone would gain a more rounded deeper understanding of the issue.

In the next section, I propose seven criteria the group could use for making meaning of any proposed changes to the design and technology curriculum. The group's first step needs to be talking and agreeing on the criteria

2.3 *The Criteria*

Each proposal for new content would need to be evaluated against a set of criteria. Some criteria may be more relevant than others to different proposed content [wc]. The criteria are a guide for the panel to evaluate the proposed new content, not a straitjacket that each proposal should meet the criteria. It is about an awareness of what adding new content might mean to the curriculum and the nature of the subject.

Hornbake (in Benjamin, 2004, p. XI) poses four questions which form the basis for these criteria:

1. What knowledge is of most worth?
2. What in our culture must be reserved and deliberately renewed in both the core and ethos of our future civilisation?

3. What is dysfunctional, likely to be increasingly dysfunctional, and therefore should be cast aside?
4. What roles of educating everyone are our schools to play in this process of transcendence?

The first five criteria focus on addressing the first three questions (content), and the last one is addressed using the final three criteria that focus on teachers, pupils, and the learning context.

2.4 *It's Fit with the Subject's Epistemology*

The first criterion is a contentious topic – the subject's forms of knowledge, which have been much debated. For example, McCormick's (1997) view was that the debate was between a process or content model for design and technology curriculum; later, Gibson (2008) proposed a conceptual framework showing the relationship between knowledge and skills; then more recently, Morrison-Love (2017) has presented an epistemology of technology education to define its curricular position. Morrison-Love's focus on technology leads to Broens and de Vries's (2003) classification of technological knowledge, which like Vincenti's (1990) research draws from the professions rather than the school classroom, this does not negate their contribution to defining D&T's forms of knowledge. Rather, in my opinion, these two provide the answer, which I show as a continuum in Figure 29.2 with a Y-axis that adds McCormick's view of the place of conceptual and procedural knowledge in D&T.

Now, this will be a polemical model for some but taking on board my stated position earlier, that this whole process is both dialogical and about making

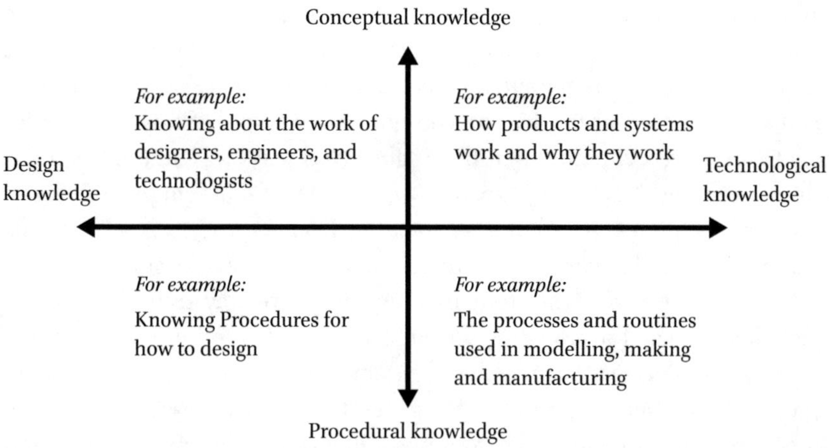

FIGURE 29.2 Forms of knowledge for D&T

sense, then it follows that there will be debate about D&T's forms of knowledge. However, it provides a basis for making sense about the proposed curriculum changes.

2.5 *Fit with the Subject's Aims*

For a new piece of content, it would need to fit with the subject's aims. Analysis of the literature shows the subject has two overarching aims: first it develops pupils' design and technology capability, and second through studying D&T pupils develop a critical awareness of the technological world. Design and technology capability is defined by Kimbell and Stables as "the power to produce change and improvement in the made world" (Kimbell & Stables, 2008, p. 18). The second aim is to develop pupils' technological literacy (Dakers, 2005; Keirl, 2007). Pre-National Curriculum in England, Harrison emphasised the value of teaching pupils about technology as they were citizens in a technological society (Black & Harrison, 1985). In terms of the philosophy of technology, this second aim could be seen as teaching pupils about technology as a volition (de Vries, 2005) and the first aim focuses on technology knowledge and processes, with the fourth category, artefacts, being included in both: the first as something pupils make when they are improving the made world, the second as an object of study.

The problem comes when the curriculum leans more towards one of these aims than the other, it is the two that make the D&T curriculum a valid subject as part of a general education.

A single piece of content will not meet either of these aims in totality, so the group needs to look at the proposed change in the whole of the curriculum. In other words, if this content were added or removed from the curriculum would this positively or negatively impact pupils' ability to intervene in the made world? If it's a new technological development, would they benefit from learning about it and its potential consequences on society or is this already included in the current curriculum content? An example of this is the inclusion of emerging technologies in the English National Curriculum, Helpfully, Barlex, Given and Steeg (2015, 2020) list examples of emerging technologies and examples are given in both versions but between the two publications some examples fade away, some emerge, and some are still emerging. It is this type of situation that the group must consider. One way around this, minimising the chance of overloading the curriculum with another new emerging technology, is to place the curriculum content in a concept group, like "emerging technology", "shaping process" or "generating design ideas". But the very nature of the topics means these examples will no longer be emerging technologies, they will just be technologies. Then they move into a category of 'current' technologies and later into 'historic' or 'redundant' (like the typewriter or Sony Walkman).

2.6 *Inclusive Curriculum*

An inclusive curriculum is one which provides all pupils with "an equitable, equal, and high-quality education regardless of [they're] characteristics, including race, ethnicity, social class, language use, gender, sexual orientation, religion, ability, and other human differences" (Mukminin et al., 2019, p. 55). Although Mukminin et al focus on the pupil, the teacher also needs to provide a curriculum that is inclusive to them (Prajapat et al., 2022). This means, that instead of focussing on the "regardless" aspect, recognising and reflecting the diversity of the teacher and pupil communities is essential for an inclusive curriculum. For example, Gumbo (2022) argues that teaching indigenous contexts to both indigenous and non-indigenous pupils ensures justice for the indigenous learners decolonising the curriculum, which benefits all pupils

There are many facets to an inclusive curriculum, and the stakeholder group's dialogue would be challenging to ensure different characteristics, beyond Mukminin et al's list, were considered. Hellard (2023) draws attention to the religious-political dimensions implicit in artefacts and Axell and Boström (2021) show how the presentation of technology in young children's books reinforces gender stereotyping. There is an argument that these examples of artefacts should be used within the classroom to teach pupils about the values and biases hidden within them as this follows on from the subject's aims; therefore, by including them they provide opportunities for pupils to learn about the unintended consequences of technology (this relates to de Vries (2005) description of technology as volition).

Exploring whether proposed changes to content were inclusive would also need to consider pupils' abilities and capacity to engage with the content. If new content had potential challenges for teachers of pupils with special educational needs or young children, the stakeholder group could consider guiding teachers to make pedagogical choices that mean the content is inclusive and provides their pupils with access to a high-quality curriculum.

Before moving on, it is important to note that this criterion is under-researched in the field of design and technology education.

2.7 *Values*

Values and their place in the design and technology curriculum are a key concepts in terms of subject content and pedagogy. For example, Layton (1992b) identified different kinds of values that pupils need to learn about in D&T for use when making judgments, such as technical values, economic values and moral values; Trimingham (2008) shows the role of values in making design decisions and Prime (1993) presents a values framework for understanding new technologies. Martin (1999, p. 202) categorises these as 'values within' D&T, but my analysis (2015) revealed that there are another two categories:

Values developed through D&T: How a pupil becomes technologically literate because of studying D&T (Dakers, 2005; Keirl, 2007).

Values ascribed to D&T: Layton's (1992a, p. 3) second perspective was not as values within D&T but how the values systems of "stakeholders involved in the socio-political shaping of school technology" influence design and technological activity.

This latter category reinforces the need for a range of stakeholders involved in shaping the D&T curriculum. The second relates to pedagogical choices, which are discussed in a later criterion, but they also need to be explicitly defined concerning the proposed change. So, the group must talk about how the proposed curriculum change aligns with these first two categories (values within and values developed in).

2.8 *Real-World: Context and Feasibility*

Researchers in design and technology education have shown that pupils and students respond and engage more positively with authentic design contexts, that is ones that are real (For example Turnbull, 2002; van As, 2019; Hill, 1998). They argue that through authentic design situations, pupils are exposed to real-world contexts that are engaging and allow them to use and practice previously learnt procedural knowledge which then allows their design and technology capability to develop. This presents challenges for teachers determining a suitable context that has a suitable balance between giving pupils opportunities to make purposeful decisions with potential demands placed on teachers' time, materials, and equipment (Mawson, 2003). Another risk is teaching pupils new knowledge, usually conceptual and technological, that is still being developed by technologists or in higher education labs, or that makes it impracticable to be taught in a school setting. An additional dimension to this is teaching pupils conceptual knowledge is "watered-down", or either ignores the science (Layton, 1992a) or is too reliant on science-fiction. Although pupils do sometimes respond well to future worlds imagined in science fiction, the real-life contexts relate to the second criterion above "Fit with the subject's aims".

There is one final aspect to the need for proposed content changes to be "real" – pupils' motivation and engagement. Priniski et al. (2018, p. 12) identify the potential of personal relevance in motivating students and "energising learning". Whilst there is not sufficient space here to explore the details of their argument, it should be noted that there are three facets to "relevance" for Priniski, Hecht and Harackiewicz: personal association, personal usefulness and identity. This leads to the next two criteria: teachers have the agency to interpret the proposed content changes appropriately for their pupils and school settings. As Davies argues: "both a teacher's agency (their intended action) and school structures (the context within which they work) play a role in change" (2022, p. 152).

2.9 *Allows for Teacher Agency*

Teachers are professionals, and to be professional they need to have agency (Biesta et al., 2015). One way of defining this in the context of design and technology is that it means they can decide how to interpret the new content. Although the proposed curriculum changes may be prescriptive – for example, pupils learn about iterative design (Department of Education, 2013), it cannot be that the teacher is expected to use it, or teach in a particular way. In other words, can they decide to teach it in a way that is relevant to their context? Which leads to the next two criteria.

A teacher decides what methods to teach knowledge, whether a process or concept, however, the parameters of the teacher's decisions are criterion 2: is the pupil begin taught new knowledge to add another building block to their design and technological knowledge they can draw on to meet the aims of the subject.

2.10 *Realistic for Various Locations and Educational Settings*

The final criterion I am proposing is practical. Can the proposed curriculum change be taught in different educational settings? Design and technology is a resource-heavy subject, pupils often use a lot of "stuff" – machines, tools, materials and kits, as part of their learning and practice. Does the proposed change require new "stuff"? Can schools afford new "stuff"? Whilst exam boards consider the accessibility of new courses, curriculum designers do not always recognise this as a fundamental concern. On the ground, proposed changes are often driven by the richest schools (usually private ones), who have the time and physical resources (if needed) to experiment with new content, but state schools - particularly those for pupils with special educational needs, primary schools or specialist settings, do not. The group needs to remember the breadth of settings and access to resources (teachers, professional development and "stuff") when talking about the proposed changes. If each child is entitled to learn "the best that has been thought and said" (Department for Education and Gibb, 2015), then surely proposed changes must consider the teaching and learning location.

Again, like criteria 3, the impact, challenges and considerations of educational settings, resources and facilities on pupils' progress and teachers' agency are under-researched in design and technology education.

3 Criteria in Practice: Artificial Intelligence

Now the criteria have been established, the next thing to do is to consider how he next consideration is how would this work in practice.

Here I am going to reflect on how or whether AI meets the criteria above drawing on some published research (e.g., Axell & Boström, 2023), professional magazines (e.g., Barrow, 2024) and discussions on my own podcast (Hardy & Taylor, 2024; Hardy, 2024). But I have two notes of warning about what follows. First, there is no dialogue as it's just me here although I'm drawing on the work of others, this is my interpretation, I am not making sense of AI as proposed earlier in this chapter. Second, AI is developing so fast that parts of my analysis may be dated before this chapter is published but that is why I think this is useful, to show how curriculum changes do not always keep pace with societal developments. Artificial intelligence is talked about in design and technology education by teachers, researchers, and exam boards but careful consideration is needed about its place in the curriculum.

Analysis of case studies written by D&T teachers show how they are exploring the use of AI for reducing time on teacher administration tasks, creating resources for the classroom, and teaching text-to-image AI to pupils to generate ideas quickly (as discussed by Ringvold et al., 2023), several mention its relevance arguing its use by designers today (like Federer in Barrow, 2024). Each example relates to a different use within D&T: the first as a productivity tool for teachers exemplifying which could be used as a case study in the classroom; the second as a pedagogical tool, next as a design instrument and finally as an example of how it is begin developed as new knowledge by a design professional. Each is valid, but not all relate to the curriculum. I suggest the second does not relate to curriculum content, the first and last are examples of conceptual design knowledge, and the third is a design instrument pupils can learn and choose to use when responding to a design context. As AI fits within the subject's epistemology, this means AI fits within the subject's epistemology – once pupils learn about AI in use (conceptual knowledge) and use AI (procedural knowledge), it can help address the subject's aims (criteria 2). However, if they only learn one form and not the other, then only one aim is met.

Whether it is inclusive is another matter. Ringvold et al. (2023) highlight its limitations if pupils have restricted vocabulary when using text-to-image AI and Axell and Boström (2023) conclude that stories created using AI about technology reinforce gender stereotypes. Does this mean AI should be avoided? No, but to counter these concerns the group may need to provide a narrative highlighting these two issues, the first to ensure an inclusive curriculum, the second as a potential learning point about in-built bias around gender, race and age (the latter two explored by Ringvold et al.).

This last point addresses the third criterion: values. Guidance indicating which values pertain to AI (Using the framework in Prime, 1993, for example; or the taxonomy of disruptive technologies in Barlex et al., 2020) could help teachers plan how they teach pupils the values implicit within this new technology. But is the use of AI to generate designs (that is AI learnt as a design instrument) realistic? Might some pupils find the inability to realise their AI-generated designs a demotivator? On the other hand, it might be a motivator for pupils who can visualise their ideas in their mind's eye but do not (yet) have the capacity to sketch them. This is where the group might reflect on the still emerging aspect of AI regarding the produce-ability of designs as something to address so that for now the focus is more on case studies of AI and its potential to disrupt rather than as a design instrument. If they take this route, then where does that leave the teacher agency? As the potential and use of AI is emerging so quickly, quicker than the group can meet and talk, there may need to be some future sensemaking extrapolating what might happen with AI to think about the choices a teacher can make when teaching AI in their own settings.

Finally, can AI be taught in different settings so that every pupil has an equal opportunity to learn? AI does need web-based "stuff" rather than physical resources. But already I am assuming all schools have equal access to computers, tablets, and stable internet access. This is where teacher and senior leadership stakeholders need to represent different physical locations (city, rural and urban) together with types of education settings (mainstream and special schools, state and private, primary and secondary, and so on).

This is an example of how the criteria could be used to explore whether a proposed change to curriculum content should be added. Here I suggest some parts of the dialogue and potential ways forward if AI was put forward to the group. But I am only one voice representing one stakeholder group, dialogue would need to come from a wider group to ensure an agreement using the criteria.

4 Conclusion

This chapter responds to Irving Louis Horowitz's definition of a Festschrift as "a call to further work, effort, and energy, a call to the improvement of learning, of a discipline, a science, an artistic vision, or an intellectual position" in that it proposes a way of a collective improving the learning and discipline of design and technology education.

My proposal for the stakeholder group's membership, the framework for their dialogue and the criteria are presented here for debate. My purpose was to challenge the accepted processes that have led, in my opinion, to a diluted, overloaded and confused curriculum. My proposal would take effort from stakeholders to talk, listen, and make sense of proposed changes to the design and technology curriculum content. My aim here is to create a process that "transcends the pre-established mores that continue to dictate policy today" (Dakers, personal communication). I will leave it to the reader to decide whether it meets this aim.

References

Axell, C., & Boström, J. (2023). Unveiling Biases: An Exploration of ChatGPT-3.5-generated 'Technology Stories'. *In: The 40th International Pupils' Attitudes Towards Technology Conference Proceedings 2023.*

Axell, C., & Boström, J. (2021). Technology in children's picture books as an agent for reinforcing or challenging traditional gender stereotypes. *International Journal of Technology and Design Education, 31*, 27–39.

Barlex, D., Steeg, T., & Givens, N. (2020). Teaching about disruption: A key feature of new and emerging technologies. In A. Hardy (Ed.), *Learning to Teach Design and Technology in the Secondary School* (pp. 137–154). Routledge.

Barrow, T. (2024). The Artificial Intelligence Issue. *FutureMinds.*

Benjamin, H. R. W. (2004). *The saber-tooth curriculum.* McGraw-Hill.

Biesta, G., Priestley, M., & Robinson, S. (2015). The role of beliefs in teacher agency. *Teachers and Teaching, 21*(6), 624–640.

Black, P.J., & Harrison, G. (1985). *In place of confusion: technology and science in the school curriculum: a discussion paper.* Nuffield-Chelsea Curriculum Trust and the National Centre for School.

Broens, R. C. J. A. M., & de Vries, M.J. (2003). Classifying technological knowledge for presentation to mechanical engineering designers. *Design Studies, 24*(5), 457–471.

Bushe, G. R., & Marshak, R. J., 2015. *Dialogic organization development : the theory and practice of transformational change.* Berrett-Koehler Publishers.

Dakers, J. R. (2005). The hegemonic behaviorist cycle. *International Journal of Technology and Design Education, 15*(2), 111–126.

Davies, S. (2022). Managing curriculum change. *Debates in Design and Technology Education.*

Department for Education and Science and the Welsh Office. (1988). *National Curriculum Design and Technology Working Group Interim Report.* DES.

Department for Education, & Gibb, N. (2015). *The purpose of education (speech).* Education Reform Summit. Department for Education.

Department of Education. (2013). *The National Curriculum in England Framework Document (July 2013).* Department of Education.

Gibson, K. (2008). Technology and technological knowledge: A challenge for school curricula. *Teachers and Teaching, 14*(1), 3–15.

Gumbo, M. (2022). Teaching for technological justice: Embracing indigenous designs. In A. Hardy (Ed.), *Debates in Design and Technology Education* (2nd ed., pp. 194–208). Routledge.

Hardy, A. (2024). *AI in D&T Education: A New Design Instrument, a Disruptor and a pedagogy.* Talking D&T.

Hardy, A.L. (2015). What's D&T for? Gathering and comparing the values of design and technology academics and trainee teachers. *Design and Technology Education: An International Journal,* 20(2), 10–21.

Hardy, A., & Taylor, J. (2024). *Creativity in the Classroom with AI Technology: a conversation with Joanne Taylor about PATT40* [online]. Talking D&T. Available at: https://www.buzzsprout.com/288393/episodes [Accessed March 1, 2024].

Hellard, A. (2023). Engaging object agency: new ways of design learning and being for young people in the museum. *The 40th International Pupils' Attitudes Towards Technology Conference Proceedings 2023,* 1.

Hill, A. M. (1998). Problem Solving in Real-Life Contexts: An Alternative for Design in Technology Education. *International Journal of Technology and Design Education, 8*(3), 203–220.

Keirl, S. (2007). The politics of technology curriculum. In D. Barlex (Ed.), *Design and technology for the next generation* (pp. 60–73). Cliffeco Communications.

Kimbell, R., & Stables, K. (2008). *Researching design learning: issues and findings from two decades of research and development.* Springer.

Layton, D. (1992a). *Values and design and technology.* Loughborough University.

Layton, D. (1992b). Values in Design and Technology. In C. Budgett-Meakin (Ed.), *Make the Future Work* (pp. 36–53). Harlow, England: Longman.

Martin, M. (1999). Exploring values in design and technology. In D. Lawton, J. Cairns & R. Gardner (Eds.), *Values and the curriculum; the school context* (pp. 199–207). Curriculum Studies Academic Group.

Mawson, B. (2003). Beyond The Design Process': An alternative pedagogy for technology education. *International Journal of Technology and Design Education, 13*(2), 117–128.

McCormick, R., 1997. Conceptual and Procedural Knowledge. *International Journal of Technology and Design Education,* 7(1), 141–159.

McCormick, R. (1990). Technology and the National Curriculum: the creation of a 'subject' by committee? *The Curriculum Journal, 1*(1), 39–51.

Mitchell, R. K., Agle, B. R., & Wood, D. J. (1997). Toward a theory of stakeholder identification and salience: Defining the principle of who and what really counts. *Academy of Management Review, 22*(4), 853–886.

Mukminin, A., Habibi, A., Prasojo, L. D., Idi, A., & Hamidah, A. (2019). Curriculum reform in Indonesia: Moving from an exclusive to inclusive curriculum. *CEPS Journal, 9*(2), 53–72.

Prajapat, B., Sinclair, R., & Hardy, A. (2022). How do we do race in design and technology? In A. Hardy (Ed.), *Debates in Design and Technology Education* (2nd ed., pp. 45–61) Routledge.

Prime, G. M. (1993). Values in technology: Approaches to learning. *Design & Technology Teaching, 26*(1), 30–36.

Priniski, S. J., Hecht, C. A., & Harackiewicz, J. M. (2018). Making Learning Personally Meaningful: A New Framework for Relevance Research. *The Journal of Experimental Education, 86*(1), 11–29.

Ringvold, T. A., Strand, I., Haakonsen, P., & Strand, K. S. (2023). AI Text-to-Image Generation in Art and Design Teacher Education: A Creative Tool or a Hindrance to Future Creativity? *In: The 40th International Pupils' Attitudes Towards Technology Conference Proceedings 2023.*

Trimingham, R. (2008). The role of values in design decision-making. *Design and Technology Education: An International Journal, 13*(2), 37–52.

Turnbull, W. (2002). The place of authenticity in technology in the New Zealand curriculum. *International Journal of Technology and Design Education, 12*(1), 23–40.

van As, F. (2019). An Exploratory Evaluation of a South African Project-Based Curriculum Module Focused on Authentic Technological Practice Utilizing Student Portfolios and an Open-Ended Questionnaire. *International Journal of Technology and Design Education, 29*(1), 107–121.

Van den Akker, J. (2003). Curriculum perspectives: An introduction. In J. Van den Akker, W. Kuiper, & U. Hameyer (Eds.), *Curriculum Landscapes and Trends* (pp. 1–10). Springer.

Vincenti, W.G. (1990). *What engineers know and how they know it: Analytical studies from aeronautical history.* Johns Hopkins University Press.

de Vries, Marc J. (2005). *Teaching about technology: An introduction to the philosophy of technology for non-philosophers.* Springer.

Weick, K. E. (1995). *Sensemaking in organizations.* Thousand Oaks.

Williams, P. J. (2007). Stakeholders in Technology Education. In J. de Vries Marc, R. Custer, & J. R. Dakers (Eds.), *Analyzing Best Practices in Technology Education* (pp. 179–190). Sense Publishers.

CHAPTER 30

Nurturing Design and Technological Thinking

A Core Goal in Technology Education

David Mioduser

1 Introduction

In our world, technology is omnipresent. Technology pervades all venues of life (e.g., work, education, health, economy, interpersonal relations, dwelling in both our personal and expanded ecosystem). There is no doubt that nowadays our postindustrial "knowledge society" (Bell, 1977) is preeminently, the "technological society" (Later in this chapter we will develop the idea that this has been always the case for every stage and historical period in human evolution since the dawn of our species – but concerning our times this claim is consensual).

Against this reality, stands the enormous gap between the omnipresence of technology in our lives, and the shallow-presence (if at all) of technological knowledge and skills in education and schooling (with the obvious exclusion of specialisation programs in high-school or higher education). In most countries, technology as a specific and knowledge-dense subject is not present at all in the curriculum, teachers are not seriously prepared for teaching, and in many cases, loosely defined curricular substitutes (more on this later) are implemented attempting to patch the hole. We should note that there is but another gap among many between schools' inner and outer environments, characteristic of the last decades' conflictive status of educational systems and schools concerning changing realities in the world – a general discussion that is beyond the scope of this chapter.

In this chapter, we will discuss a view of Design, Technology and Engineering Education (D&T&E) from a cognitive/epistemological perspective, which we have been reflecting on over many years of research, curricular development and actual educational practice. In the following we will present our particular rationale for D&T&E and a proposal for the structure, content areas and configuration for a D&T&E for all curricula.

© DAVID MIODUSER, 2025 | DOI:10.1163/9789004710948_030

2 D&T&E – Cognitive/Epistemological Rationale

The need to supply all citizens with core knowledge and skills concerning technology and the artificial world is today widely consensual, as expressed in numerous policy documents and reports. This need is rooted in a particular view of modern life emphasising that: (a) for over a century, we have lived immersed in a rapidly expanding technology-saturated environment, increasingly growing in sophistication and complexity; (b) life within such an environment imposes the need to learn and implement unique technology-related capabilities, knowledge and skills; and (c) mastering the required knowledge and skills is essential not only for the individuals' functioning and growth but also for their expected contribution to the advancement and economic development of the social unit (community, nation) of which they are members. From this perspective – a socio/economic/technological perspective – learning technology is highly important for acquiring the knowledge and skills considered necessary for functioning appropriately (e.g., as citizens, workers, students, members of a community) in 21st-century technology-saturated societies (Hanushek & Woessmann, 2020; OECD, 2019).

In contrast, without neglecting or diminishing the importance of the socio/economic/utilitarian perspective, our rationale about the need to teach and learn technology lays on a different fulcrum: a cognitive/epistemological perspective. In the following, we will briefly refer to the two main components of this perspective, i.e., the phylogeny and ontogeny of human cognition as closely related to technological thinking and doing, i.e., to humans' ability to envision and construct the artificial world.

2.1 *Technological Thinking and Human Evolution*

The first component of our rationale concerns the developmental process of traits that define our being human. Our main claim is that we should encourage technology learning because the making of the artificial world has been a defining characteristic in human evolution and intellectual development since the dawn of our species (Kirlik, 2005; Sternberg & Preiss, 2005). This line of thought emphasises the crucial role of technology as humans' way to continue the biological evolutionary path through cultural evolution – that is, by growing exosomatic instead of endosomatic organs, by inventing cultural organs and accommodating the natural world to humans' needs (Ortega y Gasset, 1941; Medawar & Medawar, 1977; Overmann & Wynn, 2019). In this evolutionary path technology stands at the interface between the 'inventor's mind' and the 'invented world', serving both the phylogeny and ontogeny of human knowledge, perceptions, beliefs, thinking processes, and skills (Cole & Derry, 2005; Simon, 1996).

Over the last century, scholars have stressed the contribution of cultural evolution to our species' development. Essential constituents of culture are tools, artefacts, methods and processes aimed e.g., satisfying necessities, changing the environment, and facilitating interpersonal communication. According to the Niche Construction Theory "Humans modify their selective environments through cultural activity, thus feeding back to affect selection" (Odling-Smee, 2003) – "when organisms affect their environment, that change can cause a shift in what traits are being selected for" (Steiner, 2020).

It is suggested that for about the last 60 millennia, since the Upper Paleolithic, Humans generated impressive cultural changes at increasing frequency at a pace that biological evolution cannot catch up – entering an era of an evolutionary mode centred exclusively on culture. At that turning point,

> Hominids went on from using rudimentary tools [...] to making them and keeping them for future use. That discovery that simply lunged into technological foresight, released the brake on evolution which the environment imposes on other animals. [since then] the evolution of man has always been culture-driven, and the driving component was technology. (Bronowsky, 1972, p. 201)

2.2 *Technological Thinking's Role in Ontogenesis*

> Human cognition [...] is one form of primate cognition, but it seems totally unique as people go around talking and writing and playing symphonies and doing math and building buildings and engaging in rituals and paying bills and surfing the web and creating governments and on and on. [...] the biological adaptation we are looking for is one that is rooted in primate cognition but then provides humans with the cognitive tools and motivations to create artefacts and practices collectively with members of their social group – that then structure their and their offspring's cognitive interactions with the world. We are thus looking for a small difference that, by creating the possibility of culture and cultural evolution, made a big difference in human cognition. (Tomasello et al., 2005, p. 690)

The second argument in support of the need for nurturing technological thinking (via explicit technology learning) refers to the ontogenetic layer.

The development of an individual's cognition [affording e.g., the intelligent deciphering, understanding and artificially enriching the world; the gradual construction of complex knowledge and skills; the development of social capabilities] emerges as a function of the interplay between maturity and experience (Tomasello, 2019). Integrating Piagetian and Vigotskyan perspectives, Tomasello proposes a neo-Vigotskian approach towards individual ontogeny, claiming that

> In much of modern psychology, maturation and experience (or genes and environment, or nature and nurture, or innate and learned) are often treated as mutually exclusive alternatives. But in actual fact maturation and experience are complementary parts of a single process. (ibid, p. 312)

Modern human cognition is thus a result not just of biological evolution, but also of cultural processes that human biological evolution made possible in both cultural-historical time and ontogenetic time. From this standpoint, acting within and interacting with the human-generated environment is critical for the ontogenetic processes to emerge. And given that a fundamental component of the environment is the *artefacts* created by humans (e.g., symbol systems, technologies, material culture, social institutions, social regulation systems), and the thinking processes involved in their creation – interacting with and within the artefactual world plays a crucial ontogenetic role.

Particularly interesting studies deal with hypotheses about apprenticeship processes within prehistoric communities. These are processes of social learning, in which the contents are mainly related to the subsistence needs of the hunter-gatherer community, as well as to the gradual building of the spiritual life and culture of the community. For example, findings regarding chisels of hand tools were produced from flint using the flaking method. The findings indicate different levels of skill – expertise level as well as immature levels of ability which may be the work of apprentices. The main pedagogical method is learning by doing in the space of life itself, for the needs of life itself. There is no separation between the living space and the learning space (Högberg, 2008; Assaf et al., 2016; Maloney, 2019).

However, living in an environment increasingly saturated by human cultural inventions, and specifically by scientific and technological intellectual and material artefacts, is highly challenging. Pinker (2002) suggests that this is due to a mismatch between the purposes for which our cognitive faculties evolved (biological evolution completed for a given reality tens of thousands of years ago) and the purposes to which we put them today – the results of the continuing cultural evolution. Thus,

> *the obvious cure for the tragic shortcomings of human intuition in a high-tech world is education* [...] *to provide students with the cognitive tools that are most important for grasping the modern world and that are most unlike the cognitive tools they are born with.* (p. 235)

Learning is essential; hence, explicit education is essential; hence, explicit technological thinking education is essential.

2.3 *Conclusion*

Nurturing technological thinking implies nurturing the development of a highly important *segment* of the cognitive, affective and moral apparatus that characterises us humans as a species (phylogeny) and as individuals (ontogeny). Learning technology is learning about thinking about the (outer) invented world, vis-à-vis the (inner) inventors' world, and about the results of the co-evolution of the designed and the designers' worlds. In Cole & Derry's (2005) words: "We have met technology and it is us" (p. 209).

3 D&T&E Knowledge and Skill

So far, we presented our rationale for the need to nurture technological thinking in the context of technology learning as an *explicit subject*. In this section of the chapter we elaborate on the translation of the above rationale into three legs of a conceptual framework for the development of a D&T&E-as-subject curriculum: (a) the cognitive/affective/moral aspects characterising a range of "technological personas" – the target learners of D&T&E education; (b) the specific constituents of the construct "technological thinking"; and (c), the corpora of content knowledge to be addressed in the curriculum.

3.1 *Technological Personas: Essence and Cognitive Toolbox*

In our framework, we will define first the diapason of technology-related individual 'technological personas' and the cognitive and behavioural features bundled in each.

Table 30.1 (adapted and expanded from Mioduser, 1998), presents a general depiction of the range of technological personas. Concerning the vast ecosystem of artefacts, technological concepts, processes and phenomena surrounding us, people act and function at varied levels, as a function of their knowledge, skills, and even motivation to deepen her or his involvement with it.

In the Table 30.1 each persona is characterised, concerning the level of involvement and/or interaction with the technological surround, by the cognitive goal

TABLE 30.1 Mapping the "technological personas" functioning and knowledge

Technological persona	Cognitive goal	Knowledge	Internal representation
Knowledgeable interactor	to be aware	– basic facts – perceived functions – social/economical/ moral implications	– declarative knowledge base – values system
	to understand	– overall structure and functions	– phenomenal model
	to use	– operating scripts – maintenance scripts	– indexed scripts-repertoire
Problem solver (naive to expert)	to repair	– device topography – functional configuration – assumptions about causal relationships	– qualitative causal model – buggy-but-working model – diagnostic model
	to adapt to devise	– necessary/sufficient functions – planning skills – debugging procedures – representational skills	– anticipatory model – solution-evaluation log
DIY-Practitioner	to construct	– tooling skills – contextualized disciplinary knowledge – "how-to" scripts	– pragmatic process-models – indexed local-solutions repertoire
Artisan	to craft	– ad-hoc design methods – generic disciplinary knowledge – normative knowledge	– proto-systemic runnable model – indexed generic-solutions repertoire
Designer/ Engineer	to design	– systematic design methodology – formal representational notations – formalized multi-disciplinary knowledge – knowledge-based heuristics	– systemic multiple-layer runnable model (declarative/ procedural/ qualitative) – formal anticipatory models

addressed, the type of knowledge required/applied and the type of internal representation of the technological entities.

We suggest several readings for this depiction. We can see it as a path of gradual specialisation concerning people's involvement with the technological world: from the level of basic interaction and convivence with technologies to the level of active creation of entities in the artificial world.

In a different reading, we can see the depicted space as a gross indication of the distribution of the various personas within the population, in a sort of pyramidal construct. We can easily assume that most people pertain to the categories "knowledgeable interactor" and possibly to the naïve "problem solver". Less people, but still many (due to nowadays social/cultural status of DIY practices), belong to the "practitioner" category. Finally, only a limited group belong to the category of formally trained expert designers and engineers.

A third reading is highly relevant to technology education, our focus in this chapter. We can refer to the mapping in the Table as (paraphrasing Bronowski) "The ontological ascent of the technological personas". We assume a gradual evolution, maturation and consolidation of conceptions, knowledge and skills concerning the material world starting from the early years and continuing as the interaction with the technological ecosystem grows in density. It is upon this spontaneous, immediate and social-driven learning that formal education processes should be built to support learners' increasing construction of the sophisticated layers of technology-related knowledge and skills.

We should note here that the contents under the "knowledge" and "internal representation" columns are in no way exhaustive, and continuous research should contribute to enrich and expand them. As well, each subsequent persona in the scale integrates the characteristics of the previous ones, thus affording different levels of behaviour depending on the technology-related challenge or context.

3.2 *Skills Toolbox – Technological Thinking Primitives*

The second leg in our framework is the technological-thinking-primitives repertoire. In this component, we refer to the cognitive building blocks, i.e., the reusable and re-configurable skills and knowledge units and chunks, involved in technological thinking and doing.

The primitives were classed into four categories (Table 30.2): Technological thinking and doing Rudiments (T-ruds), Mental Models (T-Mmodels), Methods (T-methods) and Metaknowledge (T-metas).

3.3 *T-Rudiments*

These are the basic building blocks of technology-related performance. They consist of a wide range of knowledge units which are activated in different

TABLE 30.2 Technological thinking and doing primitives

T-ruds	*Understanding and doing rudiments (building blocks)* – Declarative knowledge units (or chunks) – Procedural knowledge units (or compounds) – Skilled-doing units
T-models	*Mental Models* – Structural/functional/behavioural cognitive models (descriptive and/or runnable) of technological systems or processes
T-methods	*Technological methods and procedures* – Problem-solving processes – Design methods – Evaluation and troubleshooting procedures
T-meta	*Meta-level knowledge* – Design process control procedures – Mental model development, refinement and usage procedures – Regulation considerations (e.g., ethical, cost-effectiveness, resources management)

situations, e.g., for operating devices, using tools, building models, or programming a computer. Rudiments may be single pieces of knowledge, as well as chunks or composite units built out of these single pieces for use in solving technology problems. Rudiments sets include *declarative* units of knowledge (e.g., properties of materials, measurement units, or types and functions of Lego building blocks such as the repertoire of movement transmission blocks – gears, worm-gears, gear-racks, differential gear-house, linear-movement transmission compound, circular-movement transmission compound, differential compound); *procedural* units of knowledge (e.g., how to operate a toaster oven, how to assemble a set of transmission gears); and the comprehensive *skills-toolbox* involved in doing technology (e.g., tooling, manipulation of materials, techniques).

3.4 *T-Models*

The mental model of a target technological system or a problem situation is a key factor in any design or technological problem-solving process, whether the person is trying to understand a system, predict its behaviour, operate it, repair it or design a new one.

It was suggested that a person models a technological system through the following process: (a) the creation of an initial internal representation of the structural configuration of the system (the "device topology"); (b) complemented by a process called "envisioning"- by which the system components' functions and their functional configuration and interactions are inferred; (c) resulting in a causal model or "runnable mental model" of the system (de Kleer and Brown, 1983; Mioduser, 1998; Saba et al., 2023). At the device topology level, the person holds an internal representation of several constituents: parts (e.g., energy source, valves, clapper), conduits connecting the parts (e.g., pipes, wires), "stuff" flowing through the conduits (e.g., oil, electrons, water, information), and an overall functional map of the system or process.

Mental models serve as primitive constructs in technological thinking and are assumed to evolve through repeated experience, thus becoming more powerful tools e.g., generating explanations of a process or a system's functioning, predicting its functioning in changing conditions, optimising an existing solution to a problem, or searching for alternative solutions to it.

3.5 *T-Methods*

Primitives at this level consist of operational scripts for conducting processes implied in the creation of technological solutions. The methodological toolbox' comprises procedures which evolve and become increasingly systematic with practice, experience and formal learning. Among these, the methodological construct most referenced in the theoretical and research literature, as well as in curricular materials, is "the design process" (e.g., McCormick et al., 1994; Paans, 2022.).

Within the professional community dealing with design, a certain degree of consensus exists regarding the generic phases of the design process: from the identification of problems or diagnosis of needs leading to the formulation of the goals, requirements and constraints for designing; through a series of loops at which solutions are conceived, explored and evaluated; up to the stage in which a suitable design-solution is consolidated and instantiated (Bucciarelli, 1996). Besides that consensus the process is usually interpreted in flexible terms, leaving space for alternative configurations of stages and activities according to varied parameters, e.g., the nature of the problem or the technological field, the composition of the design team, the weight of the constraints, the availability of required resources, or level of novelty of the problem or the desired solution. (e.g., see Bucciarelli, 1996, for detailed descriptions of problem-solving processes by engineers; or Buchanan, 1992, and deVries, 2020, discussing the complexity and demands of design processes and design thinking while dealing with challenges defined as "wicked problems").

Regrettably, this flexibility is mostly lost when the design process model is redefined for educational purposes. Its reformulation by educators, in many curricular proposals as well as in actual materials, has somehow resulted in a rigid linear sequence (e.g., DES, 1990). A generic description of this sequence may include all or part of the following steps: (a) identifying needs and problems, (b) investigating and then composing a design brief, (c) generating and exploring alternative solutions, (d) choosing a solution (and eventually modelling it), (e) building the chosen solution, and (f) evaluating the outcome. The process proceeds in a stage-wise fashion and a "return path" from the evaluation stage to previous stages (revision loop in light of outcomes evaluation) is in some cases suggested.

Does this clearly defined sequential model correspond to real-life design processes? Or does this type of educational model fit real-life design processes by students? Research results on children's design activities suggest that they design in many different ways (Mioduser & Kiperman, 2002; Mioduser & Dagan, 2007).

In our curricular work for elementary school learners, we implemented a highly flexible model of the design process (see Figure 30.1). It comprises three operative phases and three practical functions. The *operational phases*, by no means in an ordered and linear manner, include a stage in which the design problem is defined along with the requirements and constraints for generating the solution; a stage in which a pool of ideas and alternatives is generated and an optimal option is chosen; and a stage in which an idea is implemented (e.g., either as a prototype or a working solution). The *practical functions* are implemented in any of the phases (e.g., 'investigation' can be conducted in the

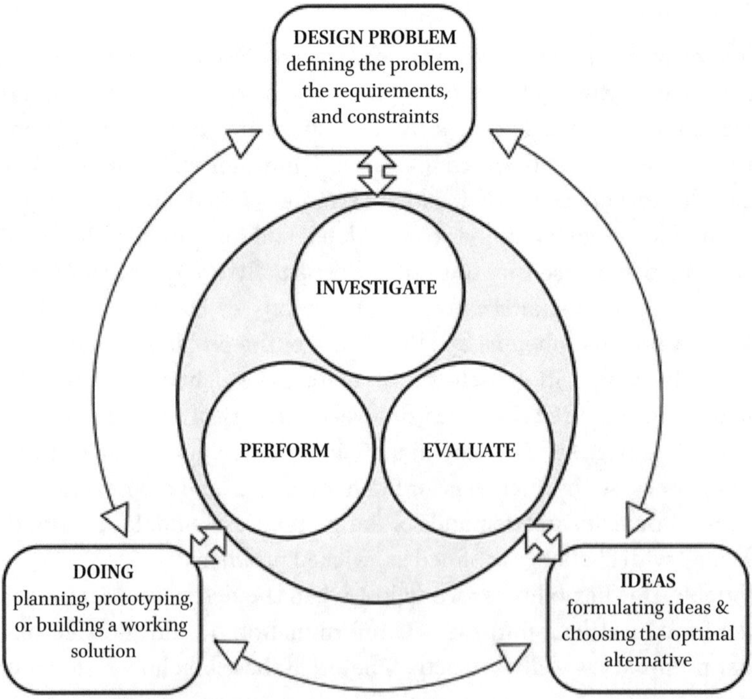

FIGURE 30.1 Design process phases and functions space

problem definition phase, or for generating ideas and alternative solutions; 'evaluation' is not a stage reserved for the end of the process but can be implemented to analyse alternative ideas or the set of constraints to be considered while generating a solution). The process depicted in the learning materials and implemented by the learners is by no means a rigid and linear configuration of the phases (and functions). As in real life, all these are interconnected and may take place in diverse configurations and, including recurrent paths among the phases (for example, starting from a preliminary formulation of a design problem; then prototyping a possible solution, reflecting on it and evaluating it; returning to the definition of requirements and constraints now with the evaluation results; investigating and evaluating other possible ideas and solutions; returning to the performing phase to implement the chosen ones).

At this point, this *design-process navigation tool* has been implemented for years in teacher training courses and actual curricular materials learnt by students in the vast majority of elementary schools in Israel (Bemabat Chadash, 2010–2024).

3.6 *T-Metas*

Meta-level (metacognitive) primitives concern the reasons and ways the learner decides to use primitives from the previous levels (e.g., rudiments, models, methods) and control the design and technological problem-solving processes. Metacognition refers to the learner's knowledge about her/his cognition and to the control and regulation of her/his cognitive actions (Flavell, 1976; Bryce & Whitebread, 2012). In our model, two sets of primitives are of relevance: those related to the meta-level features of technological thinking, and those related to the control and regulation of the design process.

On the first meta-level set, theorists and researchers refer to technological thinking as being, e.g.: analytic; synthetic; goal-oriented; anticipatory; exploratory; reflective; case-based; inventive; serendipitous; systemic (e.g., Mitcham, 1994; Chen & Strupp, 1993). For example, regarding serendipity ("the art of finding things you were not looking for"), Norbert Wiener (1994) suggests that in light of "surplus" results of a systematic research process, the question "*Now that I have come to a result, what problems have I solved?*" (a meta-level primitive in our model) is of great value when it comes to engineering problems (pp. 21–22).

The second set of primitives relates to the learners' control over the design process. For example, exploration activities may proceed in a tree-like fashion branching from one issue to another. In this case, the procedure [*stop exploration-routine?*] is highly relevant for deciding when enough investigation has been done, so that is possible now to proceed to the next design phase or function.

3.7 *The Knowledge–Space of Technology Learning*

The third leg in our framework for defining D&T&E as a learning subject focuses on the specific corpora of (disciplinary) knowledge found under its epistemological umbrella.

The terms exact sciences, life sciences, social sciences, or behavioural sciences are umbrella denominations encompassing diverse specialised disciplinary corpora of knowledge. Similarly, the term D&T&E comprehends a variety of sub-disciplines. Examples are the bodies of specialised knowledge and skills involved in the design of, e.g., buildings or bridges, computer programs, electronic artefacts, or bio-medical devices.

It is evident that the knowledge space of technology is broad and deep, demanding explicit learning. In this section, we discuss first several curricular issues, followed by a depiction of a principled configuration of content areas for a D&T&E curriculum.

3.8 *The Status of D&T&E as a Curricular Corpus*

The first issue to be discussed concerns the curricular status of technology studies. We suggest that aiming to materialise the principles of our rationale, *technology should be learnt as a deep, comprehensive and self-standing subject* in systematic form during the schooling years.

In the last decades models relying on a "literacy approach" have been developed (e.g., STS, Technological literacy, STEM). Analysis of curricular products and actual practices in educational systems, show that T and E concepts and skills appear only as shallow content layers, circumstantial to the specific topic of the activity, and attached to well-established subjects that continue to be learnt as self-standing curricular corpora (i.e., Science, Mathematics). The lack of serious T&E curricular treatment means in practice: no profound learning materials, no timeslot assigned, no teacher formation, no learning spaces (e.g., workshops), and no specific resources – all sine qua non-requirements pursuing the goal of nurturing design and technological thinking and doing.

At this point, we want to note that a broader discussion on the literacy-oriented approaches in technology education (for all) is beyond the scope of this chapter and deserves a comprehensive space by itself (see e.g., DeVries, 2021).

3.9 *A D&T&E Curriculum Precursor*

The next question to be elucidated is related to the epistemological and pedagogical characterisation of the contents and configuration of the D&T&E-as-subject curriculum.

As an inspiring start, we will look at the conceptual guidelines implemented in the German Bauhaus, considered a historical turning point in design education, despite its short life and even changing approaches along its three periods of existence (Gropius, 1919; Itten, 1975).

At the conceptual level, the teachers (renowned artists, designers and architects) developed the Bauhaus' unique pedagogical philosophy and teaching practices. The tasks and assignments aimed to develop the students' minds as well as hands, and their gradual mastery of concepts as well as skills. A principle embedded in the formation process related to the social and historical context within which the work (a design, a building, a piece of art) is performed. The product of the formation process – the craftsperson, the designer or the artist – should interact intimately with the community, aware of the social, economic, ethical and historical implications of her or his work. Correspondingly, the tasks faced were authentic (real-life-oriented), covering a whole spectrum of aspects – from pure creativity to industrial manufacturing and commercialisation issues. In sum, a school fostering the formation of professional designers and artists succeeded in developing a high-quality and complex "constructivist" (in nowadays terms) pedagogical system, that had a great impact since then on the way the holistic formation of professional designers is conceived and planned.

At the curricular level, the school's curriculum is depicted as a series of concentric circles instantiating the school's philosophy. The outer circle refers to the basic or preliminary course, addressing the essentials, e.g., the fundamental principles of design, the laws of form and colour, acquaintance with materials and textures and mastery of materials manipulation skills, and above all – aims to free the creative talents of the students as well as to free them from dead conventions (Itten, 1975).

The following inner circle comprise workshops aimed to foster minds- and hands-on experiences with a wide range of materials and technologies.

The innermost circle affords, and demands, the holistic integration of all learnt components into meaningful design tasks: new (and innovative) artefacts and ultimately – the building.

The learning journey across the curriculum integrates three main components: crafting capabilities, representational means (e.g., drawing, painting), and theoretical knowledge in the natural sciences and technology. Above all, stands the main addressee of the design activities: the community, the social ecosystem.

Upon the above inspiring approach, we will depict a preliminary configuration of contents for a D&T&E curriculum.

3.10 *D&T&E Curriculum Configuration and Contents*

Technology curricula (aimed at all students) have been successfully developed and implemented in educational systems over the years (e.g., in the UK, New Zealand, Singapore, Australia, Ireland, and Sweden). The suggested contents in the following draw in part from existing curricula, but more essentially, are configurated along the principles discussed above. The list is by no means complete or exhaustive but aims to illustrate the practical translation of the discussed principles into curricular building blocks (see Table 30.3).

Three curricular layers were defined: (a) Foundational knowledge layer; (b) Specialised packages-of-knowledge layer; and (c) Integrative authentic tasks and projects layer.

The foundational layer addresses types of infrastructural knowledge required for the development of a solid technological stance and worldview concerning the human-mind-made world: its constituents, creation process, and cognitive, affective and moral resources involved in the process.

TABLE 30.3 Curricular content areas and configuration

Foundational knowledge layer
– The world of artefacts [purposes; functional analyses; alternative solutions to similar problems; history/development]
– The design process of technological solutions [the centrality of the concepts "requirements" and "constraints"; ethical, moral, social issues]
– Manufacturing [materials; tools and machines; processes; artisanal/mass-production]
– Social and economic issues [work-life and professions; the role of technologies in history and social/cultural development; indigenous knowledge; regulations; marketing]
Specialised knowledge / Technological fields layer
– Dwelling, habitats, public spaces (civil engineering, architecture)
– Commutation (artefacts, physical infrastructure, social support systems)
– Energy infrastructures (production, distribution, alternative resources)
– Computation/coding – hardware and software engineering
– Control and robotic systems – mechatronics
– Media and communications
Looking ahead
– Biotech/Med-Tech
– Food-tech/Agri-tech/Safe-Water
– High-Tec – the digital universe, AI
– Space technologies
Authentic-integrative projects layer
– Authentic problems, contexts, reality-based requirements and constraints
– Integrative projects interlinked with knowledge from varied disciplines
– Prototyping or even implementation of designed solutions

The specialised knowledge layer refers to the range of human-activity fields in which D&T&E processes are implemented – corresponding to the different D&T&E disciplines.

The authentic projects layer addresses the curricular principle that fosters the integration of knowledge from varied areas of D&T&E with knowledge from other disciplines as well as social and value considerations while coping with authentic problems and needs.

We suggest that the three layers should be implemented in a spiral way along the schooling years increasing in depth and complexity vis-à-vis students' developmental growth.

4 D&T&E Integrated Curricular Packages

Following the above description of the legs of our D&T&E knowledge/skills framework, we suggest that the development of learning activities should be based on the integration of constituents from all three legs. In addition, real-life and contextual inputs are required to ensure the authentic character of the activities (at any level, from a single concept or phenomenon up to a complex project). And obviously – learning technology means first and foremost doing technology (Mioduser, 2015).

A depiction of the inputs feeding the development of any activity/module is shown in Figure 30.2. The small circles in the sample indicate the specific components from the different categories to be addressed in a given activity.

We envision the curriculum as an ever-growing network of activities/modules gradually covering all the constituents of the knowledge/skills space.

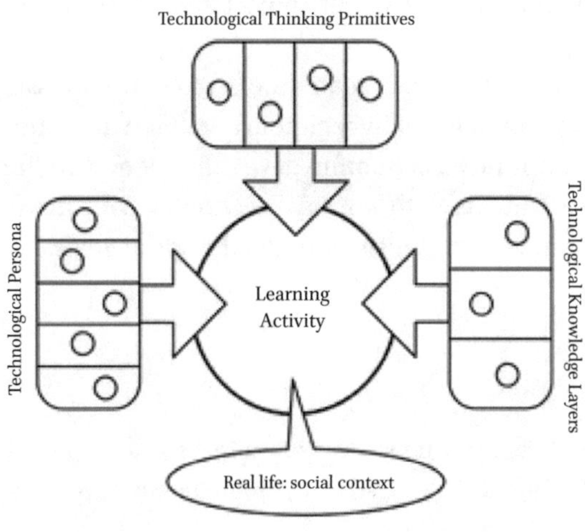

FIGURE 30.2
Inputs feeding the development of a D&T&E activity

Further research on the development of technological thinking primitives, as well as new developments in the continuously growing knowledge space of technology, will continue to enrich the categories in our framework – hence, the scope and volume of the curricular activities network.

5 Coda

> Our guiding principle was that design is neither an intellectual nor a material affair, but simply an integral part of the stuff of life, necessary for everyone in a civilized society. (Gropius, 1970, pp. 19–20)

In this chapter, I have elaborated on the need to teach and learn technology from a cognitive/epistemological perspective (in contrast with the economic/socio-utilitarian perspective). By this perspective, the designed (outer) world and the designer's (inner) world are co-constructed and evolve as a result of the recurrent and mutually affecting interactions between them.

Learning technology in this way is learning about the wide range of human assets involved in inventing the artificial world – knowledge, skills, methods and strategies, epistemic perspectives, beliefs, and moral judgements. Learning about technology in context (e.g., at the historical, social, cultural, moral, scientific and technological, economic layers) is learning about how the generation of human ideas drives our cultural evolution.

Given the crucial phylogenetic and ontogenetic role of technology, we suggest that there is a strong need to: (a) learn technology as a fundamental way for nurturing technological thinking, an essential segment of our cognitive apparatus, and (b) learn technology as an explicit subject systematically and deeply starting already in early childhood.

Closing with a remark concerning the socio/economic perspective, stressing the need to motivate students to learn formally technology-related disciplines and then contribute to the community's economic development and welfare – we believe that precisely individuals with a robust technological thinking formation will be optimally equipped to realise naturally the social goals.

References

Assaf, E., Barkai, R., & Gopher, A. (2016). Knowledge transmission and apprentice flint-knappers in the Acheulo-Yabrudian: A case study from Qesem Cave, Israel. *Quaternary International, 398*(4), 70–85.

Bell, D. (1977). *The coming of post-industrial society*. Basic Books.

Bemabat Chadash. (2010–2024). *K-6 Science and Technology curriculum*. Tel Aviv University, RAMOT.

Bronowski, J. (1972). Technology and culture in evolution. *The American Scholar, 41*(2), 197–211.

Bryce, D., & Whitebread, D. (2012). The development of metacognitive skills: Evidence from observational analysis of young children's behavior during problem-solving. *Metacognition and Learning, 7*, 197–217.

Bucciarelli, L. (1996). *Designing engineers*. MIT Press.

Buchanan, R. (1992). Wicked problems in design thinking. *Design Issues, 8*(2), 5–21.

Chen, D., & Strupp, W. (1993). General Systems Theory: Towards a Conceptual Framework for Science and Technology Education for All. *Journal of Science Education and Technology, 2*(3), 447–459.

Cole, M., & Derry, J. (2005). We have met technology and it is us. In R. Sternberg & D. Preiss (Eds.), *Intelligence and technology* (pp. 209–227). Erlbaum.

DeKleer, J., & Brown, J. S. (1981). Mental models of physical mechanisms and their acquisition. In J. Anderson (Ed.), *Cognitive skills and their acquisition* (pp. 285–310). Erlbaum.

de Vries, M. J. (2020). Wicked problems in a technological world. *Philosophia Reformata, 85*(2), 125–137.

de Vries, M. (2021). Design-based learning in science and technology as integrated STEM. In *Design-based concept learning in science and technology education*. Brill.

DES. (1990). *Technology in the national curriculum*. Department of Education and Science and the Welsh Office, HMSO.

Flavell, J. (1976). Metacognitive Aspects of Problem Solving. In L. Resnick (Ed.), *The Nature of Intelligence*. Erlbaum.

Gardner, H. (1999). *Intelligence reframed: Multiple intelligences for the 21st century*. Basic Books.

Gropius, W. (1970). *Scope of Total Architecture*. Collier Books.

Hanushek, E., & Woessmann, L. (2020): Education, Knowledge capital and economic growth. In S. Bradley & C. Green (Eds.), *The economics of education*. Academic Press.

Högberg, A. (2008). Playing with flint: Tracing a child's imitation of adult work in a lithic assemblage. *Journal of Archaeological Method and Theory, 15*(1), 112–131.

Itten, J. (1975). *Design and form*. John Wiley & Sons, Inc.

Kirlik, A. (2005). Work in progress: Reinventing intelligence for an invented world. In R. Sternberg & D. Preiss (Eds.), *Intelligence and technology* (pp. 105–134). Erlbaum.

Lemke, J. L. (2000). Across the scales of time: Artifacts, activities, and meanings in ecosocial systems. *Mind, Culture, and Activity, 7*(4), 273–290.

Maloney, T. (2019). Towards quantifying teaching and learning in prehistory using stone artifact reduction sequences. *Lithic Technology, 44*(1), 36–51.

Medawar P., & Medawar, J. (1977). *The life science*. Wildwood House.

Mioduser, D. (1998). Framework for the study of the cognitive nature and architecture of technological problem solving. *International Journal of Technology and Design Education, 8*(2), 167–184.

Mioduser, D. (2015). The ecological pedagogy of technology education – agenda for future R&D. In A. Jones & J. Williams (Eds.), *Agenda for the future of technology education*. Springer.

Mioduser, D., & Dagan, O. (2007). The effect of alternative approaches to design instruction (structural or functional) on students' mental models of technological design processes. *International Journal of Technology and Design Education, 17*(2), 135–148.

Mioduser D., & Kipperman, D. (2002). Evaluation/modification cycles in junior high students' technological problem solving. *International Journal of Technology and Design Education, 12*(2), 123–138.

Mitcham, C. (1994). *Thinking Through Technology – The Path Between Engineering and Philosophy*. The University of Chicago Press.

Odling-Smee, F. J. (2003). Niche construction. Princeton University Press.

OECD. (2019). *OECD future of education and skills 2030*. OECD.

Ortega y Gasset, J. (1941). *Toward a philosophy of history*. W.W. Norton.

Overmann, K., & Wynn, T. (2019). On tool-making minds: an archeological perspective on human cognitive evolution. *Journal of Cognition and Culture, 19*, 39–58.

Paans, O. (2022). Ontogenesis as a model for design processes. In Lockton, D., Lenzi, S., Hekkert, P ., Oak, A., Sádaba, J., & Lloyd, P. (Eds.), *Proceedings of DRS 2022*. Bilbao.

Pinker, S. (2002). *The blank slate: The modern denial of human nature*. Viking.

Saba, J., Langbeheim, E., Hel-Or, H., & Levy, S. T. (2023). Identifying aspects of complex and technological systems in the mental models of students who constructed computational models of electric circuits. *Journal of Research in Science Teaching, 60*(4), 681–723.

Simon, H. (1996). *The sciences of the artificial*. MIT Press.

Steiner, G. (2020). Calibrating vs. ideating reality: A cognitive assessment of Paleolithic abstractions and illustrations. In *NeanderART 2018 – Proceedings*. CeSMAP.

Sternberg R., & Preiss, D. (2005). *Intelligence and technology*. Erlbaum.

Tomasello, M. (2019). *Becoming Human: A Theory of Ontogeny*. Harvard University Press.

Tomasello, M., Carpenter, M., Call, J., Behne, T., & Moll, H. (2005). Understanding and sharing intentions: The origins of cultural cognition. *Behavioral and Brain Sciences, 28*, 675–735.

Wiener, N. (1994). *Invention – The Care and Feeding of Ideas*. MIT Press.

CHAPTER 31

Opportunities Won

Unexpected Encounters: Narratives from Practitioner-Based Research

Eva Hartell

1 Introduction

Hey, haven't you been to my school!? Suddenly, amidst the chatter of the crowd, a teen voice breaks out loud, grabbing my attention. I turn to see a lanky teenager standing before me in the middle of a large gathering of people on a remote island in the archipelago. Surprised, I respond, "Um, yes". I recognise him vaguely but struggle to place him, especially in this context – an event aimed at saving the Baltic Sea, attended by policymakers, philanthropists, and researchers. I tried to hide my struggle but failed. The boy tried his best to rescue me by continuing "We made a book and tried some programming stuff! It was great fun!" he says even louder for everyone to hear with a big smile, his voice full of enthusiasm and joy! Taking no notice of the crowd around him. As we begin to chat, and he tells me how much he enjoyed programming, and his new classmates joins the conversation, asking questions on programming, the project, and the like. Their chatter blending with the background noise of the event. Here I am, in a completely different setting–an island in the outer archipelago, surrounded by adults only policymakers and researcher–engaging with a former student from one of our projects. This teenager did not only remember the project after all these years, but he also recognises me and confront me with his cheers. I am stuck with awe at this unexpected encounter. What had just happened and why? The student had broken away from his group and the teacher to reconnect with me, a woman he had only met a couple of times years ago. In that moment, I felt genuinely touched by the memory of which continues to warm me to this day. Our conversation comes to an end as his current teacher, whom I recognise from before, approaches. We share with her the positive details about the project we had collaborated on years ago, much to her surprise.

Reflecting later, the boy was obviously genuinely involved in the research and development project we worked on together, where we introduced programming and, among other things, made a book together, so it is possible to capture his interest and commitment, something that seemed somewhat

© EVA HARTELL, 2025 | DOI:10.1163/9789004710948_031

surprising for his current teacher. Memories flood back, recalling him as the curious kid determined to find the best spot on his body for the pedometer he had made. I reached out to his former teacher, who confirms his identity and enthusiasm, solidifying the bond but with a bit surprise that he remembered. I ponder the significance of this encounter. What if he had been given more opportunities like this, not just by chance due to our research project? It's stories like this that fuel my passion on working with practitioner-based research with students and teachers. I am constantly reminded of the importance of such moments– of connection, inspiration, and the potential to make a lasting impact. In this Festschrift, I seize the opportunity to shine a light on some overlooked narratives. These narratives stem from practitioner-based research projects that, for one reason or another, never found their way into our previous publications. While research often drills deep into specific aims and purposes disclosed in stringent research questions, there is a whole world of experiences beyond those narrow confines–especially in educational research. Through the various interventions I have conducted alongside teachers and fellow researchers, I have witnessed some truly remarkable stories where students and teachers blossoming. Many of these narratives- despite being both rich and vibrant– were omitted from our academic papers due to the stringent focus of our research questions. Nonetheless, when collaborating with teachers and students, a wealth of other experiences unfolds. Yet, they speak volumes about the transformative power of research, offering a glimpse into the lives of young people in ways we had not anticipated. So, with this text, I aim to paint a fuller pictures of research projects, highlighting the unexpected impacts they can have on the lives of those they touch. This chapter has collected some telling examples of bonuses that happen in connection with a selection of practitioner-based research projects, in which I have been involved. It is a narrative of exploration, collaborations, and the profound resonance of narratives within the realm of academic inquiry. Hence, by providing these narratives I will attempt to highlight the multifaceted nature of research projects and the profound impact on the lives of young peoples it may have– some expected and some of them unexpected.

2 The PATT Community and the Wonderful Mark

It is impossible to contribute to this Festschrift for Marc without highlighting his significant contributions to the Pupils' Attitudes Towards Technology (PATT) conferences. Marc's contribution to this cannot be emphasised enough across all publications he has contributed to, whether as a writer, editor, or source of inspiration. A heartfelt thank you is for Marc's invaluable role in shaping our work. However, this contribution goes beyond academic achievements, as the PATT conferences or rather the PATT family, foster deep friendships and collaborations on a global scale, leading to numerous publications and presentations that extend far beyond the confines of peer-reviewed papers and conference days. The PATT community is a strong voice in the technology education domain.

The PATT conference transcends being merely the premier conference in technology education; it can be more aptly characterised as a driving force within the technology research field. New knowledge and ways of thinking regarding teaching and learning in Technology Education. Despite the name-the focus on attitudes has rather grown to include a much wider perspective and role. The PATT community is pushing boundaries, fostering both new knowledge and deeper insights, and facilitating collaboration across the globe. The PATT conferences and its community influence on policy and students' learning opportunities is profound.

Many people have contributed to all this but one person more than others – the great Marc de Vries. He, to the best of my knowledge, would be too humble to admit and not say that himself, but his contribution goes beyond recognition. His contributions have propelled the field forward, leading to growth, maturity, and increased research activity. Personally, I have endeavoured to contribute to this collective effort. During my time as a doctoral student and subsequently, as a researcher, I have been involved in numerous research projects, some of which have been documented in scientific publications while awaiting recognition.

3 Prioritising Student Access to Education: An Imperative Mission

While "PATT" stands for *Pupils Attitudes for Technology* the community has embraced a wider scope. While recognising the significance of students' interests and attitudes towards technology education, it is imperative to acknowledge that an exclusive focus on attitudes overlooks a fundamental aspect. Attitudes are inherently linked to knowledge acquisition: thus, prioritising the provision of learning opportunities becomes essential. Despite the intuitive nation of this assertion, its realisation remains elusive for a significant portion of the student population. Consequently, the relevance of the Convention on the Rights of the Child and the UN Global Goals Agenda 2030, which advocate for universal access to quality education, including technology education, remains profoundly pertinent in safeguarding students' entitlement to knowledge. Access to knowledge-rich curriculum embracing not just facts but the entire nature of technology education. Moreover, the diverse nature of the technology education landscape globally exacerbates the issue of students' access. In this regard, Sweden stands out as an exemplar, mandating technology education as a compulsory subject and even instituting a national timetable for its inclusion from year 1–9 including in special education schools. Nonetheless, despite specific learning criteria which students are entitled to achieve within the Swedish national syllabus, empirical evidence from numerous studies and reports consistently indicates a disparity between prescribed entitlements and actualised opportunities for students (c.f. e.g. Swedish School Inspectorate, 2014).

Research can serve as a crucial leverage point here. Not exclusively covering for all downsides but still offering potential for improving student learning outcomes and advancing educational research and practices. Different approaches could apply, for example by implementing existing evidence in empirical studies, and or exploring further innovative instructional strategies–both to advance the field. While research findings alone cannot prescribe exact actions, an evidence-informed approach is more likely to be effective. Still, it is vital to approach this endeavor with caution. Collaborating with teachers, who possess valuable insights into their students' needs and learning environments, is essential. Recognising the pivotal role of teachers, researchers should involve them in the research process, acknowledging their expertise. To maximise the impact of research projects, fostering closer collaboration between researchers and educators than traditional top-down approaches is imperative. The movements towards practitioner-based research, where researchers and practitioners work together, exemplifies this preferred inclusive approach.

Growing movement towards a more inclusive approach in educational research to bridge the gap between academia and practice. This chapter has collected some telling examples of bonuses that happen in connection with a couple of practitioner-based research projects, in which I have been involved. There are so many stories to be told but this chapter covers some of the valuable narratives where students were given a stronger voice and agency through the research interventions and teachers gained new perspective regarding their students.

This chapter highlights examples of the benefits derived from collaborations, particularly emphasising instances where students gained a stronger voice and agency through research interventions. It collects some telling examples of bonuses that occurred in connection with practitioner-based research projects, covering valuable stories where students were empowered through these interventions.

4 Unexpected Outcomes from a Case Study Aiming Exploring the Impact of Digital Assessment Tool for Teacher Peer Assessment in Classroom Practices

This example illustrates how interventions with one particular purpose unintentionally can inadvertently reshape relationships and expectations among home, school, teachers, and students. It highlights the ethical implications arising from teachers' biases and the presence of a visiting researcher, as well as the potential consequences for both the research project and the individuals involved.

The aim of this study was to investigate the effectiveness of utilising a digital tool for facilitating teachers' peer assessment of student work. The objective is to assess the tool's viability for potential integration into future digitalisation efforts of national tests. In Sweden, teachers assess students' achievements based on criteria based on national curriculum. Teachers design their own assessment practices and some moderation tools such as national tests are available in some subjects. However, in the subject of technology, there are no standardised national tests and teachers rarely design their own written tests, instead they primarily rely on classroom observations and interactions.

The idea behind this intervention was to reflect real classroom practices as much as possible. Product development is a commonly used learning activity. This intervention, therefore, involved students independently designing a tea strainer model for a self-selected target group, justifying their choices, with both the process and the product documented and assessed according to standard practice. To track the entire process, digital multimodal portfolios were employed. Collaboration among students was permitted, and so was asking teachers for help, mirroring regular classroom activities. Students collected and recorded their evidence using drawings, spoken word, photographs, and video recordings within digital portfolios. Subsequently, multiple teachers evaluated the student projects using the comparative judgement method (Pollitt, 2012). The results were summarised in a report for the client. The study included approximately 100 students from grades 7 and 9, along with their two teachers and a handful of external teachers as assessors.

In this evaluative study, I collaborated with the regular teachers without prior acquaintance with the students. During the planning the teachers provided a brief overview of the context including their four classes, highlighting one student primarily taught within a specialised setting (Special educational needs). They kindly advised me not to intervene if she became agitated or exited the classroom; instead, I was to allow her to leave and avoid potential conflicts. They expressed concern and even offered me the option of removing her, which I declined for ethical considerations. I did not want to be responsible for exclusion and cause trouble just for the sake of data collection. While I expressed my gratitude for their trust, I kindly asked them not to disclose the student's identity to me. As a temporary visitor, I believe it unnecessary to delve too deeply into individual student details.

The work was carried out over one full day per student group, four full days in total. The students created their own solutions and documented their work through images, text, sound, and video within digital portfolios. Given that this study was undertaken a few years back, neither the students nor the teachers were accustomed to using computers for instructions. Moreover, despite – or perhaps because of – the limited access to devices, with students sharing one laptop per pair and only a small number of cameras, the project proceeded remarkably smoothly.

It was a lengthy school day, uninterrupted. The students exhibited remarkable enthusiasm and engagement all through the day, showing no desire to take breaks or leave the classroom activities. This was true for all students, without exception. Notably, even the student of concern remains present throughout. The teachers expressed their delight and some surprise at the success of the day, particularly with the student they had been apprehensive about. Their satisfaction was reinforced when the girl's guardians contacted them later that day, expressing their gratitude for her daughter's enthusiastic account of the fantastic project work. The mother, who herself worked in product development and innovation, was especially thankful that her daughter had been genuinely challenged and returned home very content and happy. The teachers reiterated their surprise at the positive outcome of the day's activities.

Despite them not revealing the identity of the student. Given my teaching background, I could likely have guessed whom they were referring to. However, if they had not mentioned her earlier, I would not have noticed her for the same reason they briefed me about. Rather, I would have recognised her as one of the most engaged and capable students and stopped there. Do we even dare to think about what would have happened if the student had been excluded from the activity? Maybe nothing had happened but still. Maybe that is exactly what the risk is. That nothing happens. This prospect is particularly concerning within the education context, where one purpose is to foster new opportunities for all students to stretch their horizons. Nothing happens is too much of what happens. Our bias or preconceived notions should not hinder students' potential for success. One thing is certain: we fulfilled the initial purpose of the study but even more so the relationship and expectations between home, school and teacher to student were altered by this intervention, which was initially intended to evaluate a digital peer assessment tool. Evaluating digital assessment tools seems a bit insignificant in this context where the true outcome had significant consequences.

5 Unanticipated Perspective Changes: A Valuable Bonus from an Intervention Focused on Programming with Students with Dyslexia

This project was named "Tillit till IT", a Swedish anagram translated into English as *Trust in ICT*, took its starting point from teachers' concerns about the introduction of programming in the technology subject national curricula. Programming was not clearly spelled out in writing within the National curricula, yet its introduction created considerable confusion among teachers who were expected to teach the technology subject. Despite programming's non-ambitious inclusion within the curriculum, its role and place were inflated through various channels and stakeholders (aka vendors). This compounded the already daunting task of integrating yet another subject area into an already tightly packed technology curriculum. One concern raised by the teachers was the consequences for children with reading and writing disabilities. *What happens if they misspell the code?* was one of their concerns. This initiative to embark on collaborative efforts was not solely driven by teachers concerns for students with specific learning difficulties. Rather, it stemmed from teachers' own apprehensions regarding their ability to effectively teach a subject area that was largely unfamiliar to them and at the same time very much hyped. Despite this uncertainty, there was an underlying sense of excitement among them as they ventured into this new territory.

Self-efficacy refers to an individual's belief in their capacity to effectively plan, organise, and pursue actions necessary to achieve desired outcomes. Within the realm of education, teachers' self-efficacy pertains to their personal belief in their ability to plan, organise, and pursue activities related to teaching and learning. Hence, teachers' self-efficacy is of great importance for if and how teaching is expressed and enacted, and thus direct and indirectly affects students' opportunities to learn (Hartell, 2018). It is important to note that self-efficacy is contingent upon specific contexts. Teachers who exhibit low self-efficacy in a particular area are often less inclined to teach that area to the same extent, depth, and enthusiasm as those with higher self-efficacy. The overarching goal of the project was therefore to not only augment the presence of programming education in primary schools, but also supporting teachers' self-efficacy in teaching programming, with a specific focus on student encountering reading difficulties. Therefore, the project aimed to design and implement innovative, and research informed learning activities tailored to these objectives.

Here researchers and teachers designed learning activities collaboratively. It was challenging to say the least for the teacher who was not used to teaching programming herself and was almost completely unfamiliar with Micro bit, which had been chosen for its feasibility for the project. Similarly, the research team encountered similar challenges due to their lack of experience and knowledge about the students in the group. Consequently, the team complemented each other's expertise and perspectives to navigate these challenges effectively. To get started the research team took a role as instructor instead of the teacher. Students were introduced to programming in general, followed by a specific focus on Micro bit. They engaged in various exercises led by the doctoral student who took the role as prime instructor, aimed at familiarising them with both block programming on the screen and the physical Micro bit device, including its connections to the power source battery. The students demonstrated enthusiasm and found the experience exciting. Subsequent exercises gradually increased in complexity, encouraging a more exploratory approach. transferring into the second phase of the intervention where for example, students created pedometers and investigated optimal body placements for accuracy. They also designed transmitters and receivers, experimenting with different distances to assess the functionality of various solutions. Everyone was new to programming, but they harbored numerous ideas. However, it was not always the case that the students could do what they wanted the Micro bit to do. Code errors, glitches during transfer to the physical Micro bit, and occasional battery issues were encountered. In short, there were quite a few challenges, to say the least. However, the students were bursting with great enthusiasm and commitment.

One student particularly stood out from the crowd as he was able to easily help both his peers and us. He was quick to identify the problem itself. What caused the Micro bit not to achieve the desired result, e.g. flashing dice that did not flash. Find the code error or identify the gap between battery and device and or the incorrect or missing step in the code. Transfer problem from screen to device. I remember being fascinated by his speed and began to study him a bit more closely. Above all, he could easily identify errors in the code. He looked at the block programming code but most often he looked at the Python code behind. Something that the tutor only mentioned briefly. He often looked at the source code that he himself found how to produce and was able to find the errors his peers had encountered. I asked him if he spent a lot of time programming in his leisure time. He responded that this was his first experience with programming. However, he added with emphasis that he enjoyed it a lot!

When I later discussed what I had observed with his teacher, she expressed happiness and, above all, her positive surprise at his positive attitude during the lessons. How he had taken up positive space in the room helping his peers. Her fascination was evident as she continued briefing me about his situation. According to her he was a challenging student, especially with reading due to his being diagnosed with severe dyslexia. Furthermore, she told a heartbreaking story about his low self-esteem which was rooted in his great challenges that made him seek attention in other ways. This was something we as visitors had not noticed him for. We took notice of him but not as a messy kid instead rather the opposite. Instead, we had noticed him for his speed to solve tasks, and how supportive he had been when helping his peers when stuck. Taking responsibility for his peers learning. The teacher testified that trough this intervention she saw him from a different perspective and saw with pride how he had been given the opportunity to grow not only knowledge-wise but also how he assumed a different role in the group. His status was already high in the class but had now gained a more positive one.

Unfortunately, I am unaware of how his ongoing exploration of programming unfolded. I can only hope he continued his progress and got better self-esteem. Nonetheless, from this intervention his teacher was afforded the opportunity to witness a different side of her students, and perhaps even discover new aspects of herself in the process. Allowing teachers to take a step back and observer, rather than being in the midst of the action, is valuable but often too infrequent. Learning is a change in long-term memory and requires thinking hard and time to reflect, this is true for teachers as well. By affording teachers this opportunity, we may also offer their students the chance to be seen in a different light.

> Everything we see, hear, and think about is critically dependent on and influenced by our long-term memory. (Kirschner et al., 2006, p. 77)

6 The Imperative of Domain-Specific Knowledge in an Evolving Educational Landscape

Our perceptions are inevitably shaped by our preexisting knowledge and beliefs. Periodically, it proves beneficial to engage in endeavors that afford deeper insights into both us and our students, thereby fostering the acquisition of fresh perspectives. In previous mentioned example, the teacher was afforded opportunity to engage in such self-reflection. These two examples of opportunities emerged here, in part, due to the students' exploration of a novel subject area– programming – and, in part, because of the primal investigator's assumption of instructional duties. However, occurrences of this nature are regrettably infrequent. It warrants scholarly consideration whether researcher ought to take on instructional roles more frequently. While this discussion is pertinent, its comprehensive exploration must be deferred to another occasion.

In the rapidly evolving landscape of the world, domain-specific knowledge emerges as indispensable, particularly in an era dominated by information sources and the growing inclusion of artificial intelligence (AI). Access to education transcends a mere component of Agenda 2030 or the Convention of the Rights of the Child; it is an essential prerequisite for equipping students with the tools they need for the future. While specific competencies and skills demanded for the future remain speculative, one certainty is the growing importance of knowledge and the ability to critically analyse and scrutinise statements in our information-rich world.

As noted by Kirschner et al. (2006) that how we view the world is critically dependent on and influenced by our long-term memory, underscores the significance of domain-specific knowledge. This is important in the educational context in terms of teachers' knowledge including biases about their students but also on general terms for students, which has become more important than ever especially with the advancements of AI. While AI excels at data processing and synthesis, it lacks the nuanced understanding and creative inherent in human cognition and interaction (at least as far as I know now in the beginning of the year 2024). Thus, domain specific knowledge not only underpins creative thinking but also distinguishes human contribution from AI-generated outputs. Consequently, as AI continues to advance and perform various cognitive functions, domain-specific knowledge becomes even more essential for driving innovation and maintaining a competitive edge in our rapidly evolving world.

In this context we are situated in, education assumes heightened importance. This is probably true for all domains and all educational settings. However, I would particularly emphasise the importance in and of disciplines like technology education, which aims to prepare learners for the future. Technology education being both practical and theoretical, both contemporary and dynamic, evolves rapidly in tandem with societal shifts while fostering an appreciation for historical contexts. To adequately prepare our youth for the future; technology education must assume a more prominent role in schools, offering every student the opportunity to thrive, regardless of background or gender, thus benefiting both individuals and society at large. However, realising this goal poses significant challenges for schools, educators, and researchers alike. Central to addressing these challenges is the development of effective instructional strategies. This is absolutely necessary and at the same time very challenging. Fostering closer collaboration between educational research and practices holds the potential to enhance our chances of success. Therefore, I hereby recommend that further research be undertaken to optimise the delivery of a knowledge rich curriculum, encompassing not only facts but also fostering creativity and innovation. How this should be undertaken might be written in the stars but fostering enhanced collaboration between researchers and educators is advocated to achieve this endeavor.

While researchers often focus primarily on scientific articles for their merit, it is important to broaden our perspective and reflect on the broader impact we have beyond the academic sphere. Through our research projects, we not only influence the research field and, perhaps, long-term processes within the education system, but we also impact those who participate in the projects. We may provide them with opportunities for learning that they otherwise would not have had. Therefore, it is crucial to think deeply, especially when dealing with young students. We bear a significant responsibility towards both current and future generations, and it is important that we acknowledge and embrace this responsibility.

As researchers, it is our duty to deepen knowledge and propel scientific advancement, while also being willing to advocate for change. Rather than merely bluntly criticizing learning activities like spaghetti bridges, I suggest we should take more proactive steps. The time is ripe for us to suggest alternative approaches to teachers instead of dwelling on what they should not do. However, we must acknowledge that it is not our role to judge or take over teaching from those who are responsible for students' education and have intimate knowledge of them. Moving forward the key is to collaborate with teachers in conducting research that addresses pertinent issues. Ready to dive in?

While maintaining high scientific quality and publishing results is imperative, it is also valuable to be open to unexpected findings and explore alternative ways of sharing progress with active teachers and stakeholders. Collaborating with teachers on empirical studies can contribute to the school's compensatory mission, allowing researchers to implement and refine solutions based on identified practical problems in schools. Undertaking research projects involves a huge workload and planning. In addition to this, perhaps something as simple as organising study visits to your university for the school classes you work with. I have organised a couple of study visits within the K-ULF consortium[1]. Even though the impact of these study visits may be difficult to measure, they can inspire students with new perspectives and aspirations, and from that consider different pathways. I particularly remember the students saying *When I grow up- I am going to study at KTH* (my Uni). This warms me of course.

However, the unexpected outcome from the student who had already pre-decided to become an artist and said he would never consider going to KTH but learned from the study visit about the presence of public art. The study visit provided him with opportunity to gain access to the public art decorating the university buildings, which truly inspired him to even more wanting to become an artist. Another unexpected, and perhaps the most valuable outcome of these study visits is the provision of possibilities for teachers to view their students in another context.

It is imperative to engage in collaborative research with teachers as it facilitates a reciprocal exchange of knowledge instead of just transfer. Without such collaborations, researchers may overlook valuable insights gain from teachers' practical experiences in the classroom. By actively involving educators in research endeavors research can tap into their firsthand understanding of instructional dynamics and students needs this collaborative approach foster a more holistic perspective on effective teaching methods, allowing researcher to glean invaluable insights from teachers' expertise. Thus, the synergy between researchers and teachers in collaborative research endeavors enriches both parties' understanding of educational practices and promotes the development of evidence-informed strategies that truly benefits students in one way or another.

Note

1 K-ULF consortium is a partnership around practitioner-based research run from KTH Royal Insitute of Technology together with school partners including Haninge municipality in Sweden c.f. https://www.kth.se/en/k-ulf

References

Hartell, E. (2018). Teachers' Self-Efficacy in Assessment in Technology Education. In M. J. de Vries (Ed.), *Springer International Handbook of Education: Handbook of Technology Education* (1st ed., pp. 785–800). Springer. https://doi.org/10.1007/978-3-319-44687-5_56

Kirschner, P.A., Sweller, J., & Clark, R. E. (2006) Why minimal guidance during instruction does not work: An analysis of the failure of constructivist, discovery, problem-based, experiential, and inquiry-based teaching. *Educational Psychologist, 41*(2), 75–86, DOI: 10.1207/s15326985ep4102_1

Pollitt, A. (2012). Comparative judgement for assessment. *International Journal of Technology and Design Education*, 22(2), 157–170. https://doi.org/10.1007/s10798-011-9189-x

Skolinspektionen. (2014). Teknik – gör det osynliga synligt. Om kvaliteten i grundskolans teknikundervisning. Stockholm. http://www.skolinspektionen.se/Documents/publikationssok/granskningsrapporter/kvalitetsgranskningar/2014/teknik/kvalgr-teknik-slutrapport.pdf

K-ULF project

CHAPTER 32

Did You Know That Marc Is a Language Genius in Swedish?

Lena Gumaelius and Inga-Britt Skogh

We don't believe Marc has spent much time in Sweden, nor do we think he has studied Swedish extensively. Though, of course, we know he has visited technology educators in Sweden over the past decades. In Christmas 2013, KTH Royal Institute of Technology gave Marc a book about Christopher Polhem, a historical Swedish inventor. When Marc visited KTH a few months later, he had read the entire book and understood its contents, and yes, the book was written in Swedish.

Marc has many strengths, and perhaps it's not primarily his language talent that we're going to talk about here, but this experience reflects much of who Marc is. He is genuinely interested in everything related to technology, the history of technology, philosophy and books. He graciously takes on all sorts of challenges without complaining.

A festschrift offers everyone attempting to pen a few lines to a notable individual an opportunity to reflect on encounters and events that, at the time of their occurrence, might not have seemed significant, but in the light of time passed, emerge as pivotal, both professionally and personally. In this section, we share a selection of reflections spanning from the early 2000s to the present day.

Inga-Britt and Lena have had the pleasure of working closely with Marc for several years. As we were responsible in various ways for building up the field of technology didactics research at KTH in Stockholm, Sweden (Inga-Britt as a professor and Lena as a manager and researcher), Marc became a strong pillar and mentor for us in this work.

Marc's significance and role in PATT as a testing ground for students' attitudes towards technology and as a gathering point for researchers in the field of technology education are undoubtedly documented elsewhere in this book. Here, we wish to underscore the importance it holds, particularly for emerging scholars in the field, to have the chance to participate in PATT conferences, present their own research, meet and learn from more experienced researchers, and, notably, establish new national and international connections. Under Marc's leadership, a delicate balance has been maintained between, on the one hand, critical scrutiny and quality standards for the research presented, and, on the other hand, openness towards novices.

© LENA GUMAELIUS AND INGA-BRITT SKOGH, 2025 | DOI:10.1163/9789004710948_032

My (Inga-Britt's) first PATT conference took place in Glasgow in 2003. It was a nerve-wracking experience, I must admit, but also highly instructive. As the director of the research school Technology Education for the Future (TUFF, 2008–2012), I was accompanied by several of the research school's doctoral students to PATT27 in Brisbane in 2010. They and I were warmly received! Through you, Marc, supervisors and doctoral students in the TUFF research school were able to publish for the first time in an internationally published anthology (Technology Teachers as Researchers, Sense Publisher, 2013). It was a laborious, enlightening, but also immensely enjoyable period where the doctoral students (and I) gained hands-on experience of all the phases of writing.

In 2012, we came up with the brilliant idea to affiliate Marc as a professor at KTH. He then became part of KTH's faculty and supported us tirelessly. We wrote articles and books together, like the anthology Pre-university Engineering Education Research at a University of Technology (2016, Sense). Marc supported our doctoral students by participating in seminars and reading their texts. During the time of the QUEST research school, Marc was appointed as a scientific expert, which meant he participated in a number of meetings and seminars for doctoral students and stakeholders. We hope Marc remembers the reading sessions and dinners at Sigtuna Foundation, an old monastery now converted into a fantastic study place for those seeking peace and quiet for their research – a place that many should visit.

Marc is a humble man. When we met at seminars and conferences, he often said – it's Lena who is my boss now … but, of course he knew it was actually him who was in charge of us ☺. We loved when Marc decided to travel to KTH, but we also remembered that it meant we had some work to do. Then things got moving, we needed to write a bit more on the book, we needed to make sure the doctoral students had something to present.

One of our best memories from our work together is when the entire research group from KTH travelled to Delft and heard about how they worked with technology didactics – meeting like-minded people and understanding others' challenges and strengths is always inspiring.

So, Marc helped us build up the unit for technology didactics with his solid knowledge and commitment to the field. I'm sure he inspired many of those who worked in the unit.

Lena wants to wrap up this chapter with her own story about her journey with Marc over the years. As a chemical engineer, I confess I'd never heard of the design process. But Marc took me to so many different places where I met teachers from all over the world, and finally, it clicked. Today, I'm all about designing engineering education, pushing more engineers to tackle our big societal challenges with innovative solutions. So, here's to Marc for not just opening my eyes, but for giving us all a design-driven perspective on life!

CHAPTER 33

Personal Gratitude Letter to Marc de Vries

Dove Kipperman

Dear Marc

I am writing this letter to express my sincere appreciation for the inspiration and influence you have provided on our work as curriculum developers for technology education in ISRAEL.

In the mid-1990s, as part of our work (Dr. Osnat Dagan and myself) as curriculum developers for science and technology education in Israel, we were exposed to a few papers published by you. Those papers influenced and directed our professional work during the next 30 years as curriculum developers. Over the years we met you both at PATT conferences and during your many visits to Israel. Your contribution to technology education everywhere serves as a milestone for us, as well as for many other colleagues.

I would like to refer to two papers from the past that largely shaped our professional vision and reflect the curriculum materials we developed.

1 The first Article

Shaping Concepts of Technology: What Concepts and How to Shape Them

In this paper, you set the academic and theoretical basis for technology education based on concepts in technology and concepts of technology.

In short, concepts of technology are broad and abstract, while concepts in technology are more specific and concrete.

Here is an example to illustrate the difference between concepts of technology and concepts in technology.

- Concept of technology: Systems
- *Concept in technology: Electrical circuits*

This paper paved the way for us to develop technology education as a body of knowledge that helps pupils to understand and use technology concepts.

Teaching technology concepts like systems helps students organise their thinking about technology. By understanding these concepts and using them,

© DOVE KIPPERMAN, 2025 | DOI:10.1163/9789004710948_033

students will learn to see the broad patterns that are across all technology fields.

Based on the paper, and based both on 'concepts of technology' and 'concepts in technology', we in Israel developed books, teacher guides and internet sites (systems in action) for students.

One of the ways to deal with complexity is by "breaking" the system into simple sub-systems and analysing the structural and functional aspects regarding the interrelations between the elements.

Example: analysing the microwave system (Dagan & Kipperman, 2001)

2 The Second Article

Technology education beyond the 'technology is applied science' paradigm.

Within this paper, you challenge the paradigm that technology is just applied science. You analyse the relations between science and technology and state that " Science Does Play a role but not the only one". The paper highlights the "design methodology" discipline as part of the philosophy of technology that can serve as a resource for determining how we should give pupils a realistic image and experience of design.

The "design methodology" as you described it plays a pivotal concept within technology and science curricula in Israeli junior high schools (grades 7–9)

This paper serves as a reference for CPD of Israeli technology and science teachers.

Those two papers stand as a milestone for me [as well as for our team in ORT Israel] shaping the technology education in practice. For that, we are tremendously grateful.

So, keep leading the way and keep on being a source of inspiration.

References

Dagan, O., & Kipperman, D. (2001). Systems in Action": A multimedia environment for learning technology systems. In *PATT-11 conference proceedings* (pp. 59–66).

de Vries, M. J. (1996). Technology education beyond the 'technology is applied science' paradigm. *Journal of Technology Education, 8*(1), 7–15.

de Vries, M. J., & Tamir, A. (1997). *Shaping concepts of technology: What concepts and how to shape them* (pp. 3–10). Springer Netherlands.

CHAPTER 34

A Journey of Global Impact in Technology Education

Leading by Example

Alister Jones

In academic space, we rarely write about the impact that an individual academic has on a country or, indeed, globally in developing a culture of technology education research and development, including curriculum, policy, teacher education, professional learning, research culture and organisational development. In this small tribute, I want to briefly highlight the impact that an individual academic can have globally through their research approach, connecting with others and creating a research culture which has impacted policy, curriculum and teaching and learning in technology education.

I have had the privilege of being associated with and learning from Marc de Vries for the last 35 or more years. Marc and I completed our master's and doctorates in the same years each other in physics education but from opposite sides of the globe. This parallel timeline then created an early engagement. It was our common interest in the relationship between science and technology and what this might mean for teaching and learning in science and then to consider technology education that provided opportunities for collaboration and wider impact.

In the late 1980s and early 1990s, there was the introduction of technology and its role in the teaching of science. We had the science, technology and society debates and attempts to introduce technological applications to introduce physics concepts. Key moments in these times were the separation of science and technology as curriculum areas. These debates were undertaken in the context of the science research area and so *were* dominated through that lens; that is, technology was applied science. Although this was useful in the context of thinking about the teaching of learning about science, it did little to help create technology education on its own as a school subject separate from science.

With the curriculum reforms of the early 1990s, there was a greater exploration of students' and teachers' views and attitudes to technology and what this might mean for curriculum development and teaching and learning in the classroom. Teacher and student conceptualisation of technology is a complex

© ALISTER JONES, 2025 | DOI:10.1163/9789004710948_034

issue and requires an understanding of the many factors that influence it. Student learning of technological concepts and processes will be influenced by approaches supported by particular curricula, including at the national, school and classroom levels. Marc's earlier and continuing work with Pupils' Attitudes to Technology (PATT) plays a key role in promoting this research endeavour, but also creating a legacy of projects and conferences that are still relevant today.

Marc led the development of the conferences, which provided a key platform for bringing researchers together and creating a supportive community to develop the field. These were unique conferences held globally and were often linked to other conferences at the national level. These conferences also shifted the focus from what might be in a curriculum and debates about views of technology to building a more fundamental research agenda. Often, these conferences were aligned with the timing of professional conferences and helped link research with practice. When new areas of research are developing a research culture supporting emerging researchers is key. Marc built this community and sustained it through his leadership and individual engagement in this community flourished, and much is owed to Marc for this.

As countries began to develop their curricula and researchers explored classrooms in depth, there was a need to define the body of knowledge in technology and this key to curriculum development and ongoing work. Marc played a key role in this, and his work in philosophy and examination of how people in technology work was foundational. The philosophy of technology was a relative newcomer to philosophy compared with science. By studying the way philosophers write about technology we can identify the sorts of issues that determine how people think about technology. As Marc indicated, there are two types of philosophical critique: an analysis of what is meant by technology and related terms like technological knowledge, technological design and technological literacy and an analysis of the way technology and culture interact. This aimed at providing a language and terminology for having a framework for discussion on technology, and through Marc's contribution to the area of philosophy of technology, a systematic analysis of the structure and content of technology education was able to be developed and used for curriculum and research progression in learning in technology.

The developing and editing of the International *Handbook on Research and Development in Technology Education* (Jones & de Vries, 2009) was a key collaboration which sought to bring colleagues from across the world together on a joint project, included 56 chapters written by 53 authors from 13 different countries representing hundreds of studies. In this volume, Marc and I, together with Cathy Buntting, attempted to bring together 9 key areas in technology education. It included country overviews from developed and developing countries, philosophical discussions of the discipline of technology, relationships between science and technology, as well as considerations of teaching, learning, assessment, teacher education and research. It is impossible to separate the development of a field from its historical, philosophical and social setting.

In doing this, what was interesting was that in every key aspect of the areas focused on, Marc's research or the outcomes of his work contributed to that area of development. Emerging through all of this is a greater recognition of the theoretical approaches underlying the emerging discipline as well as the need to develop research that informs its ongoing development including a focus on student learning and teacher knowledge of technology in the classroom in its broadest sense

In the paper by Marc, Cathy Buntting and myself, we set out a research agenda for the next few years. In that, we are argued, that If technology education is to advance as a curriculum area of worth and as a focus of research, then we must focus on student and teacher learning and building the notion of a robust academic field in technology education. Much of the early work in technology education spent much time defining the field and the curriculum, and Marc was key to this. More recently, there has been an increase in the quanta of research that has the potential to inform practice, evidenced by the increased number of classroom-based studies and published through key journals such as *International Journal of Technology and Design Education* of which Marc was the editor and key driver.

When we set out on the journey of creating a greater emphasis and space for technology in schools, we understood the political, educational, societal and professional challenges that lay ahead. As we reflect back over the last 35 or so years, we have made huge strides while at the same time, even with an increase in STEM focus in education, we have not come as far as we would have envisioned. Yet a strong foundation has been laid, and it is crucial that the agenda set those decades ago are still relevant.

We set out on the journey, understanding what it might take. We are still not there, but through the leadership of people like Marc, we have made significant progress over the last 30 or so years. The impact of research at a system level is often an untold story. I hope through this brief narrative, I have been able to highlight what the impact of Marc's contribution has been to educational reform and the key characteristics of what made this possible. It is important to acknowledge the legacy that has been created, but it is important that the next generation is developed to continue to challenge the boundaries of what is possible for change in education and particularly to continue to enhance the possibilities for effective learning outcomes for students that will contribute to a better understanding of the world we live in and how to make it a better place.

In the construction of a research area such as technology education it is crucial that there are leaders to create space for the subject, assist in the development of the knowledge base and drive change. Marc, through his research excellence, leadership, collegial nature, and inclusiveness, made a significant impact on technology education globally.

Reference

Jones, A. T., & de Vries, M. J. (2009). *International handbook of research and development in technology education.* Sense

CHAPTER 35

Converging on Interdisciplinarity and Transdisciplinarity

The Path Forward in Technology and Design Education Within an Overwhelmingly Technological World

Dennis Cheek and Hongwu Dai

1 Introduction

Technology and Design Education (TDE) has continually struggled for legitimacy, respectability, and primacy in both the curriculum of schooling in countries worldwide as well as within general discourse within human societies. The scope, sequence, and depth of fundamental knowledge and skills regarding TDE have been continuously reshaped by generations of people purporting to make it relevant, useful, practical, applicable, and worthy of thought and reflective actions by all, not just the chosen few. The "few" frequently meant those whom society deemed suited to a particular future employment path where intellectual skills were seen as non-essentials but manual dexterity and technical skills as prerequisite.

Is it possible to conceive of a path towards the future of TDE that will encapsulate the best of its past while making it an essential subject for ALL citizens both now and in the foreseeable future? This essay will argue that interdisciplinarity and transdisciplinarity, already very evident in research & development activities, will force needed changes in TDE while at the same time considerably raising its profile as a necessary critical thinking and socially responsible preparation for all citizens of all countries for life and living in an overwhelmingly technological society.

The many threads that run through the life and academic pursuits of Marc de Vries speak to the necessity and viability of such a multidimensional approach to the new TDE advocated in this chapter. His intellectual journey through multiple disciplines, facilitates means of making sense out of disparate disciplinary knowledge, and conceptualisation and articulation of both philosophical understandings of technology and practical implications of such understandings in terms of practitioner and societal practices serve as pointers for what is being advocated.

© DENNIS CHEEK AND HONGWU DAI, 2025 | DOI:10.1163/9789004710948_035

2 Immense Challenges Present Immense Opportunities

Continuous news cycles chronicle the many challenges facing humanity in the 21st century. The United Nations' *2030 Agenda for Sustainable Development*, adopted by all UN member states in 2015, delineates the immense challenges across the 17 Sustainable Development Goals (SDGs) facing the world. Just six years away from 2030, it now seems quite clear that most of the goals will remain elusive, despite all the activities and actions that can be applauded (United Nations, 2023). We are reminded just how difficult reaching these goals will be. An international movement is growing to add "culture" as an additional goal even though the word appears in most of the SDGs. Cultures around the globe are under pervasive threats from government-led and other efforts to minimise human diversity, homogenise cultures and languages, disrespecting local customs, artefacts, and territorial claims, and defund or deliberately ignore the need to value, preserve, protect, and promote different human ways of being in the world (Culture Action Europe, 2019; Ottone, 2023). Cultural entrepreneurship has also emerged to remind societies that culture is a human creation and must continually be innovated and evolved for its viability but also for what it contributes to the quality of human life and spiritual, emotional, and ethical sensibilities, beliefs, and feelings that it evokes and sustains (Lockwood & Soublière, 2022; Ratten, 2022; April et al., 2023).

The pervasive presence of human technologies around the world, even in quite remote locations, brings with it promises of human redemption but also imposes upon the world human suffering of an unimaginable scale never achievable during the lifespan of prior generations among the world's 107 billion or so total estimated inhabitants since the dawn of prehistory (Haub, 2011). Concerns about the diminishment of human abilities, capacities, and contributions in the face of the growing powers and transformational potentials inherent within technologies prompt growing debates and gnawing fears – with artificial intelligence (AI) in its varied forms being just the most recent cause for increasing concern (Sinnreich & Gilbert, 2024).

If we take China as an example, we can see that research topics related to the future form of human existence and social development are spreading in more and more disciplines and professional fields (Luo, 2024). Two recent areas of concern found within Chinese academic literature are food security (including safety) and environmental protection.

According to the *Global Food Crisis Report 2023* (Food Security Information Network, 2023), over 258 million people in 58 countries and territories were facing acute food insecurity at crisis levels or worse in 2022. The serious impact of multiple factors such as economic crises, climate disasters, geopolitical conflicts, wars, and major infectious diseases on the global food system has undoubtedly significantly increased the instability and uncertainty of global food security. Considering that food security is a comprehensive and complex system, it involves a series of interwoven, interactive processes such as agricultural production, food supply, regional and global supply chains, household food acquisition, food security evaluation, and response to food shortages strategies (Behnassi et al., 2022).

Chinese scholar Xin (2024) pointed out that the current food security evaluation indicator system is imperfect and insufficient. Most indicators suffer from problems such as inadequate quantification, inconsistent standards, difficulty in making informed comparisons, and the nagging persistence of inaccurate data. Chinese and other geography and agricultural researchers increasingly are focusing their research on the impact mechanisms of food supply, food supply chains, and the food security evaluation index system (Sarmiento & Frolich, 2020).

The deterioration of the ecological environment is a common challenge facing life on earth and China has sought to make major strides forward in facing, ameliorating, and restoring ecological systems that have suffered extensive deterioration due to human actions. Chinese scholars Zhang and Li (2024) call for transcending anthropocentrism to implement environmental protection. Many people care about environmental and ecological issues around them while being less concerned about issues some distance from them – even when those regions are indirectly affecting them in a myriad of ways. The problem of environmental protection poses a challenge to human survival and development. Successful remedies inevitably require stoking people's interests and enhancing the value perceptions they might attach to these issues.

In environmental protection, domestic and international interest demands and rights claims are sometimes contradictory. There is ongoing debate within China and elsewhere over which should take precedence (Zhang & Li, 2024). Chinese scholars believe that when solving international environmental protection issues, the legitimacy of the allocation of environmental protection responsibilities must not only be based on the basic assumption of universal human equality but also consider the social development levels and people's actual needs among different countries and regions where customisation to these circumstances must be given due consideration rather than imposing from the outside uniform solutions across all nations (Zhang & Li, 2024). In China and elsewhere, topics such as toxic pollution, deforestation, global warming, and biodiversity loss have become major ecological and environmental issues, giving rise to multidisciplinary research fields such as environmental ethics, green development theory, environmental protection theory, and deep ecology (Dryzek et al., 2008; Bainbridge & Roco, 2016).

3 The Ever-Changing Landscape of Research and Development

Human endeavours to understand more fully both the natural world and the human-designed world have served as an excellent stimulus to the development of ever-growing research approaches to investigate the universe we inhabit and to speculate about even other universes and dimensions of reality. The history of science and technology, humanities, the arts, and social sciences demonstrate that knowledge is indeed a progressive discovery of our ignorance as we falsify our prior conceptions and understandings and replace them with new explanatory and prediction frameworks. Multiplying methods of investigation multiply understandings, generate new questions, and evoke new perspectives – some of which prove to be revolutionary in their extent and ramifications. The role of technologies themselves in opening new avenues of research and making possible the investigations of various phenomena at different levels of scale has proven especially fruitful while also simultaneously reshaping our perceptions and ways of being in the world (Chatfield, 2024).

Accompanying these technological advances is a growing recognition of the vast terra incognito of human knowing, including both the knowable and evolving understanding of what likely is unknowable. Increasingly as we move through the centuries, a very interesting development has occurred. There was a time in human history when it was potentially possible for a single person to know everything that was known at that time (Peters, 2017; Robinson & Rees, 2023). The 1789 first edition of the Encyclopedia Britannica published in America was three volumes with large print containing everything an educated person should know (Arner, 2016).

Natural philosophers explored reality with no limitations on what they could think about or investigate. In modern terms, we would say they intentionally crossed many disciplines as they sought to make sense of phenomena they encountered in the world (cf. Siraisi, 2013; Sgarbi, 2023). The English phrase "Renaissance person" is still used to denote someone in the present who is recognised as cognisant of work in multiple disciplines and who works across multiple fields with some degree of success. Professor Marc de Vries certainly merits this moniker for his professional practice moving across multiple fields in pursuit of new formulations and understanding – a very good physics teacher who also became a well-known philosopher of technology, a global expert and an intellectual leader within TDE, and who moved into the realm of researching communicating science and technology knowledge and viewpoints to varied audiences, including Protestant church congregations and leaders. This volume recognises his thought leadership, diverse research activities, and continuing impact in multiple areas of human endeavour.

As professionalisation and specialisation arose from the 16th century onwards, we seemed to be reaching a state where not only could a single person not know everything that was known; they could not even keep up with what was known within a particular discipline or even a sub-speciality within a discipline. One of the hallmark institutions of the world, the modern university with research labs (initially limited to chemistry), was purposefully created in Germany and quickly copied in the United States and elsewhere in the 19th century. It was organised into specialised faculties within discrete disciplines where an individual specialised in a given topic. The work of thousands of these specialists within different disciplines caused human knowledge to rapidly expand and specialised academic literature continued to the point where even a specialist in a narrow area found it difficult to remain abreast of everything that was occurring that seemed relevant (Rüegg & Ridder-Symoens, 2003–2010; van Berkel & Homburg, 2023). This period was also accompanied by the rise of professionals – occupations where specialised knowledge and skills were taught and externally certified by universities and professional bodies within distinct areas such as medicine, law, social work, clergy, and dentistry and no longer obtained via an apprenticeship form of education (Bledstein, 1976).

The multiplicative and exponential growth of human knowledge in various arenas has increasingly made such organisational structures look almost quaint. The OpenAlex database, for example, tracks over 100,000 different journals and forms the basis for Leiden University's Centre for Science & Technology Studies new OpenEdition Ranking 2024 of the world's universities.[1] Specialists find it hard to even locate relevant literature in so many disparate sources akin to the proverbial finding the needle in a haystack. Intelligent tools are starting to make some progress in aiding (or further confusing?) the researcher but ten years from now we will likely view current attempts as primitive and not very illuminative.

One early solution to deal with this problem of knowledge proliferation/separation was the rise of multidisciplinary research teams and entire departments organised in universities around problems and issues. Frequently these multidisciplinary efforts led to what came to be recognised as entirely new disciplines, whose academic program of study for this new degree blended in varying amounts of studies across multiple disciplines (e.g., materials science, biogeography, biochemistry, earth sciences). Well-recognised scholars working in these and other "newer" disciplines that evolved in this way, as well as people interested in research policy and its development, came to understand that many problems, issues, and situations could not be investigated fruitfully or well when explored by specialists in a single discipline using the set tools, techniques and knowledge routinely used within that arena. We see this trend reflected also at the international level as seen in the UNESCO thesaurus which classifies interdisciplinary approach, interdisciplinary research, and integrated curricula all under "scientific methods" as the broader concept. Transdisciplinary and interdisciplinary projects, programs, initiatives, and endowed chairs in UN institutions are spread across the entire structure and functions of the United Nations, although one can note that interdisciplinary notations remain more prevalent than transdisciplinary (www.un.org and its many UN organisational links).

Renaissance scholars roamed across what we would today call multidisciplinary, cross-disciplinary, and interdisciplinary approaches to knowledge creation, problem-solving, and decision-making as seemed appropriate to the problems on which they were working (Dekka, 2020). Today we return to this approach to growing human knowledge within contemporary scholarly communities but with much greater depth, more powerful techniques, more elaborate explanatory frameworks, and highly specialised knowledge being co-mingled in new ways, resulting often in powerful effects over time (Gibbons et al., 1994; Nowotny et al., 2001). Many developments were driven by the ongoing complexities associated with human impact on the environment, the development of new military weapons and systems for their use, the rise of global monitoring and reporting systems, and the creation of new multidisciplinary approaches like systems science, STS (both science, technology, & society studies and science and technology studies) and services science investigating these kinds of changes and their many ramifications (Dai et al., in press). The National Science Foundation in the United States 2016 highlighted the unique characteristics of this type of research coining the term "convergence research" which it defined as:

> a means of solving vexing research problems, in particular, complex problems focusing on societal needs. It entails integrating knowledge, methods, and expertise from different disciplines and forming novel frameworks to catalyse scientific discovery and innovation. Convergence research is related to other forms of research that span disciplines – transdisciplinarity, interdisciplinarity, and multidisciplinarity. It is the closest to transdisciplinary research which was historically viewed as the pinnacle of evolutionary integration across disciplines". (EOPUS, 2022, p. 9)[2]

The importance of interdisciplinary and transdisciplinary research continues to elevate in academic circles from the 1960s to the present time (e.g., Dai et al., in press, for a brief history and examples such as Klein, et al., 2001 and Keitsch & Vermeulen, 2021). Convergence has been a clear theme in recent scientific and technological research and development which has highlighted examples such as biocognitive evolution, collective intelligence systems, consilience, convergent-divergent processes, self-organisation, the emergence of dynamic systems, global climate change, systemic interdependencies, complex biological systems, and techniques for investigating such phenomena such as modeling and simulation, science & technology forecasting, and whole Earth monitoring (Brainbridge & Roco, 2016; cf. National Academies of Sciences, Engineering and Medicine, 2014, 2018, 2019, 2023).

Guru Madhaven is a biomedical engineer and a senior program officer at the National Academy of Engineering in the United States. In his provocative new book, Wicked Problems: How to Engineer a Better World (Madhaven, 2024), he distinguishes between hard problems, soft problems, and messy problems – delineating how engineers approach each of these types of problems differently. He describes what we can learn from them to deal with wicked policy challenges and national and global problems more generally starting from the systems sensibility that engineers exhibit. He correctly notes that contrary to popular stereotypes, engineers since the earliest civilizations have always applied engineering concepts within civic contexts as they sought to improve standards of living, raise human know-how, and expand human thought. He gives engineers high marks for their ability to recognise sensitivities, shape synergies, and account for side effects – at least in their better moments. However, he reminds us that “wicked problems” are a collision among hard problems, soft problems, and messy problems in the real world as they are intertwined in ways that make solutions extraordinarily difficult to come by no matter how many different disciplines seek to attack them.

Convergence is just the latest in a series of approaches that have been invented by human beings as the complexities of life multiply and the ambitions of humans to ameliorate them grow unabated. Steve Jobs once remarked concerning the work of Apple that “It’s technology married with liberal arts, married with the humanities, that yields us the result that makes our heart sing” (Cited by Joseph DeSimone in National Academies of Sciences, Engineering and Medicine, 2021, p. 8). Humans, always in search of better solutions and breakthroughs, are increasingly collaborating in new ways that promise new insights, the invention of new tools and techniques and disposed towards a general hopefulness about the future of the human race – despite its many challenges (National Academies of Sciences, Engineering, and Medicine, 2023).

Measuring the worth and impact of convergent research and development turns out to be difficult given the varied ways in which it is conceptualised, actualised in practical R & D activities, funded, and more generally seen as vital to making further progress in both understanding our places of existence but also in dealing with the many issues which face humans and the flora and fauna of this planet which we all call home. Expert participants at a recent National Academies of Sciences, Engineering, and Medicine workshop in the United States provide an excellent summary of the many dimensions that must be considered, questions that might be asked and answered, and issues that must be adequately thought through, experimented, and addressed (National Academies of Sciences, Engineering and Medicine, 2021). Fortunately, it appears that the general public in America and elsewhere in the world still has considerable trust in the work that scientists do. That confidence has declined in recent years but is similar to, or higher than, the confidence the public has in civic, governmental, or cultural institutions (Lupia et al., 2024). While many in the public think scientists and technologists are the same, we have less clear information about the level of trust the public has in engineers vs. scientists vs. technologists who are not engineers and whether the public even understands such distinctions.

4 Looming Changes in General Education

Commensurate with changes going on in the wider world and the spectre of some incredible predicaments facing humankind, there is a growing consensus that the educational systems we currently possess from primary through secondary to post-secondary education need further redesign. The Organisation for Economic Cooperation and Development (OECD) is comprised of 18 member nations at present and along with its partners accounts for 80% of the world's trade and investment. It has spearheaded an effort across its member states and partners to take an in-depth look at upper secondary education to evaluate the adequacy of current secondary education offerings, policies, procedures, and results with a particular focus on its members. A recent report by Stronati (2023) looked carefully at the design of upper secondary education in terms of student choices, coherence, and specialisation and revealed just how dramatically different the structures, choices, and pathways are across member states. It advocated more consistent policies in line with bodies of accumulating research about the effects of various current policies and overall achievement results, with special attention to historical gaps in achievement regarding subgroups within each nation. A companion report looked specifically at how student transitions are managed in upper-secondary pathways (Santos, 2023). It found much to recommend from selected countries but much to criticise, in the light of research and results achieved within each nation. It found that much more work is needed to align research, policy, and practice across the OECD (and one would suspect, elsewhere in the world). The OECD (2024) recently released its 500+ pages PISA Vocational Education & Training (VET) Assessment & Analytical Framework which considers career-focused education in automotive technology, electrician, business & administration, healthcare, tourism & hospitality, and employability skills in general. It seems that a similar type of OECD approach for TDE would be quite worthwhile in serving as a prestigious and influential venue for the formulation of the very kinds of changes for which this chapter and much of this volume advocates.

Louder and ever-more influential voices are also calling for the redesign of educational systems to confront the global challenge of climate change (Nusche et al., 2024) and use this evolving crisis as leverage points for change. They argue for rethinking both STEM education and cross-curricular learning, utilising the power of place-based approaches to climate change, and making school systems themselves climate change resilient. The United Nations recognises that the SDGs can never be fully addressed unless education at all levels focuses on the 17 current SDGs in more predictable, engaging, sustained, and fruitful ways (United Nations, 2023). Wessels (2022) has recently argued that the societal challenges of the present time require all educators to adopt a "pedagogy of entanglement" where we consciously complexify our students' understanding of contemporary issues and actively engage them in the quest for solutions. The education arm of the Smithsonian Institution in Washington, DC, formulated and has been promoting in its programs and throughout its extensive network of museums and research organisations a five-stage Global Goals Action Program (GAP) progression which consists of 1) questioning, 2) investigating, 3) critical reasoning and systemic understanding, 4) synthesising, and 5) acting. GAP promotes the purposeful introduction in educational settings of cultural values, cultural identities, different epistemological beliefs, varying cultural contexts and the specificity of "place". It seeks to create students at all grade levels who are ready and willing to take (local) actions consistent with their learning and engagement (Gibson, 2021).

A broad-based coalition that works with the Center for Curriculum Redesign in the United States led by the founder of the CISCO Academies, has argued that in the age of artificial intelligence (AI), humans will always need human interactions to develop expertise and transfer since education is broad and transfer involves accounting for contexts and making appropriate adaptations in response to what is found. They also rightly note that the "T" in STEM usually just means computing applications and rarely invokes or involves investigation of the technologically pervasive, human-designed world we inhabit where innovations of even the long-distant past have become so commonplace in human experience that we scarcely consider where they came from and who was first responsible for formulating these innovations in historical or prehistorical times (Fadel et al., 2024; Chatfield, 2024). Despite all the buzz about AI eating the world, the keenest observers and leading AI advocates themselves, are quick to point out both its usefulness and its limitations – the concept of tradeoffs that are familiar to everyone teaching and learning TDE (Sinnreich & Gilbert, 2024; Mainzer & Kahle, 2024; Aoun, 2017).

5 Convergence Education

Consistent with the interdisciplinarity and transdisciplinarity research across federal agencies within the US government regarding STEM convergence research, this naturally leads to the identified need for "convergence education" for all Americans across ALL federal agencies which provide funding for educational efforts as portions of their annual operational budgets. The Executive Office of the President of the United States in its report (EOPUS, 2022, p. 7) defines convergence education as "driven by compelling or complex socio-scientific problems or topics, where learners apply knowledge and skills using a blended approach across multiple disciplines (i.e., transdisciplinary) to create and innovate new solutions". Long-time advocates of both STS education and TDE might rightly quip, "where have you been all this time?" Many sources in environmental education from the 1960s to the present as well as STEM education have advocated for interdisciplinary and transdisciplinary education efforts within primary, secondary, and higher education (e.g., Sengupta et al., 2019; Anderson & Li, 2020; Bodzin et al., 2010; Scott & Vare, 2021; Häggström & Schmidt, 2022).

The good news is that this new consensus on convergence education across all US government agencies if it holds, will portend potentially good implications for TDE efforts in the United States and as well as providing additional conceptual and programmatic justifications for TDE efforts elsewhere in the world; although we must recognise that there is a certain impracticality embedded in how we think about change informed by social science and other types of research, given its history and uses (cf. Popkewitz, 2020). This prominent executive branch document lays out a framework to understand the relative contributions of different types of approaches to disciplines using an ice cream analogy (Figure 35.1) and a helpful table that looks at what STEM learning, teaching and assessment look like at the various levels of integration (Table 35.1).

TABLE 35.1 Elements of STEM learning, teaching, and assessment at progressive levels of integration (EOPUS, 2022, p. 14f.)

	Disciplinary	Multi-disciplinary	Inter-disciplinary	Trans-disciplinary
Learning	Students directed by teacher to focus on a single discipline or topic within a content area.	Students directed by teacher to focus on more than one discipline in a unit of study, but the disciplines are not connected as part of the learning experience for the student.	Students focus on more than one discipline in a unit of study and the disciplines are connected to each other but don't have to be fully integrated.	Students focus on broad real-world constructs drawn from our increasingly interconnected world for their relevance and student interest; integrates values, ethics, worldviews, and student action.
Teaching	Teacher identifies an isolated science, technology, engineering, or math concept (fact, idea, or practice) and does not approach it in an integrated way.	Teacher identifies a STEM concept involving multiple STEM disciplines addressed independently on different aspects of the same concept.	Students and teachers collaborate to identify a STEM concept involving multiple STEM disciplines (and even other disciplines such as art, social studies, etc.) in an integrated way that makes the concept authentic and real-world.	Teachers work together with those of other disciplines to support students to identify a real-world problem and work together to create a shared conceptual framework and draw together theories, concepts and practices that transcend individual disciplinary boundaries.

(*continued*)

TABLE 35.1 Elements of STEM learning, teaching, and assessment at progressive levels of integration (EOPUS, 2022, p. 14f.) (*cont.*)

	Disciplinary	Multi-disciplinary	Inter-disciplinary	Trans-disciplinary
Assessment	Assessment items address only a singular discipline.	Assessment items address multiple disciplines related to the STEM concept but they are not integrated.	Assessment item is problem or phenomenon based and integrates multiple STEM disciplines and other disciplines.	Assessment item is scenario-based, measuring content knowledge along with creative and critical thinking frameworks, and focused on 21st century learning.
Example	Students examine the properties of soil. *Assessment*: define soil.	Students examine the properties of soil in one lesson, then they calculate the porosity of soil in another. *Assessment*: one item assesses science, another math, etc.	Students study the porosity of soil, assess its impact on erosion, and engineer a solution to a soil erosion problem. *Assessment*: students present their engineered design proposal to solve defined problem.	Students identify an erosion issue in the school community (e.g., drainage related to local construction and propose solutions while exploring related issues such as infectious disease, biohazards, and community relations. *Assessment*: rubric includes identifying issue and scientific evidence, knowledge, and skills learned.

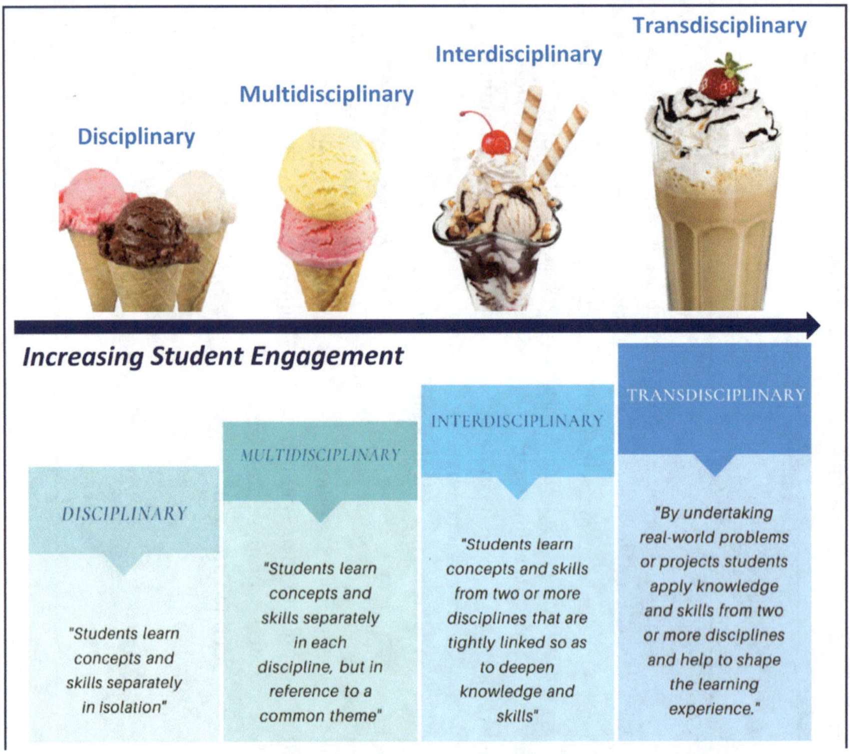

FIGURE 35.1 Increasing levels of integration, (EOPUS, 2022, p.19)

The EOPUS document (pp. 17–18) goes on to delineate a series of what it designates as "near-term opportunities" that include academic teaming, applied learning, CTE [career and tech education], engineering design, inquiry-based learning, invention education, phenomenon-based learning, place-based learning, problem-based learning, project-based learning, scientific modelling, and work-based learning experience. TDE specialists around the world can powerfully relate to these near-term opportunities, many with which they are already experienced, and immediately appropriate them in their endeavours in terms of curricula, teaching, and assessment.

6 Supply-Focused TDE Responses

A futuristic supply-focused TDE features current world problems and issues within a framework that draws teachers and students to consider multidisciplinary, interdisciplinary, and transdisciplinary responses. The subject will be freed from its many strictures of the past while taking forward the best complex problem-solving, scenario building, design evaluation, and rigorous testing from TDE's rich history of innovations. There are, of course, numerous barriers to such an imagined outcome, some of which the Executive Office of the President of the United States enumerate including: 1) convergence is a new and undefined idea in education, 2) alignment in implementation across grade bands, 3) existing educational systems organised around discrete disciplines with attendant standards and assessments, 4) time and resources allocations to enable relearning by teachers in ways consistent with what convergence education would require, and 5) attitudes of teachers and students towards convergence education (EOPUS, 2022, p. 16).

The history of education in the 20th and 21st centuries is replete with examples of these very same barriers being faced by other educational innovations, some of which have proven reasonably successful and others perhaps as abysmal failures (cf. Cole, 2009; Levine & Van Pelt, 2021; Spring, 2021). Reorganising higher education at the entire organisation level in this manner has encountered broad resistance. However, Arizona State University in the US under the 20-year leadership of President Michael Crow seems to have successfully overcome opposition and very significantly redesigned the entire university, and its operations, and greatly elevated its academic reputation due to its strong emphasis on interdisciplinary and transdisciplinary research and teaching efforts (Crow & Daber, 2018, 2020). Wayne Clough (2021) performed a similar feat when President of the Georgia Institute of Technology from 1994–2008. Meanwhile, another higher education private nonprofit innovator launched in 2007, the former Quest University in Squamish, BC, Canada, designed its entire curriculum around one-month block scheduling and granted a single degree which combined liberal arts disciplines in unique ways organised around problems and issues. However, it failed to find a large enough clientele and achieve stable funding and ultimately had to close its doors in 2023 (Kozelj, 2023). A new American-led university called Minerva was intentionally created to be quite different as a model for the future of global higher education with its students living in major cities all over the world (40 countries as of this date) and studying using an online environment coupled with local experiences that fit with the design of the curriculum. As of this date, it is still in business as a thriving private, nonprofit university (Kosslyn & Nelson, 2017).

Teaching and guiding learners within this reoriented school subject will require teachers considerably different than what was required of prior generations of teachers (National Academies of Sciences, Engineering, and Medicine, 2020a, 2020b, 2020c; Davis & Stephens, 2021; Floden et al., 2020; Wilson et al., 2016). It will also require thinking more deeply about how we contextualise learning more fully, and deliberately engage in learning experience design, i.e., creating an environment where we design for the types of learning we want to see occurring as students collaboratively work together and the "teacher's" role varies throughout the week as guide, mentor, provocateur, recorder, observer, listener, consultant, resident "expert", and co-collaborator. TDE must seek to develop students' expertise, and understanding in differing contexts, and learn how to transfer and transform what they know and can do in ways that match the problems and issues they are seeking to address (Fadel, et al., 2024; Aoun, 2017; Gibson, 2021; Nusche et al., 2024). We need to assume that students already have latent abilities in the realms of imagination, creativity, innovation, invention, design, and entrepreneurial orientation that need to be regularly unleashed in school-based learning and come more into line with what they routinely are capable of doing in their everyday lives (Hunter, 2023; Sawyer & Henriksen, 2024; Moulding et al., 2019, Liu & Noppe-Brandon, 2011; James & Brookfield, 2014). This requires specific attention to these abilities and introducing to students a wide tool kit of techniques that enable them to enhance their abilities in each of these areas and move across them in a manner consistent with the principles of interdisciplinarity and transdisciplinarity in a personal quest for expanding competencies and a growth mindset that can be essentially continuous across their lifetimes (Cheek, 2021; see Riveros et al., 2022 for higher education considerations and the role of institutionalisation). Transdisciplinary education calls for a broadening of the average teacher's instructional techniques repertoire, and the first explicit handbook to aid this transition has now appeared under a Creative Commons Attribution-ShareAlike 4.0 licensed document from DeGruyter which authorises others to reproduce and/or adapt handbook materials that they find useful, so long as they continue to authorise uses of their own adopted or adapted materials similarly (Philipp & Schmohl, 2023).

7 The Danger of Thinking the Battle Is Won Over TDE

The TDE community dare not relax its vigilance in advocating for TDE and its proper place in the school curriculum throughout the various grade bands of primary/elementary and secondary school and undergraduate and graduate education (Baptista & Klein, 2022). If we use the USA as an example, we can say that the current situation in schools and universities is one where instead of STEM we have SteM, i.e., students study far more years of the sciences and mathematics than they study about TDE and engineering. Much of the literature from the National Academies in America appears to subsume TDE completely into engineering when history, experience, and the very essence of what occurs in these teaching areas, while manifesting some overlaps, suggest that TDE can be easily distinguished from engineering. Technologists are not some lesser version (type) of engineers; rather we might think of engineers as a specialised form of technologists in the same way that invention can be thought of as a specialised form of activity within the broader category of innovation.

Technology remains the absent presence in the American school curriculum – students inhabit a pervasively technological world their entire lives, yet they spend the rough equivalent of a single full year learning anything about it and engaging it purposefully within their formal studies in most US states and our 19,000 or so public-school districts. Vicariously, they may pick up some meagre insights about technology via the many tools and affordances they utilise in schools and outside of them, but this is not formal learning that enables insightful mastery and transfer (Fadel et al., 2024). The various national standards documents produced in America across many subject areas have been altered in their state-specific versions to the point where some essentials have been lost. These products of hard work and orchestration at the national level remain decidedly NOT national in everyday teacher understandings, instruction, assessment, and school-level practices. The central role that a high-quality curriculum plays in effective teaching and learning cannot be minimised. Research has demonstrated that the use of high-quality curriculum materials must be: 1) appropriately linked to standards, 2) provide a clear exposition of the material to be learned, 3) offer sound and interesting activities linked to those same standards, and, 4) be accompanied by professional development focused on these materials and their use. If all of these things happen, the result is higher and more uniform learner achievement (Petrilli, 2023).

If we apply the standard definition and practices associated with a legitimate area of academic study to TDE we can see that it is indeed legitimate (Cheek, 2015). It has all the trappings of a legitimate area of study, including bodies of research and professional literature, master and doctoral programs, specialised journals, forums for professional meetings and discussions, jobs, professional associations, accreditations, and certifications and is present all over the globe. The problem is that TDE is insufficiently valued, promoted, required, and tested across the USA.

The engineering community has done much better at organising itself and advocating for engineering in primary and secondary education in the United States – although even engineering education at pre-collegiate levels is still a long way from widespread implementation across all public schools, let alone private schools in the US. It has been rightly said that "a vision without resources is a hallucination" (DeSimone in National Academies of Sciences, Engineering, and Medicine, 2021, p. 10) and the old phrase, "show me the money", clearly indicates that both the "t" and the "e" of SteM are underresourced in elementary and secondary education in the USA compared to their two multidisciplinary cousins.

Attitudes and beliefs must be altered in the USA but also in many other countries. If we take China as an Asian exemplar, we note that compared with attitudes towards science and technology in general, Chinese scholars have conducted more empirical studies on attitudes towards science and technology in specific fields, such as the public's attitude towards climate change (e.g., Huang et al., 2016), nanotechnology (e.g., Liu, et al., 2016), and nuclear energy applications (e.g., Zhang et al., 2016; Deng et al., 2016). There are even more abundant studies on attitudes toward the applications of biotechnology, especially genetically modified (GM) food, as the influence of traditional Chinese medicine and respect for the inherent worth of nature, have made many Chinese citizens wary of GM foodstuffs (e.g., Chen & Jin, 2019; Pan et al., 2020; Jia, 2020). When explaining the influencing factors of attitudes towards science and technology in these specific fields, most studies are still based on research ideas of "public understanding of science" and "risk governance", focusing on individual-level scientific knowledge (literacy), risk-benefit perception, and emotional influence (such as Deng et al., 2016; Jia, 2020). Little attention has been paid to the influence of scientific and technological information acquisition channels or media (cf. You & Jin, 2020). Attention to social factors mainly focuses on social or institutional trust and social networks (Qiu et al., 2007). Few involve discussion of other social structural factors. All research is problem-oriented, either theoretical or practical.

When Chinese scholars study technological attitudes, some people will mention PATT in reviews (Xu, 2020; Xu & Gu, 2021). However, judging from the literature found, very few people in mainland China pay much attention to PATT. PATT has better acceptance in Hong Kong and Taiwan (Yu et al., 2005; Volk & Yip, 1999; Volk et al., 2003). Volk and Yip (1999) Chineseised and localised the PATT items and developed the PATT-Hong Kong scale.

In order to present, verify and explain the relationship between technology understanding and technology attitude, Xu and Gu (2021) conducted an empirical investigation of 835 high school general technology teachers in seven geographical regions in mainland China. They believe that the formation of attitudes among these teachers is controlled by both rational and irrational factors.

Attitudes formed on a rational level are decisions and judgments based on information. For example, when a high school general technology teacher believes that the subject is in a weak position, it is easy to have ideas that are not supported by external resources. The formation of this attitude combines personal experience and rational judgment. Technical attitudes formed on an irrational level may be guided by personal interests or a certain ideology. For example, when high school technology teachers generally believe that high school general technology courses are complex and difficult to teach compared to other subjects, the expression of this emotion may be to attract the attention of relevant national and local departments to high school general technology courses. The formation of attitudes is not entirely based on objective technical experience and may only depend on personal motivations and interests (Xu & Gu, 2021).

Without a shift in attitudes and beliefs among teachers, administrators, parents, policymakers, and students about the technological world and its impacts, tradeoffs, and issues, we will make little progress in seeing substantial change in the USA, China, or many other nations regarding TDE. Potential allies for the TDE community can be found in communities everywhere in the world. But we need major and more effective marketing, public education, outreach efforts, and leaders who can articulate a powerful case for the full and equitable inclusion and quantum increase of TDE in curricula and programs worldwide.

8 Conclusion

Professor de Vries and many others have provided a solid intellectual foundation for the TDE movement. There are many good examples of curricula and learning activities in TDE that lead to important changes in students' knowledge, skills, attitudes, and actions in the sphere within which TDE focuses. But we must go further, faster. Evolving environments in R & D, education, and societies provide a platform into which a better organised, cohesive, and more strategic TDE can speak truth to power. Discussions about the importance of TDE conceptualised, designed, and taught within a convergence education framework might enable TDE to ultimately become THE subject that the best and brightest teaching candidates are equipped and eager to teach well. It could become the mid-21st century equivalent of physics and calculus teaching in high schools in the 20th and early 21st centuries. However, instead of being a subject limited to what is widely (but perhaps wrongly?) perceived as the best and brightest students, TDE will be the new integrator of the school curriculum through its adept interdisciplinary and transdisciplinary learning experiences and the new culminating verification for eligibility to graduate from upper secondary school. SteM has the potential to become STEM where TDE takes it rightful place as an equal partner to reshape the school curriculum in ways commensurate with contemporary life, work, and the challenges of modernity. We owe it to every student to prepare them to understand, navigate, evaluate, value, critique, and make a difference in this pervasively technological world that they will inhabit their entire lives. Their stake in their future has already commenced. They need to be involved in designing the innovations, inventions, systems, processes, and sociopolitical solutions that enable humankind not just to survive but to thrive.

Notes

1 www.openleidenranking.com
2 https://www.nsf.gov/od/oia/convergence/index.jsp

References

Anderson, J., & Li, Y. (Eds.). (2020). *Integrated approaches to STEM education: An international perspective*. Springer.

Aoun, J. E. (2017). *Robot-proof: Higher education in the age of artificial intelligence*. The MIT Press.

April, W. I., Adeyanju, A. A., & Tafirenyika, B. (2023). *Context, policy, and practices in Indigenous and Cultural Entrepreneurship*. IGI Global Business Science Reference.

Arner, R. D. (2016). *Dobson's 'encyclopedia': The publisher, text, and publication of America's first Britannica* (pp. 1789–1803). University of Pennsylvania Press.

Bainbridge, W. S., & Roco, M. C. (Eds.). (2016). *Handbook of science and technology convergence*. Springer. https://link.springer.com/referencework/10.1007/978-3-319-07052-0

Baptista, B. V., & Klein, J. T. (Eds.). (2022). *Institutionalizing interdisciplinarity and transdisciplinarity: Collaboration across cultures and communities*. Routledge.

Behnassi, M., Baig, M. B., Sraïri, M. T., Alsheikh, A. A., & Abu Risheh, W. A., (Eds.). (2022). *Food security and climate-smart food resilience in the global south*. Springer.

Bledstein, B. (1976). *The culture of professionalism: The middle class and the development of higher education in America*. W. W. Norton.

Bodzin, A. M., Klein, B. S., & Weaver, S. (Eds.). (2010). *The inclusion of environmental education in science teacher education*. Springer.

Chatfield, T. (2024). *Wise animals: How technology has made us what we are*. Picador.

Cheek, D. (2015). Legitimization strategies across the various guises of entrepreneurship: Implications for cultural entrepreneurship educators. In O. Kuhlke, A. Schramm, & R. Kooyman (Eds.), *Creating cultural capital: Cultural entrepreneurship in theory, pedagogy, and practice* (pp. 22–31). Eburon Academic Publishers.

Cheek, D. (2021). Guiding principles for integrating disciplines and practices in pursuit of complex and diverse learning outcomes. In B. Hokanson, M. Exter, A. Grincewicz, M. Schmidt, & A. Tawfik (Eds.), *Intersections across disciplines: Interdisciplinarity and learning design* (pp. 13–25). AECT Education Communications and Technology: Issues and Innovations series. Springer.

Chen, X., & Jin, J. (2019). Gong Zhong Dui Zhuan Ji Yin Ji Shu de Tai Du Xing Cheng: Ji Yu Wen Xian de Li Lun Kuang Jia Shu Li [The formation of public attitudes towards genetically modified technology: Review of theoretical framework based on literature]. *Global Media Journal, 6*(4), 18–43.

Clough, G. Wayne. (2021). *The technological university reimagined: Georgia Institute of Technology, 1994–2008*. Memoirs of a university president. Mercer University Press.

Cole, J. R. (2009). *The great American university: Its rise to preeminence, its indispensable national role, why it must be protected*. Public Affairs.

Crow, M., & Dabars, W. B. (2018). *Designing the new American university*. The Johns Hopkins University Press.

Crow, M., & Dabars, W. B. (2020). *The fifth wave: The evolution of American higher education*. The Johns Hopkins University Press.

Culture Action Europe. (2019). *Implementing culture within the sustainable development goals: The role of culture in Agenda 2030*. Author.

Dai, H., Cheek, D., & Li, T. (in press). Western and Chinese dialogue about the origins, evolution, and meaning of transdisciplinarity and some implications for learning experience designers. In B. Hokanson, M. Exter, A. Grincewicz, M. Schmidt, A. Tawfik, & Y. Earnshaw (Eds.), Transdisciplinary learning experience design: Futures, synergies, and innovations. AECT Education Communications and Technology: Issues and Innovations series. Springer.

Davis, E. A., & Stephens, A., (Eds.). (2021). Science and engineering in preschool through elementary grades: The brilliance of children and the strengths of educators. *National Academies of Sciences, Engineering and Medicine*. The National Academies Press. https://doi.org/10.17226/26215

Dekka, J. J. H., (Ed.). (2020). *A cultural history of education in the Renaissance*. Bloomsbury Academic.

Deng, L., Zhen, X., & Zhou, Z. (2016). Ke Guan Zhi Shi Yu Zhu Guan Zhi Shi: Qing Nian Xue Sheng de He Dian Zhi Shi Shui Ping Ji Dui He Dian Tai Du De Ying Xiang

[Objective knowledge and subjective knowledge: the influence of young students' nuclear power knowledge level and attitude towards nuclear power]. *Ke Xue Yu She Hui, 6*(2), 85–109.

Dryzek, J. S., Honig, B., & Phillips, A. (2008). *The Oxford Handbook of Political Theory*. Oxford University Press, USA. https://doi.org/10.1093/oxfordhb/9780199548439.001.0001

EOPUS – Executive Office of the President of the United States. (2022). Convergence education: A guide to transdisciplinarity STEM learning and teaching. A report by the Interagency Working Group on Convergence, Federal Coordination in STEM Education Subcommittee, Committee on STEM Education of the National Science and Technology Council. November 2022. ERIC ED 630556.

Fadel, C., Black, A., Taylor, R., Slesinski, J., & Dunn, K. (2024). *Education for the age of AI*. Center for Curriculum Redesign.

Floden, R., Stephens, A., & Scherer, L. (Eds.). (2020). Changing expectations for the K-12 teacher workforce: Policies, preservice education, professional development, and the workplace. *National Academies of Sciences, Engineering and Medicine*. The National Academies Press. https://doi.org/10.17226/25603

Food Security Information Network. (2023). *Global food crisis report 2023: Joint analysis for better decisions*. Author.

Gibbons, M., Limoges, C., Nowotny, H., Schwartzman, S., Scott, P., & Trow, M. (1994). *The new production of knowledge: The dynamics of science and research in contemporary societies*. SAGE.

Gibson, H. (2021). *From ideas to action: Transforming learning to inspire action on critical global issues*. Smithsonian Institution Scholarly Press, Open Monographs. https://scholarlypress.si.edu/store/life-sciences-biodiversity/ideas-action-transforming-learning-inspire-action/

Häggström, M., & Schmidt, C., Eds. (2022). *Relational and critical perspectives on education for sustainable development: Belonging and sensing in a vanishing world*. Springer.

Haub, C. (2011). *How many people have ever lived on Earth? Population Reference Bureau*. Downloaded 04.05.2024 from www.prb.org.

Huang, L., Ren, L., & He, W. (2016). Zhong Guo Gong Min Dui Quan Qiu Qi Hua Bian Hua de Ren Zhi Ji Tai Du [Chinese public's understanding and attitude towards global climate change]. *Ke Pu Yan Jiu, 11*(3), 45–52.

Hunter, D. (2023). Learning and teaching creativity. *You can only imagine* … Ithasca Books.

James, A., & Brookfield, S. D. (2014). *Engaging imagination: Helping students become creative and reflective thinkers*. Jossey-Bass.

Jia, H. (2020). Zhi shi Ke yi Gai Bian dui Zhuan Ji Yin Shi Pin de Tai Du ma [Can knowledge change the public attitude to GM food] Hua Zhong Nong Ye Da Xue Xue Bao (Social Sciences Edition), (1), 135–142.

Keitsch, M. M., & Vermeulen, W. J. V. (Eds.). (2021). *Transdisciplinarity for sustainability: Aligning diverse practices*. Routledge.

Klein, J. T., Grossenbacher-Mansuy, W., Häberli, R., Bill, A., Scholtz, R. W., & Welti, M. (Eds.). (2001). *Transdisciplinarity: Joint problem solving among science, technology and society: An effective way for managing complexity*. Birkhäuser.

Kosslyn, S., & Nelson, B. (2017). *Building the intentional university: Minerva and the future of higher education*. The MIT Press.

Kozelj, J. (2023). Inside the shutdown of Quest University. University affairs/affaires universitaires. 16 August 2023. https://universityaffairs.ca/news/news-articles/inside-the-shutdown-of-quest-university. Accessed 04.13.2024.

Levine, A., & Van Pelt, S. (2021). *The great upheaval: Higher education's past, present, and uncertain future*. The Johns Hopkins University Press.

Liu, E., & Noppe-Brandon, S. (2011). *Imagination first: Unlocking the power of possibility*. Wiley.

Liu, Z., Wang, X., Luo, C., Huang, F., & Hou, H. (2016). Gong Zhong Can Yu Na Mi Ji Shu Feng Xian Gou Tong Yi Xiang Ren Zhi Yin Su Shi Zheng Yan Jiu [An empirical study on the cognitive factors of public participation in nanotechnology risk communication intention]. *Ke Ji Jin Bu Yu Dui Ce, 33*(8), 22–29.

Lockwood, C., & Soublière, J-F. (Eds.). (2022). *Advances in cultural entrepreneurship*. Emerald Publishing Ltd.

Luo, Y. (2024). Wei Lai jiao yu fa zhan cheng xian yuan yu zhou he zhi wang kong jian zhan de sheng ming xing tai tan xi [Analysis of the life form of metaverse smart network space station in future education development]. *Ji Xu Jiao Yu Yan Jiu, 5*, 66–71.

Lupia, A., Allison, D. B., Jamieson, K. H., Heimberg, J., Skipper, M., & Wolf, S. M. (2024). Trends in U.S. public confidence in science and opportunities for progress. *PNAS, 121*(11), e2319488121. https://doi.org/10.1073/pnas.2319488121.

Madhaven, G. (2024). *Wicked problems and how to engineer a better world*. W. W. Norton.

Mainzer, K., & Kahle, R. (2024). *Limits of AI – theoretical, practical, ethical*. Technik im Fokus series, Springer.

Moulding, B., Songer, N., & Brenner, K. (Eds.). (2019). *Science and engineering for grades 6–12: Investigation and design at the center*. National Academies of Sciences, Engineering, and Medicine. The National Academies Press. https://doi.org/10.17226/25216

National Academies of Sciences, Engineering and Medicine. (2014). *Convergence: Facilitating transdisciplinary integration of life sciences, physical sciences, engineering, and beyond*. The National Academies Press. https://doi.org/10.17226/18722

National Academies of Sciences, Engineering and Medicine. (2018). *The integration of the humanities and arts with science, engineering, and medicine in higher education: Branches from the same tree*. The National Academies Press. https://doi.org/10.17226/24988

National Academies of Sciences, Engineering and Medicine. (2019). *Fostering the culture of convergence in research: Proceedings of a workshop*. The National Academies Press. https://doi.org/10.17226/25271

National Academies of Sciences, Engineering and Medicine. (2020a). *Changing expectations for the K-12 teacher workforce: Policies, education, professional development, and the workplace*. The National Academies Press. https://doi.org/10.17226/25603

National Academies of Sciences, Engineering and Medicine. (2020b). *Building capacity for teaching engineering in K-12 education*. The National Academies Press. https://doi.org/10.17226/25612

National Academies of Sciences, Engineering, and Medicine. (2020c). *Teaching K-12 science and engineering during a crisis*. The National Academies Press. https://doi.org/10.17226/25909

National Academies of Sciences, Engineering and Medicine. (2021). *Measuring convergence in science and engineering: Proceedings of a workshop*. The National Academies Press. https://doi.org/10.17226/26040.

National Academies of Sciences, Engineering and Medicine. (2023). *Transforming research and higher education institutions in the next 75 years: Proceedings of the 2022 endless frontier symposium*. The National Academies Press. https://doi.org/10.17226/26863.

Nowotny, H., Scott, P., & Gibbons M. (2001). *Rethinking science: Knowledge and the public in an age of uncertainty*. Polity.

Nusche, D., Rabella, M. F., & Lauterbach. (2024). *Rethinking education in the context of climate change: Leverage points for transformative change*. OECD Education Working Papers No. 307. OECD Publishing, JT03537057. https://doi.org/10.1787/f14c8a81-en

OECD. (2024). *PISA vocational education and training (VET): Assessment and analytical framework*. OECD Publishing. https://www.oecd.org/en/publications/pisa-vocational-education-and-training-vet_b0d5aaf9-en.html

Ottone, E. R. (2023). Culture, a transformative force for sustainable development: Paving the way for a standalone goal in the post-2030 agenda. *The ASEAN*, 32–33. Retrieved April 5, 2024, from https://theaseanmagazine.asean.org/

Pan, Z., Wang, Y., & Zhao, C. (2020). Ji Yu Xin Xi Yi Lai He Li Xing Gui Fan de Zhuan Ji Yin Ji Shu Gong Zhong Tai Du Xing Cheng Lu Jing Yan Jiu [Research on the formation path of public attitude towards genetically modified technology based on information dependence and rational norms] *Hanzhou Nong Ye Da Xue Xue Bao*, (6), 145–168.

Peters, M. (2017). *The man who knew everything: The strange life of Athanasius Kircher*. Annick Press.

Philipp, T., & Schmohl, T. (Eds.). (2023). *Handbook transdisciplinary learning*. De Gruyter.

Popkewitz, T. S. (2020). *The impracticality of practical research: A history of contemporary sciences of change that conserve*. University of Michigan Press.

Ratten, V. (Ed.). (2022). *Cultural entrepreneurship: New societal trends*. Springer.

Riveros, P. S., Meriño, J., Crespo, F., & Vienni Baptista, B. (2022). Situated transdisciplinarity in university policy: Lessons for its institutionalization. *Higher Education, 84*(5), 1003–1025. https://doi.org/10.1007/s10734-021-00812-6

Robinson, A., & Rees, M. (2023). The last man who knew everything: Thomas Young, the anonymous polymath who proved Newton wrong, explained how we see, cured the sick, and deciphered the Rosetta Stone, among other feats of genius. Open Book Publishing.

Rüegg, W., & Ridder-Symoens, H. d. (Eds.). (2003–2010). *A history of the university in Europe*. Cambridge University Press, four volumes.

Santos, A. V. P. E. (2023, April 19). *Managing student transitions into upper secondary pathways*. OECD iLibrary. https://dx.doi.org/10.1787/663d6f7b-en

Sarmiento, F. O., & Frolich, L. M. (Eds.). (2020). *The Elgar companion to geography, transdisciplinarity, and sustainability*. Edward Elgar Publishing.

Sawyer, R. K., & Henriksen, D. (2024). *Explaining creativity: The science of human innovation*. Oxford University Press.

Scott, W., & Vare, P. (2021). *Learning, environment and sustainable development*. Routledge.

Sengupta, P., Shanahan, M-C., & Kim, B. (Eds.). (2019). *Critical, transdisciplinary and embodied approaches in STEM education*. Springer. https://link.springer.com/book/10.1007/978-3-030-29489-2

Sgarbi, M. (2023). *The democratization of knowledge in Renaissance Italy: The philosopher and the people*. University of Amsterdam Press.

Sinnreich, A., & Gilbert, J. (2024). *The secret life of data: Navigating hype and uncertainty in the age of algorithmic surveillance*. The MIT Press.

Siraisi, N. G. (2013). *Communities of learned experience: Epistolary medicine in the Renaissance*. The Johns Hopkins University Press.

Spring, J. (2021). *American education*. Routledge.

Stronati, C. (2023, April 5). *The design of upper secondary education across OECD countries*. OECD iLibrary. https://www.oecd-ilibrary.org/education/the-design-of-upper-secondary-education-across-oecd-countries_158101f0-en

United Nations. (2023). The sustainable development goals report 2023. *Special edition: Towards a rescue plan for people and planet*. United Nations Department of Economic and Social Affairs. https://unstats.un.org/sdgs/report/2023

Van Berkel, K., & Homburg, E. (Eds.). (2023). *The laboratory revolution and the creation of the modern university, 1830–1940*. Amsterdam University Press.

Volk, K. S., & Yip, W. M. (1999). Gender and Technology in Hong Kong: A study of pupils' attitudes toward technology. *International Journal of Technology and Design Education, 9*(1), 57–71. https://doi.org/10.1023/a:1008894006039

Volk, K., Yip, W. M., & Lo, T. K. (2003). Hong Kong pupils' attitudes toward technology: The impact of design and technology programs. *Journal of Technology Education, 15*(1), 48–63.

Wessels, K. R. (2022). *Pedagogy of entanglement: A response to the complex societal challenges that permeate our lives*. Springer.

Wilson, S., Schweingruber, H., & Nielsen, N. (Eds.). (2016). Science teachers' learning: Enhancing opportunities, creating supportive contexts. *National Academies of Sciences, Engineering, and Medicine*. The National Academies Press. https://doi.org/10.17226/21836.

Xu, M., & Gu, J. (2021). Ji shu Li Jie Yue Shen Ke Ji Shu Tai Du Yue Ji Ji?—Ji Yu wo Guo 835 Wei Gao Zhong Tong Yong Ji Shu Jiao Shi de Shi Zheng Jian Yan [Is the Deeper the Technological Understanding, the More Positive the Technological Attitude?]. *Zhong Xiao Xue Dian Jiao*, (6), 121–128.

Xu, MD. (2020). Gao Zhong Tong Yong Ji Shu Jiao Shi de Ji Shu Li Jie, Ji Shu Tai Du. Ji Shu Xing Wei Ji Qi Guan Xi Yan Jiu [A study on technology understanding, technology attitude, technology behavior and their relationship among high school general technology teachers], [Unpublished doctoral dissertation]. Nanjing Normal University.

You, C., & Jin, J. (2020). Xin Mei Ti Huan Jing Xia Ke Xue Zhi Shi Dui Zheng Yi Xing Ke Ji Tai Du De Ying Xiang—Yi Zhuan Ji Yin Wei Li [The impact of scientific knowledge on controversial scientific and technological attitudes in the new media environment—taking genetic modification as an example]. *Guo Ji Xin Wen Jie*, (5), 81–98.

Yu, K. C., Han, F. N., Hsu, I. Y., & Lin, K. Y. (2005). The development of attitude toward technology scale for junior high school students. *Kaohsiung Normal University Journal, 19*, 69–83.

Zhang, C., Huang, L., & Ren, L. (2016). Zhong Guo Gong Min Dui He Neng Li Yong de Ren Zhi Ji Tai Du [The Chinese public's understanding and attitude towards the use of nuclear energy]. *Ke Pu Yan Jiu*, (3), 53–58.

Zhang, G., & Li, Z. (2024). Human interests and dignity of life in environmental protection. *Zhe Jiang Xue Kan*, (1), 144–156. https://d.wanfangdata.com.cn/periodical/ChlQZXJpb2RpY2FsQ0hJTmV3UzIwMjMxMjI2Eg16anhrMjAyNDAxMDE2GghqZTc2NTQ5NQ%3D%3D.

CHAPTER 36

Some Selected Memories of PATT

Wendy Dow/Dakers

Prior to 2002, I must confess to never having heard of PATT, nor its founder Marc de Vries. That year, however, my husband, colleague (and editor of this book), with the customary (some would claim foolhardy!) zeal of a newcomer to the field of technology education, after a lifetime spent in architecture, set off to attend his first ITEA conference in Columbus Ohio. There, he met, and was warmly welcomed by, Marc de Vries who had organised the PATT conference as part of ITEA. After that auspicious meeting, Marc and PATT became a familiar and integral part of both our lives.

Impressed both by the quality of the presentations and the friendliness of the PATT and ITEA delegates in Columbus, John asked whether there was a possibility that a future conference might be held in Glasgow. Marc's response was that it would certainly be possible – provided John was willing to undertake its organisation. It was an early indication of the inclusive ethos promoted by Marc that he was so relaxed about placing his trust in such an unknown and, as far as he knew, inexperienced and indeed perhaps even inept, new kid on the block. John was at first exhilarated, and then immediately filled with trepidation as the enormity of the task dawned. Who on earth would want to come all the way to Glasgow with its overcast skies and persistent rain?

Nobody has ever really been able to fully explain what happened next. Some claim it was the promise of a whisky-tasting session – and they may well be right! Whatever the reason, delegates converged upon Glasgow from all corners of the world- from India, Australia, New Zealand, Japan, Africa, South Africa, the USA, South America, Israel and Russia as well as from many parts of Europe. Numbers were so great that, for the first time in the history of PATT, it became necessary to run presentations in parallel sessions. This of course had disadvantages: it was not possible for all delegates to attend all papers, which is traditionally an important part of the ethos of PATT, and there were tight constraints on time. All, however, accepted the situation with characteristic good humour and John, whose job it was to make sure that presentations didn't overrun, became widely known as "the ten-minute-ator", as he would hold up both hands, followed ten minutes later by a cut-throat gesture to inform presenters that they had to stop, which they all did.

© WENDY DOW/DAKERS, 2025 | DOI:10.1163/9789004710948_036

This was my first experience of a PATT conference, and what was evident throughout was not only the sincere respect which delegates had for Marc de Vries and for each other's work and culture, but also the sense of camaraderie and fun that permeated the serious work of presentation and discussion.

Such memorable moments were evident when the piper, who had been hired to play traditional Scottish music, allowed anyone who wanted, to try out their luck with the unwieldy bagpipes. While everyone laughed along with Franscesca from Chile as she grappled hopelessly with the numerous drones, Frank Banks from England caused astonishment and awe by demonstrating that he could actually play! Another was when Steve Keirl from Australia proudly gave the Address to the Haggis in a broad Northumberland accent.

From that time, I attended many PATT conferences in both Europe and the USA. Although technology education was not my area of expertise, I found that most topics had direct relevance to my own interest in psychology, and I always felt totally accepted as part of the community of PATT. At each conference, interesting ideas were exchanged, and important friendships and professional collaborations were forged.

By the time of the second PATT conference to be held in Glasgow in 2007, the largest room in the hotel was needed to accommodate the numbers for the conference dinner. And that was despite the fact that the Glasgow weather had lived up to its reputation so well that questions were raised as to why the tour of the town on an open-top bus had appeared to end up driving through a car-wash!

At this conference, another important element, which continued to widen the scope of research in technology education, was introduced. In recognition of the importance that Marc has always placed on providing a philosophical underpinning for technology education, other eminent philosophers in the field, such as Don Ihde, Andrew Feenberg and Joseph C. Pitt were invited to attend, along with Marc in his philosophy of technology role, to present their ideas, to answer questions and take part in formal and informal discussion groups. By these means, exciting new perspectives on technology education were presented for further development in the field.

From my involvement over the years since 2002, it has been clear that the influence of Marc de Vries on technology education is both wide-ranging and profound. It is largely through his endeavours that technology has emerged as such an important and respected area of research for educators across the world. The community he has created through PATT is one which is stimulating, interesting and above all welcoming and inclusive – even to those like me who have only a peripheral and tenuous involvement in the field. It has been a privilege to have been involved.

CHAPTER 37

Getting Inspired for the Future

Reflections on Dreams

Marc J. de Vries

1 A Variety of Dreams

This book offers a wonderful collection of dreams for the future of Technology Education. Some of the contributions, like the one by Alister Jones, are very autobiographic, which makes them very personal. Although each chapter is unique in its own way, the dreams together share some common themes. Perhaps not surprisingly, given the context of this book as a Festschrift, they are all themes that are close to my heart. I will go through them one by one and add my own reflections on what the authors have written.

2 (Ethno)Technological Literacy

Several chapters show how Technology Education has moved away from a craft-oriented, and often vocational, school subject to a curriculum component with a much broader scope. What was once a dream has now become a reality. For decades previously, the term 'technological literacy' had been used to suggest that Technology Education should offer more than simply a transfer of the knowledge and skills needed for certain professions or to survive as a consumer in a technological world. Technology Education should in fact enable future generations to act wisely and responsibly when dealing with both existing and new technologies. It can by no means be taken for granted that young people will do that. New technologies can be mesmerising, and it is often too easy to assume that technologies exist simply to be used by us, without reflection. The philosopher Albert Borgmann expressed concern about this in what he called the 'device paradigm': our world is full of buttons that invite us to push them, and we cease to wonder why we actually keep pushing them. The quest for sense and purpose fades away and the quality of our experience of reality decreases the more we give in to our inclination to push the buttons of devices. Technology Education should therefore bring back our awareness that technology should serve a purpose that goes beyond naïve consumerism, as

© MARC J. DE VRIES, 2025 | DOI:10.1163/9789004710948_037

Mike Martin addresses so well in his chapter. Or as David Spendlove expresses it: we need to get away from a capitalist approach and move towards a human-centred approach in Technology Education.

John Ritz points out that technological literacy is now an important goal of Technology Education in the USA. His dream is that the connection with other STEM subjects will enhance this because it will free Technology Education from the isolated position that has often adversely affected the reputation of the subject. Phil Reed also sees a value in breaking down the, often artificial, boundaries between subjects in the school curriculum. It is interesting to note the contrast here with John Dakers who also believes in technological literacy as a crucial aim in Technology Education, at least when it is more than mere instrumentality (as he sees it in the traditional use of the term; for him this suggests a need to introduce a new term: ethno-technological literacy). For John, however, the connections that are emphasised in STEM pose a threat rather than an opportunity, as they may mean a movement back to a more instrumental form of Technology Education. According to John, not even adding the A from Art to STEM can fix this. My personal view is that it will depend very much on how we shape STEM or STEAM. I believe that, in itself, the relation with science need not harm the ethnotechnological literacy perspective, as science and mathematics are as much human activities as technology in that we bring all our human vices and virtues with us to the subjects. This, however, is unfortunately not evident in the majority of Science Education and Math Education practice today. The 'healthy' influence of an ethnotechnological literacy-oriented Technology Education could perhaps even lead to a paradigm shift in Science and Mathematics Education. This would definitely require a very careful and conscious development of STEM Education because the pitfalls identified by John are by no means imaginary.

3 Inclusiveness

An important characteristic of a curriculum that aims for ethnotechnological literacy is inclusiveness. It is striking that this is an issue that is brought up in several of the chapters. Traditionally, inclusiveness meant striving for a gender balance in Technology Education, and in her chapter Sonja Niiranen and her colleagues address that concern. My very first international conference was a GASAT (Girls and Science and Technology) conference in Ann Arbor, USA, and from that point I have taken this as a focal point for my own research and development work. Today, however, inclusiveness means much more than this, and justly so. Minorities of all kinds should have equal opportunities to become ethnotechnologically literate. In her chapter Wendy Fox-Turnbull mentions the Maori population in New-Zealand. Gerald van Dijk and Erwin Savelsberg identify low-SES and migrant parts of the population. Personally, I have also become intrigued by the wisdom of indigenous knowledge. I remember how, during one of my teaching activities in South-Africa, I took a group of technology teachers to a local open-air museum, a Sotho settlement, to challenge them to recognise the richness of that context for teaching about technology. In essence, Technology Education should cater for special needs pupils, pupils of all races, religions, genders, gender-orientations, ethnic and socio-economic backgrounds, indeed whatever makes one human being different from another. Inclusiveness also means that we must take account of the abilities of all pupils by respecting their autonomy as learners, a point brought up by Richard Kimbell. David Mioduser in his chapter addresses the idea of technological thinking as a characteristic of our human species. In her chapter Osnat Dagan relates her experiences in teaching at the Beit Berl college in Israël. Some time ago I had the privilege of teaching a course on the ethics of technology there and was delighted to see how student teachers of different ethnic backgrounds and religions peacefully worked together in that institute. A true example for the whole country. Steve Keirl, as is typical of him, emphasises the quest for our humanity in all of Technology Education and I wholeheartedly agree with him. This is also why we should have a genuine interest in pupils' attitudes towards technology, not to 'test' whether that attitude has 'improved' but to see how education impacts the way pupils develop their own perspective on technology. Piet Ankiewicz in his chapter describes the ways in which we have been trying to work out how to investigate this over the course of time.

4 Technology as a Social Practice

The ability to hold values lies at the heart of being human. For many years, however, it was considered inappropriate to address values in Technology Education, for fear of indoctrination. That fear was certainly not imaginary, but neither was it necessary. Values and worldviews are closely connected, and both determine technology as a social practice in which normativity is crucial for understanding. This is something that my colleagues and I have worked on, in what used to be called 'reformational philosophy', but now goes under the name of 'Christian philosophy' (without, of course, claiming that there can be no other philosophy done by Christians). I was pleased, therefore, to find several chapters by my philosophy colleagues in this book. The notion of normative practices was developed primarily by Henk Jochemsen, together with

Gerrit Glas and Joan Hoogland. The value of this for education is nicely highlighted by Henk Jochemsen: all professionals (including teachers) need moral formation, in order to serve responsibly, in their normative practice. Sometimes widening the scope of the practice is required. This also holds true for teachers. It can for example, be useful for their development to get involved in research, as Eva Hartell points out, although originally that was not part of their practice as teachers. Maarten Verkerk, on the other hand, focuses on technology as a normative practice, something that is an important factor within the content of Technology Education. I also very much agree with HildaRuth Beaumont, who emphasises the importance of virtues in Technology Education.

Practices have a historical dimension. Roel Kuiper, therefore, addresses the value of tradition for understanding these. This is also at the heart of Jonas Hallström's chapter. History must be a component of Technology Education, otherwise we cannot get a proper understanding of technology. The challenge is to connect the historical and social dimensions of technology with the practicality of Technology Education, as Charlotta Nordlöff so well describes in her chapter. Food, the domain discussed by Marion Rutland, is certainly also clearly a suitable topic in this respect.

One of the consequences of technology being a social practice is its complexity and uncertainty. In his chapter, therefore, my former colleague in Delft, Maarten van der Sande, addresses the need to educate professionals to work and communicate in a complex reality with many uncertainties. That certainly also applies to technology teachers. In their chapter, Dennis Cheek and Hongwu Dai convincingly argue that the social impact of technology is what makes Technology Education so important. This also, however, necessitates it taking a transdisciplinary turn, which is also Maarten van der Sande's major concern.

5 Disruptive Technologies

It is interesting to see that a number of chapters address the issue of new, revolutionary technologies as a necessary element in Technology Education. Some experts use the term 'disruptive' for the dramatic changes that such technologies can bring about in society. Artificial Intelligence (AI) is definitely an example of one such technology. Moshe Barak discusses robotics in Technology Education as probably the earliest example of AI in Technology Education. Currently, Generative AI invades schools and of these technologies probably ChatPGT is best known. My former Ph.D. student Alice Schut, together with Christian Detweiler, discusses the use of Generative AI for generating, not texts, but images. This clearly has great potential for Technology Education. Fortunately, all authors who address the future role of AI in Technology Education point out that AI is not the same as HI (Human Intelligence). In addition, Alison Hardy writes about the ambivalence of AI as potentially useful but not without its pitfalls. I also very much agree with Kay Stables when she claims that there is an open future and that we are responsible for choosing between good and evil. That certainly applies to the way we deal with AI. It is part of ethnotechnological literacy to recognise both how it can be a blessing but also a curse, depending on how conscious we are in dealing with it.

6 The Importance of Context

A final theme I see in the chapters is the need to address the local needs of learners. Andrew Doyle demonstrates clearly how this should play out in Technology Education. This matches well with Frank Bank's claim that it can be harmful to try to push the curriculum forward in a top-down approach. Such an approach is bound not to take account of the specific circumstance of schools. The danger of this is that the new curriculum will be biased towards an urban context, with the consequence that pupils in a rural school will feel estranged from its content. In the context-concept approach, contexts are used to enhance the learning of technological concepts. I am currently working on a monograph in which this is elaborated in more detail.

7 Personal Remarks

I must admit that, originally, when John Dakers approached me with his dream of a Festschrift for me, I was not very happy with the idea. Why me? There are so many colleagues who deserve this honour more than I do. John himself is one of them. Yes, I organised PATT conferences, but so did he. Yes, I wrote and edited books but so did he. He even did something that I never accomplished, namely bring a group of the world's most famous philosophers of technology and technology education colleagues together to discuss the relevance of philosophy for education. John and I have worked together on many occasions. He and I share an interest in 'translating' insights from the Philosophy of Technology to Technology Education. Over the course of time, we have become very close friends. John was there when I gave my inaugural address as a professor in Delft. We have laughed and cried together, which is the ultimate proof of friendship as some say. Had it not been for all sorts of silly formalities he would have defended his thesis with me in Delft. Subsequently, however, that publication became his recent book on nomadic pedagogy, truly a landmark publication in our field and I was happy that I could be instrumental in getting it published by Brill. Given the history we share I simply could not refuse John's request to let him do this for me. Now, when I see the outcome, I am very happy that he persisted because the result is a book full of dreams about the future of technology as there has been no such one before. The colleagues who contributed to the book really embraced the freedom to write from their hearts and not be too concerned about references and academic exactness (which does not mean that the book has no academic qualities because it certainly does). John, thank you so much for doing this for me and for all of us.

I still feel shy about the words of praise written by colleagues such as Carl Mitcham, Kendall Starkweather, Lena Gumaelius , Dov Kipperman, Inga-Britt Skogh (my 'boss' at KTH) and Wendy Dakers. In my own eyes, I just did my little thing, like we all do our little thing. One of the features of PATT conferences is that the atmosphere is such that everyone is equally valued, from the just-starting Ph.D. student to the emeritus professor. That is why we generally have no keynote lectures and, if we do have them, they are usually presented by local outsiders. Therefore, I can sympathise with colleagues who may have been asked to contribute to this book but felt uncomfortable with the idea of a Festschrift, irrespective for whom it was meant. Perhaps that was the case with John Williams with whom I have a longstanding relation of friendship and who would also immediately come to my mind as a colleague who deserves a Festschrift more than I do. For many years John was the editor of a scholarly journal in our field and he is still the editor of an academic book series in Technology Education. I highly respect him as a colleague and friend. Likewise, there are several other contributors to this book that in my view played a more important role than I did. Among them are several journal editors, book-series editors and colleagues who had a leadership role in national and international projects which had a great impact in our field.

PATT is truly a community, and that is why I continued my involvement during the ten or so years that I had a formal appointment in philosophy and not in technology education. I still enjoy being involved in it and hope to remain so in the coming years. The inclusivity of that community has been consistently emphasised. From the beginning we made efforts to enable the participation of people who normally would not have had the chance to do so (such as colleagues from Africa and Eastern-Europe). The flipside of being part of a community, however, is that one feels the loss of colleagues who retire or pass away. The most recent losses have been Bill Dugger and Jacques Ginestié. Before that, we lost friends like Tapani Kananoja, Chitra Natarajan and Dietrich Blandow.

The loss I personally felt most deeply was that of my supervisor Jan Raat, who also initiated PATT, together with Ilja Mottier and myself. Many years ago, Ilja put together a Festschrift for Jan and I contributed to it. It is a strange experience that now I myself am the recipient of one.

Let me finish by emphasising that whatever I have been able to do, it is only thanks to the cooperation of so many others. PATT is a perfect example of that. I was involved in the initiation and am now the only one of the remaining initiators who continues to be involved. Gradually, however, others began to organise conferences and now the only thing that is left for me to do is to delegate the distribution of conferences over the years in different places and keep an eye on the preservation of the PATT 'culture' of community by the local organisers. Likewise, I owe a debt of gratitude to my Ph.D. students and other young scholars who constantly inspire me with their enthusiasm. And of course, I owe my thanks to all those colleagues with whom I have collaborated over the years, many of whom have contributed to this book. When I consider myself as an 'excuse' to bring together all the dreams and insights that are in this book, I can reconcile myself with the idea that this book is a Festschrift. I think we can see it as a Festschrift for what we have all together been able to achieve over the course of the past decades. Our common dream was, and still is, that Technology Education should get the recognition that it deserves as a learning area that can have an impact on a generation that is growing up in a world that is full of technology and at the same time a world that is full of concerns, for which technology can be part of a solution, although certainly not the only one. Technology is part of our being human, but not the only part. What our world needs is people with heart for one another. Perhaps in that respect, too, the international Technology Education community can be seen as an example of how that can work. For me personally, faith in God is also part of who I am. It is even the primary part. I thank Him that he gave me health and motivation to play my modest role in the Technology Education community and I hope He will allow me some years to come in which I can continue to play that role.

Index

Printed in the United States
by Baker & Taylor Publisher Services